Comb-Shaped Polymers and Liquid Crystals

SPECIALTY POLYMERS

Series Editor
J. M. G. Cowie, *University of Stirling, Stirling, Scotland*

ALTERNATING COPOLYMERS
Edited by J. M. G. Cowie

COMB-SHAPED POLYMERS AND LIQUID CRYSTALS
N. A. Platé and V. P. Shibaev

Comb-Shaped Polymers and Liquid Crystals

N. A. Platé
and
V. P. Shibaev
Moscow State University
Moscow, USSR

Translated from Russian by
S. L. Schnur

Translation Edited by
J. M. G. Cowie
University of Stirling
Stirling, Scotland

Plenum Press • New York and London

Library of Congress Cataloging in Publication Data

Platé, Nikolai Al'fredovich.
Comb-shaped polymers and liquid crystals.

(Specialty polymers)
Translation of: Grebneobraznye polimery i zhidkie kristally.
Bibliography: p.
Includes index.
1. Polymer liquid crystals. I. Shibaev, V. P. (Valerii Petrovich) II. Title. III. Series.
QD923.P5613 1987 547.7 87-18518

ISBN-13: 978-1-4612-9082-7 e-ISBN: 978-1-4613-1951-1
DOI: 10.1007/978-1-4613-1951-1

Softcover reprint of the hardcover 1st Edition 1987
A Division of Plenum Publishing Corporation
233 Spring Street, New York, N.Y. 10013

Preface to the American Edition

We are pleased that our modest work, published some time ago in Russian in Moscow* and which attracted the attention of polymer specialists,† will now be available to the English-speaking audience of scientists – chemists, physicists, and technologists engaged in creating new types of polymer materials for modern technology and working on the fundamental problems of the solid-state physics and structure of polymers – due to the initiative of Plenum Press.

In polymer science, the 1980s were marked by the birth of a new field and a new scientific trend related to the discovery and study of a previously unknown class of polymers – thermotropic liquid-crystalline polymers – and the further development of the fundamental theoretical concepts of the liquid-crystalline (mesomorphic) state of macromolecular compounds. This state is a phase state in thermodynamic equilibrium characterized by the anisotropy of the structure and properties as a result of one-dimensional or two-dimensional ordering. Such systems have an ordered but simultaneously labile structure which can easily be altered by mechanical, electrical, or magnetic fields; the polymer system then acquires unique physical and optical properties. These properties, which are acquired in the liquid-crystalline state, are then fixed in the solid at the operating temperatures.

*N. A. Platé and V. P. Shibaev, Comb-Shaped Polymers and Liquid Crystals [in Russian], Khimiya, Moscow (1980).
†See the review of this book by H. Mark in J. Polym. Sci. Polym. Lett. Ed., 20, 139 (1982).

This feature of liquid-crystalline polymers opens up broad possibilities for the preparation of new polymeric glasses, films, fibers, and coatings.

The class of branched macromolecules in the comb state, i.e., containing long aliphatic branches in each monomeric unit, is one of the typical representatives of such polymers which tend toward self-organization. Addition of mesogenic groups to the side branches of comb-shaped polymers can result in the formation of any of the known low-molecular-weight liquid crystals of the mesophase type: nematic, smectic, or cholesteric; on the one hand, this significantly expands the number of liquid-crystalline substances and, on the other hand, it also allows the creation of other new types of macromolecular compounds. Polymers of this type are now being synthesized and widely studied in polymer laboratories in the FRG, USA, USSR, UK, Japan, France, and other countries and, together with other liquid-crystalline polymers with a linear structure, are the subject of discussion at international conferences and symposia. Within the International Union of Pure and Applied Chemistry (IUPAC), the European Physical Society, and the American Chemical Society alone, there have been more than 20 international meetings of scientists concerned with questions relating to the examination of problems of liquid-crystalline ordering in polymers and their synthesis during 1982-1985, not to mention the many national symposia and conferences.

Many reviews and collections published in the FRG, USA, and USSR (see the references to the Introduction and to Chapter 4) have been dedicated to the chemistry and physics of polymeric liquid crystals. In addition to other questions, they examine the problems of the synthesis, structure, and properties of liquid-crystalline polymers with mesogenic side groups. Nevertheless, up to now there has not yet been any concentrated complete analysis of the field of comb-shaped macromolecules and the liquid-crystalline polymers based on them; our book is the first attempt in this direction.

In comparison to the Russian edition published in 1980, the contents of this monograph have been significantly revised. Basic changes were made in the chapters on the liquid-crystalline state, which have actually been rewritten in consideration of the important new publications which appeared in 1980-1984, and a totally up-to-date picture of the status of this field and the trends in its development is provided.

Chapters on the structure and molecular mobility of comb-shaped polymers in the solid phase and in solutions have been added, although the subject matter and exposition style have generally remained the same. At the same time, we believed it possible to exclude two sections which are not very important for understanding the entire field as a whole: those which deal with the structures of gels of linear block copolymers and problems of the radical polymerization of higher alkyl methacrylates. The discussion of the properties of gels of comb-shaped macromolecules was moved to another chapter.

We would like to thank our colleagues of many years, Dr. Raisu Tal'roze, Dr. Yakov Freidzon, and Dr. Sergei Kostromin, for their useful discussions on liquid crystals reported in this monograph, and Aleksei Platé for his assistance in preparing the graphs and illustrations.

N. A. Platé and V. P. Shibaev

Moscow
June, 1985

Contents

Chapter 3

Comb-Shaped Macromolecules in Solutions and Intramolecular Interactions

Chapter 4

Thermotropic Liquid-Crystalline Polymers

Introduction

The so-called comb-shaped polymers, macromolecules which contain relatively long side branches spaced comparatively closely along the main chain, occupy a special position among the large number of polymers sometimes used in unusual areas. They differ from ordinary branched polymers due to the fact that they have many side chains: each monomeric unit can contain such a chain (Fig. 1), and this quantitative difference results in the appearance of a new set of qualitative properties which are absent in linear or weakly branched polymers.

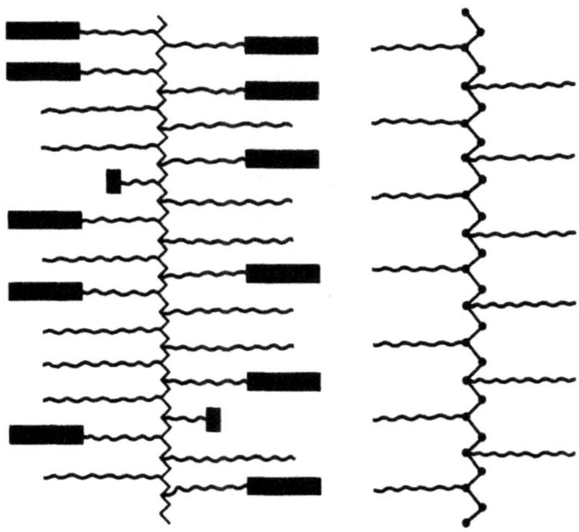

Fig. 1. Schematic diagrams of comb-shaped macromolecules.

Higher homologs of poly-α-olefins [poly(1-alkyl ethylene)s], poly(alkyl acrylate)s and poly(alkyl methacrylate)s with aliphatic groups in the side chain, poly(vinyl alkyl ether)s and polyesters, poly(acyl styrene)s, polyacrylic derivatives of amino acids, comb-shaped polypeptides, and some others are comb-shaped polymers. Although all of these homopolymers and copolymers are formally linear polymers (the side branches are incommensurably shorter than the main chain), comb-shaped polymers are actually a special class of branched polymers whose properties can differ significantly from the properties of ordinary linear polymers, as will be demonstrated below [1-3]. The chemical formulas of the basic units of some comb-shaped homopolymers are shown below:

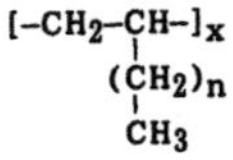

Poly(1-alkyl ethylene)s

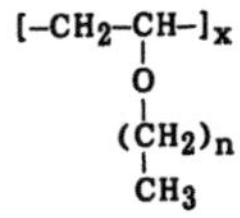

Poly(alkyl vinyl ether)s

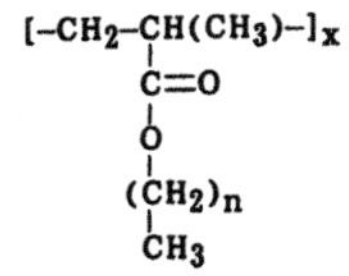

Poly(alkyl methacrylate)s

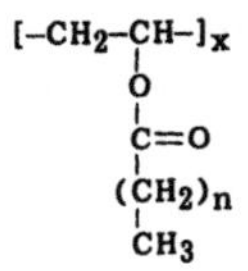

Polyvinyl esters

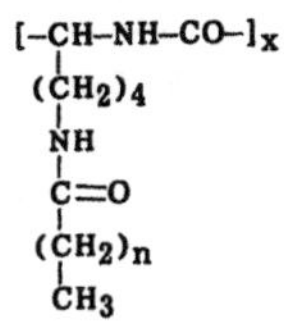

Comb-shaped polypeptides

The value of n for the aliphatic groups usually varies from 5 to 20, but the side groups in these polymers can be very different and can contain carbocyclic, heterocyclic, amide, alkylaromatic, and other moieties; however, they are most frequently aliphatic. It is important that the length of these side groups is significantly greater than their cross section. The unique set of properties of comb-shaped polymers which will be discussed below only occur under this condition.

The presence of two types of units – units which participate in the formation of the backbone of the main chain, and units from which the side branches are constructed – is the most important property of these polymers. The side branches are joined to the main chain by very different chemical bonds: carbon–carbon, ester, amide, etc. The spatial proximity of the side chains to each other within the same chain creates the conditions for their interaction, frequently very similar to what takes place for a group of small molecules of similar structure. However, the presence of the main chain on which these side branches are "strung," with its inherent flexibility and the possibilities of intermolecular interactions with the main chains of other macromolecules, is the basis for the search for properties resulting from the structure of the main chain in such a polymer (polyacrylates, for example, remain polyacrylates, and polypeptides remain polypeptides). This duality of structure of the molecules results in the dual character of the ordering and properties of the system as a whole.

Comb-shaped macromolecules are also models of graft copolymers if the structure of the side chains differs from the structure of the main chain. Polymers with sufficiently long side chains are convenient models for studying the structure and properties of copolymers, and are advantageous in that the investigator knows precisely both the length of the side branches and the frequency of their spacing. Actually, the well-known independence of behavior of the set of side chains and the main units of the macromolecules, as in many graft copolymers, is frequently observed in such systems.

The presence of sufficiently long side chains containing aliphatic, alkylaromatic, and other types of units causes comb-shaped polymers to have very high intramolecular ordering; they are even sometimes called "crystal-like macromolecules" [4]. This intramolecular ordering imparts the unusual properties of comb-shaped polymers in the solid phase and in solutions [2].

The study of the structure of these polymers and their distinct capacity for crystallization and ordering due to the presence of side chains makes it possible to clarify the basic questions on the structure of amorphous and crystalline polymers, the ratio of the amorphous and crystalline phases, and the nature of the boundary between them which passes through a segment of the chain known to belong to the same macromolecule.

The similar structure of comb-shaped polymers and low-molecular-weight triglycerides of higher aliphatic acids which form the lipid bilayers of biomembranes is also interesting for investigating the structure and physicochemical properties of these polymers, which simulate biological membranes. It is known that most of the membranes of living cells are constructed of oriented layers of hydrophilic polymers (polypeptides, polysaccharides) and complex hydrophobic (lipid) compounds, so that modeling the structure of such systems with comb-shaped polymers can also be useful in this field.

The amphiphilic character of the structure and properties of most comb-shaped macromolecules creates difficulties for the unambiguous interpretation of their behavior in solutions; for example, when the habitual concept of the θ-temperature is lost, gel formation occurs instead of the expected phase stratification of the system, etc. This is also of interest to the physical chemist, since it allows the discovery of unexpected properties of the system and enables critical experiments on a similar type of sample to be conducted in many cases.

There are two major ways of synthesizing comb-shaped polymers. One is the polymerization of unsaturated or cyclic derivatives, where the side unit is contained in the molecule of the monomer from the beginning: for example, polymerization of a monoolefin or alkyl acrylate

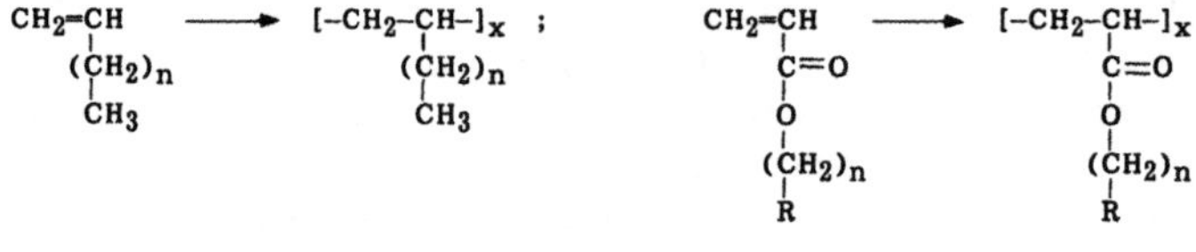

or N-carboxyanhydride. The comparative ease of conducting radical or ionic polymerization of the monomers used makes it possible to synthesize comb-shaped polymers with side branches in each unit of the main chain, and in the individual units of the main chain in copolymerization with other monomers. This is the primary method of synthesizing comb-shaped polymers.

Another method of synthesis is the reaction of polymer-analogous transformations where the side units are "strung" on a main chain previously formed by some method. The esterification of polyvinyl alcohol or polyacrylic acid with long-chain monocarboxylic acids or monohydric alcohols is an example:

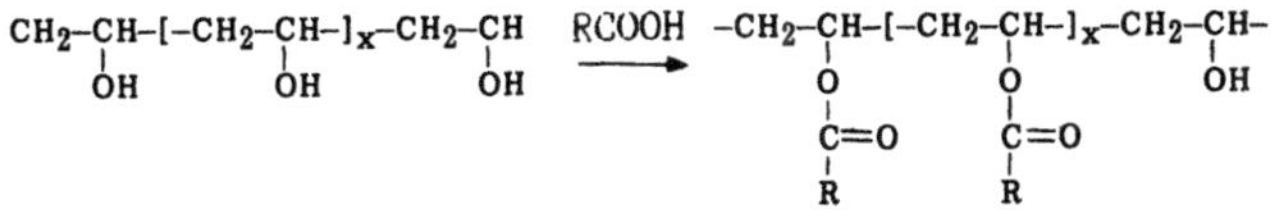

where R = C_nH_{2n+1}.

In this case, it is difficult to obtain the 100% conversion which is usual in other polymer-analogous reactions. The final product usually contains some amount of unreacted units.

Comb-shaped polymers have already been used in practice. Copolymers of ethylene and higher α-olefins containing side aliphatic groups of 8-12 carbon atoms in the chain are used industrially as one of the types of modified polyethylene. The presence of side groups causes a decrease in the melt flow index and facilitates the processing of this polymer. Copolymerization of methyl methacrylate with higher alkyl methacrylates is used for a similar purpose. Glasses which can be cast molded and which have high resistance to impact loads have been prepared in this way.

The tendency of comb-shaped carbon-chain polymers to gel led to their use as thickeners for motor oils and lubricants to preserve the required viscosity of the basic substance in a wide range of temperatures.

Comb-shaped polymers and copolymers have also been used in another area; polymers based on derivatives of styrene and dialkyl methacrylates in the form of small porous spherical particles are used as foams for absorbing liquid hydrocarbons (in laboratories or on the surface of water for petroleum, fuel, oil spills, etc.). Containers of such spherical particles placed in a water supply system absorb the oils and liquid fuel which accidentally enter the system.

One of the most important properties of comb-shaped polymers with mesogenic groups in the chains of theoretical and practical interest is their capacity to form a liquid-crystalline phase (together with the fact that they can exist in both the usual crystalline and isotropic amorphous states) [5].

In the past 10-15 years, the importance of the role of the liquid-crystalline state of a substance in the vital activity of organisms has been discovered, and broad possibilities have opened up for the use of low-molecular-weight liquid crystals in electro- and magnetooptics, microelectronics, medical diagnosis, and other fields. The chemical substances known to form a liquid-crystalline phase now number over 10,000, and each month new publications appear in this field. Until the 1970s, basic attention was focused on the synthesis, structure, properties, and the practical use of low-molecular-weight liquid crystals. However, the past 8-10 years have been marked by significantly increasing interest in polymeric liquid crystals, and the liquid-crystalline state of polymers is now attracting the interest of specialists in very different fields [5-16].

Crystallographers and crystal chemists are studying the features of the packing of chain molecules in lyotropic and thermotropic polymeric liquid crystals. Specialists engaged in the study of the structure of polymers are attempting to determine how widely distributed the liquid-crystalline state is in the world of macromolecular substances and what this means to modern polymer physics. Synthetists and physical chemists are investigating ways of creating polymeric liquid crystals and are studying the properties of these unusual compounds. Specialists in the field of chemical fiber processing have successfully used lyotropic liquid-crystalline polymer systems for creating high-modulus, high-strength fibers based on rigid-chain polymers, including aromatic polyamides and polyesters, for several years now. Specialists in optics and electronics are studying the use of thermotropic polymeric liquid crystals for the creation of sensitive light filters, temperature-sensitive elements, and other technical instruments. It is difficult to recall that only 15-20 years ago this field simply did not exist in polymer chemistry and physics, and the isolated studies on lyotropic liquid-crystalline polymers (polyglutamic acid and its derivatives, for example) aroused the interest of only a few specialists in the entire world.

Liquid-crystalline polymers with a comb-shaped structure are only some of the liquid-crystalline polymer systems, but it is possible to demonstrate how unusual the properties of this class of polymers are, using them as an example.

The features of the phase state of lyotropic polymer systems and the structure and physicochemical properties of linear rigid-chain polyamides, polypeptides, and polyesters which form a liquid-crystalline phase when mixed with solvents, have been examined in many review articles and monographs: for example, in [4, 7, 10, 16]. In the present book, most of the attention is focused on thermotropic liquid-crystalline polymers with mesogenic groups in the side chains.

The first publications on the synthesis of liquid-crystalline polymers based on comb-shaped macromolecules with mesogenic groups added to the main chain through flexible aliphatic bond were published in the USSR in 1974. Since then, this principle has been accepted in different laboratories throughout the world, and the present monograph generalizes not only the data on comb-shaped polymers available in the literature, but also shows how these polymers are related to liquid crystals. It has been found that the greatest possibilities for controlling the structure and properties of liquid-crystalline polymers are again manifested in the use of branched macromolecules with a comb-shaped structure.

We have not included material on the polymerization of monomers in the mesophase and long-chain monomers in the crystalline state. The kinetic and structural features of such reactions are the subject of special publications (see, e.g., Amerik and Krentsel [11] and the publications of Hardy et al. [13]).

The area of the physical chemistry of polymers examined here is currently developing explosively. In the habitual language of polymer chemists, a significant concentration of "active sites" has been created, and the massive "formation of propagating chains" has taken place. These sites "are propagating" at a great rate, which is resulting in the formation of a new substance: new knowledge and new materials. The "polymerizable" character of the material, due to the continuous acquisition of new information, results in the fact that many concepts have not yet been definitively established. However, if the material reported below draws the attention of polymer chemists and specialists from related fields, we will consider our task to have been completed.

REFERENCES

1. N. A. Platé and V. P. Shibaev, Vysokomol. Soedin., A13, 410-424 (1971).
2. N. A. Platé and V. P. Shibaev, J. Polym. Sci., Macromol. Rev., 8, 117-253 (1974).

3. P. L. Magagnini, Makromol. Chem., Suppl. 4, 223-238 (1981).
4. V. N. Tsvetkov, Rigid-Rod Polymeric Molecules [in Russian], Nauka, Leningrad (1985).
5. V. P. Shibaev and N. A. Platé, Vysokomol. Soedin., A19, 923-972 (1977).
6. A. Blumstein (ed.), Liquid-Crystalline Order in Polymers, Academic Press, London (1978.
7. S. P. Papkov and V. G. Kulichikhin, The Liquid-Crystalline State of Polymers [in Russian], Khimiya, Moscow (1977).
8. V. P. Shibaev, in: Advances in Liquid-Crystal Research and Applications, Vol. 2, L. Bata (ed.), Pergamon Press, Oxford; Akademiai Kiado, Budapest (1980).
9. H. Finkelmann, in: Polymer Liquid Crystals, A. Cifferi, W. Krigbaum, and R. Meyer (eds.), Academic Press, New York (1982).
10. I. Uematsu and Y. Uematsu, Adv. Polym. Sci., 59, 37-74 (1984).
11. Yu. B. Amerik and B. A. Krentsel', Chemistry of Liquid Crystals and Mesomorphic Polymer Systems [in Russian], Moscow (1981).
12. R. V. Tal'roze, V. P. Shibaev, and N. A. Platé, Vysokomol. Soedin., A25, 2467-2487 (1983); Pure Appl. Chem., 56, 403-416 (1984).
13. F. Cser, K. Nyitrai, and G. Hardy, in: Mesomorphic Order in Polymers and Polymerization in Liquid-Crystalline Media, ACS Symposium Series No. 6, A. Blumstein (ed.), Washington, D. C. (1978).
14. V. P. Shibaev and N. A. Platé, Adv. Polym. Sci., 60-61, 173-252 (1984).
15. V. P. Shibaev and N. A. Platé, Pure Appl. Chem., 57, 206-221 (1985).
16. A. Blumstein (ed.), Polymeric Liquid Crystals, Plenum Press, New York (1985).

Chapter 1

Structure of Comb-Shaped Polymers

1.1. EARLY STUDIES

In the past twenty years in polymer physics, it has been possible to construct relatively universal models of crystalline polymer structure with folded or straight linear chains and to postulate the basic conditions and features of the crystallization of linear polymers [1-8]. The number of studies on the structure of more complex macromolecular compounds, such as branched and comb-shaped polymers, statistical copolymers, block and graft copolymers, is significantly smaller.

Although the theory of the crystallization of statistical copolymers began to be successfully developed during the late 50s and early 60s by Flory [9-12], Kilian [13-16], and others [1, 17-20], there were still too few reliable experimental data which would allow any serious generalizations to be formulated at this time concerning the structure formation and crystallization of block and graft copolymers* and comb-shaped polymers.

The study by Rehberg and Fisher [26] should be mentioned for their investigation of the dependence of the brittle temperature T_{br} (actually, the glass transition temperature T_g) on the length of the side chain in the series of atactic

*Some information on structure formation in block and graft copolymers can be found in [21-25].

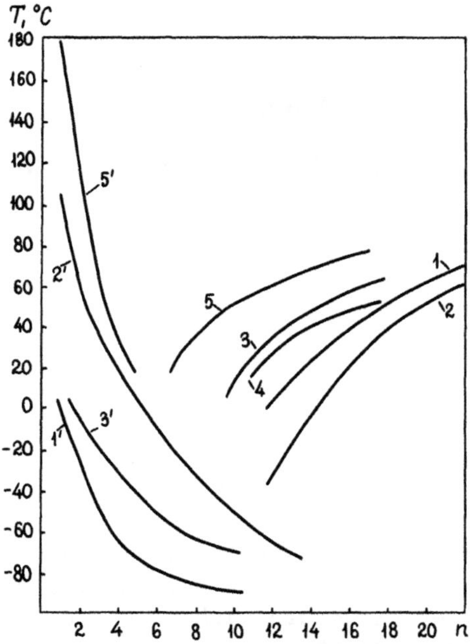

Fig. 1.1. Dependence of T_g (1'-3', 5') and T_m (1-5) of comb-shaped polymers on the number of carbon atoms in the alkyl group of the side branches: 1, 1') poly(n-alkyl acrylate)s [26]; 2, 2') poly(n-alkyl methacrylate)s [26]; 3, 3') poly(n-alkyl vinyl ether)s; 4) poly(n-alkyl vinyl ester)s; 5, 5') isotactic poly(n-alkyl ethylene)s.

poly(n-alkyl acrylate)s (PA-n)* and poly(n-alkyl methacrylate)s (PMA-n); they found that an increase in the length of the side chain to 8-10 carbon atoms is initially accompanied by a decrease and then by an increase in T_g (Fig. 1.1, curves 1 and 2).† [At the recommendation of the IUPAC Committee on Polymer Nomenclature, homologs of poly-α-olefins will subse-

*The following abbreviations will subsequently be used for designating the actual polymer: PA-n, PMA-n, PVE-n, etc., where the first letters indicate the affiliation of the polymer in a certain homologous series and n corresponds to the number of carbon atoms in an alkyl group of normal structure; for example, PA-16 indicates poly(n-hexadecyl acrylate).
†The melting points (T_m), and not the glass transition temperatures, are shown on the right in Fig. 1.1; the curves on

quently be called poly(1-alkyl ethylene)s (PE-n), where n is the number of carbon atoms in the n-alkyl group.] In [26], and based on the x-ray data of Kaufman et al. [27], this observed trend in the dependence of T_g on the number of carbon atoms in the alkyl group of PA-n and PMA-n could be attributed to crystallization of the side chains. Based on the data from x-ray analysis, Greenberg and Alfrey [28] hypothesized that crystallization of the side chains in these polymers takes place in a rhombic cell similar to crystallization of molecules of n-paraffins and polyethylene. Their conclusions were based on the presence of a strong reflection corresponding to an interplanar distance of 4.2 Å in the x-ray patterns of all compounds studied, which is usually characteristic of long-chain paraffins and their derivatives and also of polyethylene. The presence of a small-angle x-ray spacing was found for some samples, but it was not always reproducible [28, 29].*

The possibility of crystallization in the side chains of comb-shaped PA and PMA (primarily with reference to [26, 28]) is also reported in [31-36]. The conclusion on paraffin side-chain crystallization was later extended to other comb-shaped nonstereoregular polymers: poly(2-n-alkyl-1,3-butadiene)s [37], poly(n-alkylvinyl ether)s and esters [29, 38-41], poly(n-alkyl and acyl styrene)s [42, 43], cellulose esters [44], poly(n-alkyl acrylamide)s [45], and poly(n-alkyl malonamide)s [46, 47], for which a dependence of T_g and T_m on the length of the side chain similar to the one found in [26] was also observed in many cases.

The use of comb-shaped polymers for the preparation of monomolecular films based on PMA and poly(n-alkyl vinyl ether)s (PVE) and some copolymers containing a long-chain comonomer [48, 49] unfortunately did not stimulate detailed investigation of their structure. Fort and Alexander [49]

---

the left correspond to the decrease in T_g and those on the right correspond to the increase in T_m.

*The unsuccessful attempts to prepare oriented samples reported in [28, 29] are apparently due to the low degrees of polymerization of the polymers studied. It was shown [30] that highly oriented samples of comb-shaped PA and PMA can only be obtained with degrees of polymerization greater than 1500-2000.

studied polyvinyl stearate (PVE-17) films on the surface of aqueous solutions of hydrochloric acid and found that the area of the cross section per side chain was 26 Å, which indicated the relatively dense packing of the side chains.

The presence of a strong reflection in the region of 4.2 Å was the only criterion indicating that crystallization of these polymers occurred in a rhombic lattice, as hypothesized.

Brownawell and I Ming Feng [43] studied the structure of low-molecular-weight atactic poly(n-acyl styrene)s with n-acyl chains containing 10, 12, and 14 carbon atoms and atactic poly(hexadecyl styrene) (PS-16)* [the degree of polymerization was 5-9 for the poly(acyl styrene)s and from 16 to 20-25 for PS-16]. According to the findings of the small-angle x-ray study, the existence of large d spacing, whose values are reported below, was found for both isotactic poly(1-octyl ethylene) (PE-8) and poly(1-decyl ethylene) (PE-10):

Poly(n-C_{10}-acyl styrene)	26 ± 1
Poly(n-C_{12}-acyl styrene)	29 ± 1
Poly(n-C_{14}-acyl styrene)	32 ± 1
Poly(n-hexadecyl styrene) (PS-16)	29.0
Poly(1-decyl ethylene) (PE-10)	29.0
Poly(1-octyl ethylene) (PE-8)	23.8

It was suggested that the presence of a large spacing for these polymers is due to the arrangement of the main chains in parallel planes separated by hydrocarbon chains with a planar zigzag conformation (Fig. 1.2). This arrangement of the macromolecules was called mesomorphic. However, a simple geometric calculation and studies of molecular models indicate the impossibility of the structure shown in Fig. 1.2 (if it is assumed that all side chains lie in one plane), since the diameter of a side chain (4.5-4.7 Å) significantly exceeds the distance between neighboring tertiary carbon atoms (2.54 Å) with branches. Nevertheless, the attempt to correlate their appearance with the uniform arrangement of the main chains in the interpretation of the values of the d spacings undoubtedly merited attention, although this point of view also contradicted the hypothesis advanced by Reding

*The number corresponds to the number of carbon atoms in the side n-alkyl chain.

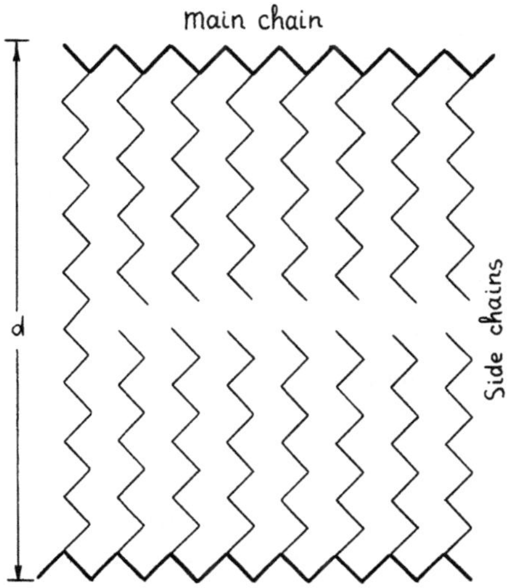

Fig. 1.2. Diagram of the packing of macromolecules of isotactic poly(1-octyl ethylene) based on the data of Brownawell and I Ming Feng [43].

[50] that the crystallization of higher homologs of isotactic poly(1-alkyl ethylene)s only takes place due to packing of the side chains with a random conformation of the main chains.

The absence of model concepts on the structure of comb-shaped polymers which take their molecular parameters into account (length of the side chains, microstructure, flexibility, and the chemical nature of the main chain and linkage, etc.) and the difficulty of estimating the molecular mobility of such chains did not allow any overall hypothesis on the structure and behavior of these macromolecules in the 1960s.

1.2. THE AMORPHOUS STATE

At least two points of view concerning the nature of macromolecular ordering in the amorphous state appeared in the literature some time ago. According to one [51-54, 56-58, 62-64], amorphous polymers were considered to be systems characterized by a distinct short-range order in the arrangement of the polymer chains with ordered aggregates of macromolecules in the form of bundles of chains constructed of molecules stacked in parallel [62, 63], or domains composed of folded polymer chains [52, 55, 64]. Each of these models hypothesized the existence of a certain structural organization of the macromolecules which was confirmed by a number of experimental data.

Another point of view which is particularly supported by the results of neutron scattering analysis in polymers [66, 68] suggests the absence of any order in the arrangement of the polymer chains in amorphous polymers. Recent studies concluded that the size of flexible-chain macromolecules is the same in bulk and in an ideal solution, and most scientists support this concept [66-73].

In view of the specific structural features of comb-shaped polymers, it is natural to assume that the short-range order should be of a slightly different nature than in linear polymers. If this order is determined in linear polymers by the intermolecular interaction of the main chains, then two types of short-range order can be realized for comb-shaped polymers due to both the interaction of the long side chains and to the intermolecular interaction of the main chains. This hypothesis has been completely confirmed experimentally. For example, let us examine the diffraction patterns of amorphous poly(1-alkyl ethylene)s (Fig. 1.3) [74, 75]. As Fig. 1.3 shows, the presence of one diffraction maximum of 4.5 Å corresponding to the distance between the macromolecules is characteristic of linear polyethylene in a melt. The value of this parameter increases to 5.4 Å in going to a poly(1-methyl ethylene) (polypropylene) melt. A further increase in the length of the side substituent in the poly(1-ethylene)–poly(1-octyl ethylene) series is accompanied by an increase in the reflection in the region of small angles of scattering and the appearance of a new maximum which coincides in value with the diffraction maximum of polyethylene.

The x-ray and electron diffraction studies of amorphous poly(n-alkyl acrylate)s (PA), poly(n-alkyl methacrylate)s (PMA), and poly(n-alkyl vinyl ester)s (PVE) with a side chain comprised of 4 to 18 carbon atoms showed [76, 77] that, in addition to the maxima corresponding to the sets of distances between the carbon atoms inside the macromolecules (1.5 and 2.5 Å), there are two more diffraction maxima, d' and d", whose corresponding values for some comb-shaped polymers* as a function of the length of the side chain are shown in Fig.

*The values of d' for most comb-shaped polymers almost coincide.

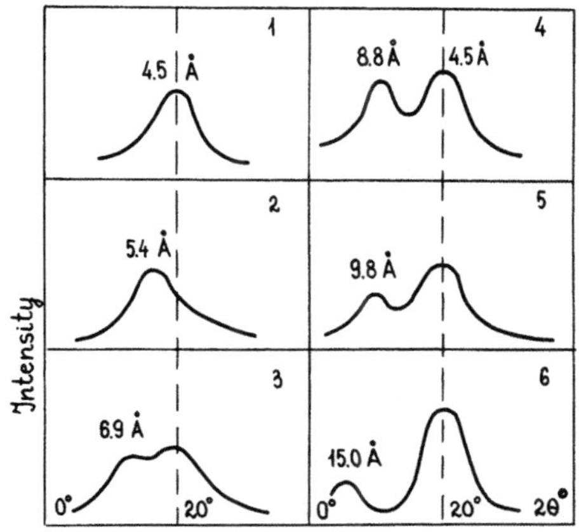

Fig. 1.3. Diffraction patterns of amorphous poly(1-alkyl ethylene)s [74]: 1) polyethylene; 2) poly(1-methyl ethylene); 3) poly(1-ethyl ethylene); 4) poly(1-propyl ethylene); 5) poly(1-butyl ethylene); 6) poly(1-octyl ethylene) [75].

1.4. The maximum corresponding to d' = 4.6-4.7 Å for poly(1-alkyl ethylene)s, PA, and PVE and d' = 4.8-5.0 Å for PMA is almost independent of the length of the side chain of the polymer in the range of n values from 4 to 18. (A slight increase in d' for the lower members of the PA-n series was observed in [78].) This value of d' is similar to the value of the diffraction maximum observed in the x-ray patterns of amorphous melts of polyethylene and n-paraffins and is due to the interaction of the side methylene chains [73, 80].

A similar maximum whose value is not dependent on n is also observed for a series of poly(phenyl methacrylic ester)s of n-alkoxybenzoic acids (PPMOB) [81, 82] with the general formula*

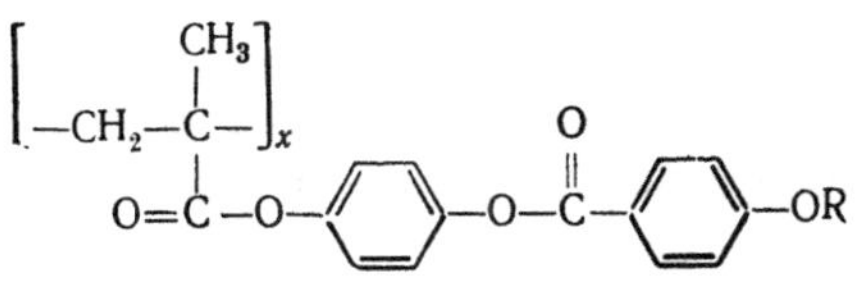

*It should be noted that one of the polymers of the PPMOB series with n = 16 can also exist in the liquid-crystalline state (see Chapter 4).

where R = C_3H_7, C_9H_{19}, $C_{16}H_{33}$, and poly(N-alkyl maleimides) (P-N-MI) [83]

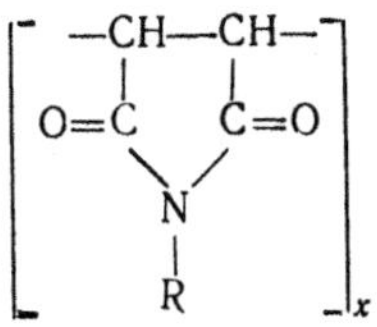

where R = C_8H_{17}, $C_{10}H_{21}$, $C_{14}H_{29}$.

The value of the second diffraction maximum α'' observed in the small-angle region, as Fig. 1.4 shows, is essentially dependent on the length of the side chain of the polymer. This type of change in d'' reflects the specific features of the structure of all comb-shaped polymers.

The increase in d'' with increase in n forms the basis for correlating the reflections observed in the x-ray patterns primarily with the length of the side chains. Since the dependence of d'' on n is linear in nature, as Fig. 1.4 shows, it is natural to consider the possibility of layered ordering of the methylene chains, i.e., parallel packing of the side chains, with their ends (methyl groups and main chains) also arranged in a certain order which causes the appearance of "spacing" in the x-ray patterns. The value of d'' can serve as a distinctive criterion of the conformational structure of the methylene side chains and the value of d_0, obtained by extrapolation of d'' to n = 0, provides information on the diameter of the main chain (including the linkage segment of the side chains to the main chain).

The slight arbitrariness and limitation of this examination should naturally be taken into consideration, since this does not involve a crystalline polymer with precisely established positioning of atoms in the crystal cell, but an amorphous substance characterized by the presence of local order alone. Nevertheless, it is possible to evaluate comparatively the conformational state of the main and side chains of comb-shaped polymers in the amorphous state.

The values of the parameters of the following equation

$$d'' = d_0 + a n \qquad (1.1)$$

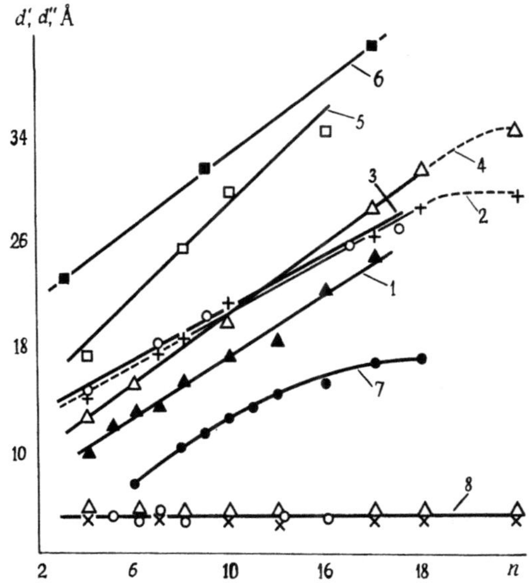

Fig. 1.4. Dependence of the maxima d' (8) and d'' (1-7) on the number of carbon atoms in the alkyl portion of the side chain for polymers in the amorphous state: 1) isotactic poly(1-alkyl ethylene)s [74]; 2) PA-n; 3) PV esters; 4) PMA-n; 5) P-N-MI-n [82]; 6) PPMOB-n; 7) brominated n-alkanes [84].

calculated for some comb-shaped polymers with n = 4-18 are reported in Table 1.1. In this equation, d_0 corresponds to the so-called effective diameter of the main chain, including the linkage of the alkyl group, a is the slope of the line corresponding to an increment in the length of the side chain per CH_2 group (a = 1.27 Å for the methylene chain in the planar zigzag conformation), and n is the number of carbon atoms in the alkyl radical.

It is important to emphasize that the calculation of the parameters of Eq. (1.1) was based on the values of small-angle reflections d'' in the entire range of n values. This means that for higher crystallizable polymer homologs, the values of d'' corresponding to the small-angle reflections observed in the x-ray patterns of their melts were used, but the change in the values of d'' with the temperature was not considered for the lower homologs. In addition, despite the

TABLE 1.1. Parameters d_0 and a in Eq. (1.1) for Amorphous Comb-Shaped Polymers

Polymers	d_0, Å [85]	d_0, Å [78]	a [85]	a [78]
Isotactic poly(1-alkyl ethylene)s	5.13	6.20	1.18	1.09
Polymethacrylates (PMA)	6.90	9.80	1.27	0.73
Polyacrylates (PA)	10.30	4.40	1.01	2.0
Polyvinyl esters (PVE)	10.68	11.90	0.98	0.89
Poly(N-alkyl maleimide)s (P-N-MI)	10.85	7.60	1.76	1.48
Poly(phenyl methacrylic ester)s of n-alkoxybenzoic acids (PPMOB)	18.80	–	1.38	–

"scattering" of the experimental points in Fig. 1.4, which makes it possible to draw two lines within one homologous series (for the lower and higher homologs), extrapolation to n = 0 was effected with one line. This is apparently the reason for the slightly "exaggerated" values of d_0 for some polymers.

The values of d_0 and a for these and other polymeric homologous series of comb-shaped polymers were calculated in [78] using the same approach. These calculated values are also reported in Table 1.1. Despite the noncoincidence of a number of values of d_0 and a for the same polymers due to estimation methods, virtually the same conclusion was arrived at in both studies, indicating the presence of some "regularity" and the existence of local order in amorphous polymers with side chains.

As Fig. 1.4 and the data in Table 1.1 indicate, layered packing in which side chains with a stretched conformation (a = 1.18 Å and 1.27 Å, respectively) are arranged parallel to each other is apparently characteristic of amorphous isotactic poly(1-alkyl ethylene)s and PMA. The main chains of the macromolecules are located at distances approximately equal to the length of the side chain (single-layer packing), and the values of d_0 obtained are only slightly higher than the corresponding calculated values. The significantly larger differences in the values of a (decreasing in comparison to the value of 1.27 Å) and the elevated values of d_0 observed for PA and PVE are apparently related to the greater "conformational freedom" in the segments of their linkage to the main chain, which causes the appearance of defects (kinks, twisting, etc.) [85]. On the one hand, this should result in an increase in the effective diameter of the main chain; on the other hand, it could cause a slight deviation in the conformation of the side chain from the zigzag conformation, which would decrease the value of a. This "conformational freedom" is not realized in the case of poly(1-alkyl ethylene)s due to the significantly higher potential barrier to rotation of the segments of the main chain around the C—C bond (in comparison to PA and PVE) and, in the case of PMA, due to steric hindrance superimposed by the α-methyl group in the main chain, which prevents free rotation around the C—O bond.

For such polymers as PPMOB and P-N-MI, which have a relatively complex and rigid linkage, the experimentally determined [using Eq. (1.1)] value of d_0 also exceeds the calculated value, and the value of a significantly exceeds 1.27 Å. This means that the increased stiffness of the linkage, which causes even stronger distortion of the conformation of the comb-shaped macromolecule in the segments adjacent to the main chain (in comparison to PA and PVE), prevents convergence of the main chains at a distance equal to the length of the side chain ($a > 1.27$ Å). Despite the layered packing of the side chains, the values of d" obtained in this case (PPMOB and P-N-MI) correspond to some average distances which are greater than the length of one side chain.

Nevertheless, in both the first and second cases, the main chains of comb-shaped polymers in the amorphous state are arranged in a certain order with respect to each other at distances close or equal to the length of the side chains, which are oriented perpendicularly to the longitudinal axis of the macromolecule.

The presence of long methylene side chains, which are analogs of low-molecular-weight paraffins, in comb-shaped macromolecules results in a similarity between the structure of a comb-shaped polymer in the amorphous state and the structure of a mixture with a melt of a low-molecular-weight compound located in the polymer matrix as one of the components.

Unfortunately, the number of studies of the structure of melts of low-molecular-weight, long-chain organic compounds is very limited [86-88]; in particular, there are no data on the structure of melts of esters of higher fatty acid homologs which are low-molecular-weight analogs of comb-shaped polymers. However, if n-alkanes are examined for comparison, we find [86-89] that the molecules of liquid n-paraffins are packed parallel to each other, similar to cylinders of rotation; the average distance between their axes is 4.6-4.8 Å, and the number of nearest neighbors is approximately six, i.e., the packing is nearly hexagonal. No small-angle spacing is observed in melts of pure n-paraffins [79], although a small-angle diffraction maximum whose value increases in going to the higher homologs [84] (cf. Fig. 1.4, curve 7) is observed in the x-ray patterns of melts of n-paraffin derivatives containing bromine atoms at the ends of the molecules. The existence of small-angle spacing is also observed for such n-paraffin derivatives as liquid and n-aliphatic alcohols [89] and acids [90].

The presence of this spacing is usually correlated with a certain order in the liquid due to the formation of cybotactic groups [79] or clusters [87, 91] bound by the same forces of interaction which bond molecules in crystals. In the case of comb-shaped polymers in the amorphous state, the packing of the side chains is apparently highly similar to the hexagonal packing of low-molecular-weight paraffins. The main chain makes an additional contribution to the formation of layered ordering observed even for poly(1-alkyl ethylene)s containing no polar groups. As Fig. 1.4 shows, the dependence of d" on n for comb-shaped polymers is almost linear, while a deviation from linearity is observed for low-molecular-weight compounds, for example brominated paraffins, when $n \geq 10$-12. This indicates the deviation of the conformation of the paraffin chains from the planar zigzag conformation and bending or kinking of the methylene chains which begins when the length exceeds 10-12 carbon atoms. This is in good agreement with the data in [92] on the study of the conformational structure of gaseous and liquid alkanes with $n \geq 5$, for which the possibility of significant bending (kinking) of the molecules has been established.

In the case of comb-shaped macromolecules, a slight change in the linear dependence of d" can be found only with significantly larger values of n ($n \geq 20$), indicated by the broken lines in Fig. 1.4. The presence of a main chain in comb-shaped polymers which ensures a relatively high level of cooperation in the interaction of the side chains places certain limitations on their conformational arrangement (in comparison to their low-molecular-weight analogs). In this sense, it can be conjectured that branched polymers are convenient models for studying the structure and conformational features of normal hydrocarbon molecules and their derivatives in which the main chain can be considered a unique restriction of the possible degrees of freedom of small molecules.

The presence of layered ordering of the methylene side chains is thus a distinctive feature of the structure of all comb-shaped polymers in the amorphous state. The existence of short-range order in the arrangement of the methylene side chains indicates that each methylene chain can vibrate in a "characteristic" cell formed by its nearest neighbors. Such ordered regions are also structural elements in which crystallization of the side chains takes place when they reach a certain critical length, as we will show in the next chapter.

1.3. THE CRYSTALLINE STATE OF ISOTACTIC POLY(1-ALKYL ETHYLENE)S, POLY(1-ALKYL ETHYLENE OXIDE)S, AND POLY(ALKYL ALDEHYDE)S

As shown above, contradictory opinions concerning the possibility and method of crystallization were often advanced in the early studies of the structure of comb-shaped polymers. Most investigators believed that crystallization was possible when the side chain was longer than a certain length and the main chain had a random-coil conformation.

The contradictions in the opinions concerning the methods of packing of comb-shaped macromolecules were significantly resolved by the detailed x-ray studies of Turner-Jones et al. of the structure of a number of isotactic poly(1-alkyl ethylene)s [93-95] and a series of studies by Kargin et al. on the structure and molecular mobility of comb-shaped polymers of the acrylic and vinylic series [85, 96-106]. The results of these studies stimulated the reexamination of the concepts of comb-shaped polymer structure and essentially resulted in new interest directed toward the study of processes of structure formation in branched polymer systems. Studies of comb-shaped polymers were subsequently most actively developed by Italian (Magagnini et al. [107-113]) and Spanish (Barrales-Rienda et al. [83, 114-116]) investigators. Let us examine the basic results of the structural studies of polymers in the homologous series of poly(1-alkyl ethylene)s [93-95, 117-122].

TABLE 1.2. Crystallographic Parameters of Isotactic Comb-Shaped Poly(1-Alkyl Ethylene)s (PE-n)*

Polymer	Type of coil†	Unit cell parameters			Type of coil	Density, g/cm³		Melting point, °C
		a, Å	b, Å	c, Å		crystal-line	amor-phous	
Poly(1-butyl ethylene) (PE-4) [94]	O or	11.7	26.9	13.7	7_2	0.91 (expt)		
	M	22.2	8.9	13.7 γ = 94.5°	7_2	0.73 (expt)	0.80 (expt)	−55
Poly(1-pentyl ethylene) (PE-5) [94]	Modif. I, M or T			6.45	3_1			17
Poly(1-hexyl ethylene) (PE-6)								10 [121]
Poly(1-heptyl ethylene) (PE-7) [94]	Modif. I							17-18 22 [121]
Poly(1-octyl ethylene) (PE-8) [94]	Modif. I			13.2				34‡ [117, 121]; 40 [95]; 29
Poly(1-nonyl ethylene) (PE-9)								36 [121]
Poly(1-decyl ethylene) (PE-10) [94]	Modif. I			13.2	3_1 [121]			49; 20, 33 [120]; 46 [117]; 49 [121]
Poly(1-dodecyl ethylene) (PE-12) [94]	Modif. I, T(?) Modif. II, O	7.5	56.0	6.7	4_1 [121]			57; 31, 54 [120]; 52-55 [117]; 3, 54 [121]

Poly(1-tridecyl ethylene) (PE-13)							13, 59 [120]
Poly(1-tetradecyl ethylene) (PE-14)	Modif. I, T(?) Modif. II, O	7.5	63.2	6.7			67.5 47, 63 [120]; 58-63 [117]; 28, 63 [121]
Poly(1-hexadecyl ethylene) (PE-16)	Modif. I, T(?) Modif. II, O	7.5	70.4	6.7	0.95 [95]	0.905 0.86 [118]**	71 [94, 117]; 41, 68 [118]
	Modif. H [123]	4.24	4.24		0.91 [118]		60, 74 [120]; 43, 71 [110]
Poly(1-octadecyl ethylene) (PE-18) [107, 110]							80, 85 [107]
Poly(1-eicosyl ethylene) (PE-20)							55, 80 [119]††; 74 [119]‡‡; 69, 92 [121]

*The number in the abbreviation of the samples of poly(1-alkyl ethylene)s (n) corresponds to the number of carbon atoms in the aide chain of the polymer.

†O: orthorhombic; M: monoclinic; T: triclinic; H: hexagonal modifications.

‡The T_m do not differ by more than 20°C in the series of PE-8 and PE-16 for the same polymer in modifications I and II according to [94]; the maximum values of these temperatures according to the data in [94] are reported in the table.

**Obtained by extrapolation of the dependence d = f(T) from 80-100°C to 20°C.

††Crystallization from melt.

‡‡Crystallization from solution.

The x-ray studies of Turner-Jones [94] in the isotactic poly(1-butyl ethylene) (PE-4) to poly(1-hexadecyl ethylene) (PE-16) series showed that when the side chain is 4-10 carbon atoms long, the polymers are rubbery substances at room temperature and only crystallize when held for a long time at low temperatures (from −40 to −20°C), and even then the degree of their crystallinity is low. The macromolecules of poly(butyl ethylene) (PE-4), which have a helical conformation of the main chains, are packed in the crystal in an orthorhombic or monoclinic cell whose parameters are reported in Table 1.2. Poly(1-pentyl ethylene) (PE-5) can be obtained in two different crystalline modifications corresponding to oriented and unoriented samples. However, based on the x-ray data obtained, it is not possible to determine the parameters of the unit cell, whose symmetry corresponds to a monoclinic or triclinic system according to [94] (see Table 1.2). Poly(1-hexyl ethylene) (PE-6) is amorphous and does not crystallize even at a low temperature, while poly-(1-heptyl ethylene) (PE-7) can be transformed into the crystalline state after holding for 1 week at −20°C.

The side chains also begin to participate in the formation of the crystalline phase together with the main chains when the length of the side chains reaches 10-12 carbon atoms [poly(1-decyl ethylene), poly(1-dodecyl ethylene)]. A further increase in the length of the side chains has a determining effect on the structure of poly(1-alkyl ethylene)s. The higher homologs of isotactic poly(1-alkyl ethylene)s form two crystalline modifications dependent upon the conditions of heat treatment. Modification I is obtained by fast cooling of melts of the polymers to 0°C with subsequent annealing for 1 h in the range between 30°C and a temperature 5°C higher than their melting point. Modification II is formed either by heating form I at a temperature slightly below their melting point, or by slow cooling of a melt of the polymer.

In the opinion of Turner-Jones, the presence of large d spacings in the x-ray patterns of poly(1-alkyl ethylene)s for both crystalline modifications, whose values increase linearly with increasing sidechain length as Fig. 1.5 shows, refutes Reding's hypothesis [50] that crystallization of the polymers takes place only due to the side chains, which was advanced to explain the behavior of poly(1-hexadecyl ethylene) (PE-16).

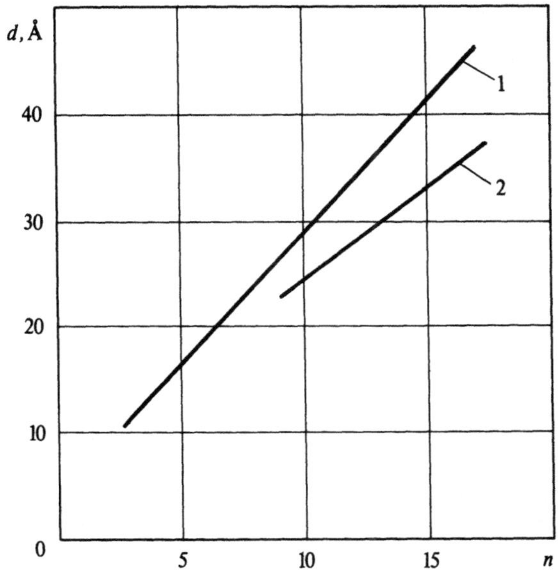

Fig. 1.5. Dependence of the d spacing of isotactic poly(1-alkyl ethylene)s in crystalline modifications I (1) and II (2) on the number of carbon atoms in the alkyl side chain [94].

Based on the results of the x-ray studies, Turner-Jones proposes the following packing schemes for comb-shaped macromolecules (Fig. 1.6). In modification I polymers, which all form isotactic poly(1-alkyl ethylene)s beginning with PE-5 (see Fig. 1.5, curve 1), the side chains are at right angles with respect to the main chain (whose conformation can be helical), and the value of the experimentally found d spacing is in good agreement with the calculated values (Table 1.3 and Fig. 1.6a). It has been proposed that the crystallization of the main chains, which are located in parallel planes due to their helical conformation, plays a major role in the formation of crystalline modification I; this type of packing to some degree corresponds to the triclinic form of polyethylene.

An orthorhombic cell in which the main chain has the conformation of a triple or quadruple helix corresponds to the polymers in modification II, which have a higher degree of crystallinity but a lower value for the d spacing (see Tables 1.2 and 1.3); the side chains are tilted with respect

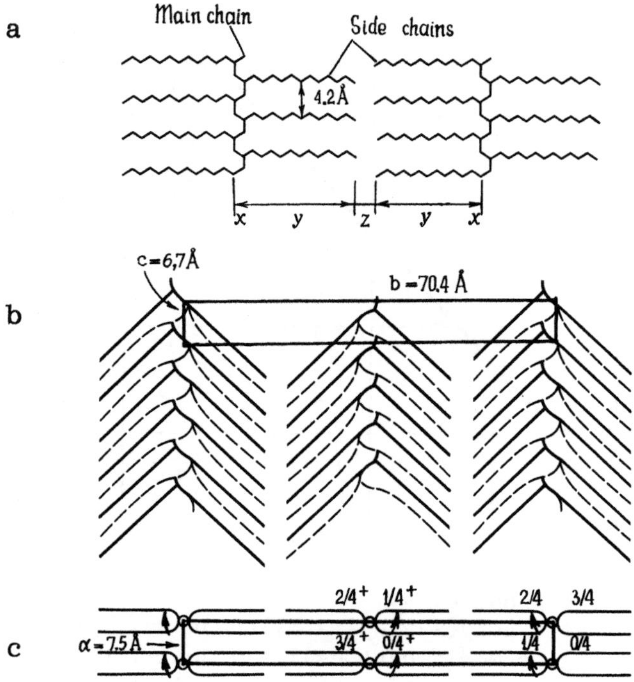

Fig. 1.6. Diagrams of the assumed packing of macromolecules of isotactic poly(1-alkyl ethylene)s in a crystal in modifications I (a) and II (b, c). Figure 1.6b shows projections of the orthorhombic cell containing two quadruple helices with tilted side chains: b) projection on axis c; c) projection on axis *a* [94].

to the axis of the helix by an angle of 130.6°, and they are packed together with the main chains in the form of the orthorhombic cell of polyethylene (see Fig. 1.6b, c).

We note that, in addition to the crystalline forms indicated for PE-14 and PE-16, a partially ordered hexagonal structure is obtained when melts of these polymers are rapidly cooled. In the opinion of Turner-Jones, the formation of this modification is due to the almost parallel arrangement of the hydrocarbon chains in the absence of any ordered arrangement of the main chains, which do not participate in crystallization. Isotactic poly(1-alkyl ethylene)s with shorter side chains do not form this modification. The hy-

TABLE 1.3. Calculated* and Experimental Values of the d Spacings of Isotactic Poly(1-Alkyl Ethylene)s in Modifications I and II [94]

Polymer	Maximum calculated value of d	Experimental value of d	
		modification I	modification II
Poly(1-hexadecyl ethylene) (PE-16)	45.3	44.5	35.2
Poly(1-tetradecyl ethylene) (PE-14)	40.3	39.5	31.6
Poly(1-dodecyl ethylene)(PE-12)	35.2	34.8	28.0
Poly(1-decyl ethylene)(PE-10)	30.1	29.5	–
Poly(1-octyl ethylene)(PE-8)	25.0	25.4	–
Poly(1-heptyl ethylene)(PE-7)	22.5	21.5	–
Poly(1-pentyl ethylene)(PE-5)	17.4	16.2	–

*The calculated value of the d spacing reported above was obtained using the equation $d = x + 2y + z$ (see Fig. 1.6a), where $x = 1.4$ Å or has a smaller value as a function of the actual conformation of the main chain, y is the length of the side chain of the polymer, and z is the distance between the ends of the side chains of neighboring macromolecules, equal to 3.3 Å.

pothesis of Turner-Jones concerning the participation of the main chain of isotactic poly(1-alkyl ethylene)s in crystallization is in agreement with the findings of Brownawell and I Ming Feng [43], who also studied the structure of isotactic PE-8 and PE-10 (see p. 13). However, in the opinion of the latter [43], the main chains cannot be assigned a helical conformation; those of the polymers studied are located in parallel planes and have a planar zigzag conformation (see Fig. 1.2). Despite the fact that the values of the d spacings obtained by Brownawell and I Ming Feng coincide almost precisely with the findings of Turner-Jones for modification I, their opinions differ with respect to the actual conformation of the main chain.

The opinion of Turner-Jones concerning the coiling of the main chain in the formation of crystalline modifications I and II is apparently correct, although the similar values of the melting points of these modifications should be noted. According to the data in [94], the melting points for the same polymer in modifications I and II in the PE-8 to PE-16 series differ by no more than 2°C; the highest melting points are reported in Table 1.2. Similar melting points of the

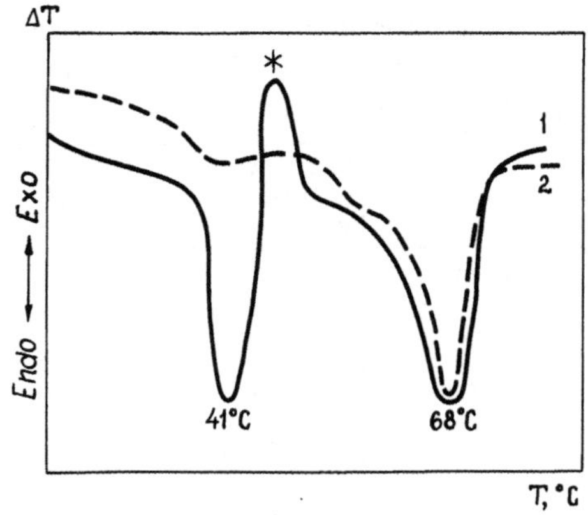

Fig. 1.7. Thermograms of poly(1-hexadecyl ethylene) (PE-16) [118]: 1) quenched sample obtained by cooling a melt from 100 to 0°C; 2) same sample after annealing for 4 h at 50°C (heating rate of 10°C/min in both cases).

corresponding isotactic poly(1-alkyl ethylene)s were also found in [117] (see Table 1.2). However, two endothermic transitions were found in subsequent thermographic studies of poly(1-hexadecyl ethylene) (PE-16) by Aubrey and Barnatt [118] and in a series of studies by Hummel and Holland-Moritz et al. [119-122] simultaneously using radiography, IR spectroscopy, and differential scanning calorimetry (DSC) for studying isotactic poly(1-alkyl ethylene)s (PE-12 to PE-20).

The hexane-soluble fraction of PE-16 has one first-order transition in the region of 41°C, while the fraction of the same polymer which is not soluble in hexane has two endothermic transitions at 41.5 and 68°C; the same transitions are also observed in the thermogram of a quenched sample which is not soluble in hexane (Fig. 1.7, curve 1) [118]. Annealing of the quenched sample results in the total disappearance of the first peak (Fig. 1.7, curve 2). It was concluded that the sample of PE-16 with a lower melting point is atactic, and only the side chains are involved in the crystallization of this polymer, while the fraction of the polymer with T_m = 68°C is a stereoregular polymer and crystallizes with the participation of both the main and side chains. However, it is difficult to explain the transition from thermogram 1 to thermogram 2 in Fig. 1.7 with this hypothesis, since the microstructure of the polymer does not

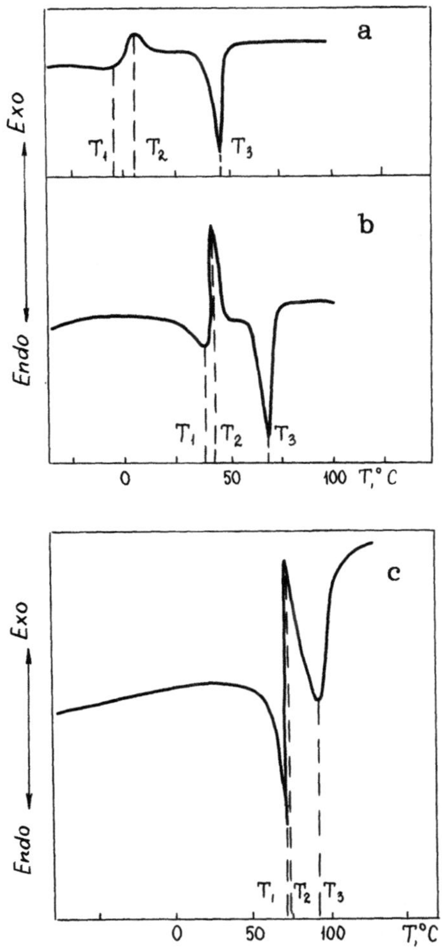

Fig. 1.8. Thermograms of quenched samples of isotactic poly(1-alkylethylene)s [122]: a) PE-10; b) PE-14; c) PE-20.

change during annealing. Unfortunately, there are no additional experimental data in the study to confirm the hypothesized existence of fractions with different microstructures in the samples studied.

The events observed in [118] can be interpreted in another way in light of the results obtained by Hummel et al. [120-122]. There are also two endothermic peaks T_1 and T_3 and one exothermic peak T_2 in the thermograms of all quenched samples of isotactic poly(1-alkyl ethylene)s (PE-

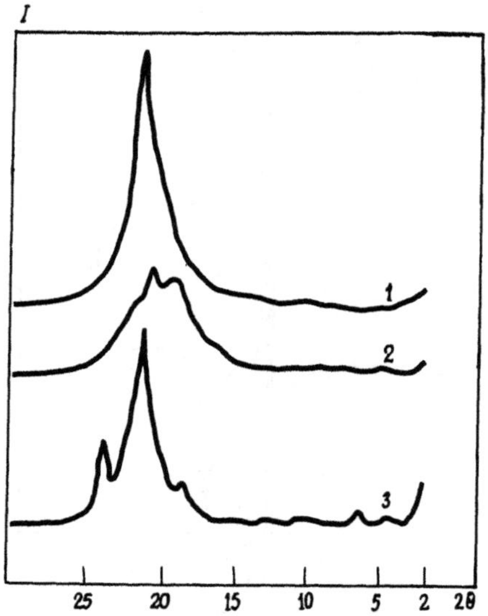

Fig. 1.9. Diffraction patterns of a sample of PE-20 at different temperatures [122]: 1) quenched sample, 17°C; 2) same sample at 82°C; 3) same sample after annealing for 4 h at 82°C.

10, PE-11, PE-12, PE-14, and PE-20) (Fig. 1.8) ([122], also observed in [118]; see Fig. 1.7),* but these were not taken into consideration in the discussion of the results [107]. In studying the x-ray patterns of samples of poly(1-alkyl ethylene)s made at temperatures corresponding to different sections of the thermogram, the mechanism of the structural transitions was convincingly revealed [122]. The initial structure of the quenched samples of poly(1-alkyl ethylene)s shown in Fig. 1.9 (curve 1) on the example of PE-20 in all probability corresponds to the partially ordered structure of the polymer observed by Turner-Jones for PE-14 and PE-16 in sudden cooling of their melts.

The presence of one maximum in Fig. 1.9 corresponding to $2\theta = 21.1°$ (4.21 Å) suggests that crystallization of the polymer in a hexagonal cell takes place during quenching due to the dense packing of the methylene side chains positioned parallel to each other. The first endothermic peak in the

*This exothermic peak is indicated in the thermogram by an asterisk.

thermograms, T_1 (see Fig. 1.8), corresponds to the melting of this crystalline modification. The polymer is subsequently transformed into a more stable form, to which the exothermic peak in the thermogram at temperature T_2 corresponds. The diffraction patterns of the samples made in the region of T_2-T_3 are characterized by the appearance of new crystal reflections, in particular a reflection of 3.75 Å (see Fig. 1.9, curve 2), which corresponds to crystallization of the side chains of the polymer in an orthorhombic cell with a helical conformation of the main chain [120]. The detailed study of the IR and Raman spectra [119-121, 124] of isotactic poly(1-alkyl ethylene)s also indicates that, beginning with PE-12, crystallization of the side chains in form II (according to the Turner-Jones classification) takes place, similar to the orthorhombic structure of PE, with two chains in the cell. Melting of this stable modification takes place at temperature T_3, which is close to the melting points of samples of poly(1-alkyl ethylene)s determined in [94, 95] (see Table 1.2).*

To determine the relationship between the microtacticity, structure, and thermal properties of samples of poly-(1-alkyl ethylene)s, Italian scientists headed by Magagnini compared the data examined above with the results of their detailed studies of three samples of polymers from this homologous series: poly(1-decyl ethylene), poly(1-tetradecyl ethylene), and poly(1-octadecyl ethylene) [107-111]. The results of these studies are most clearly shown with poly(1-octadecyl ethylene) (PE-18). After synthesizing and carefully separating a number of samples of PE-18 of varying microtacticity, it was convincingly shown in [108, 109-111] that totally atactic and isotactic samples have a different structure and different melting points (Fig. 1.10, curves 1 and 5); they do not combine when stirred, and they melt differently.

Since macromolecules with different degrees of tacticity are usually formed during synthesis, a polymer which is essentially a mixture of atactic and isotactic fractions (and possibly with the same stereoblock structure) is formed as a result.

---

*When a polymer has two melting points, the first value reported in Table 1.2 corresponds to T_1 and the second corresponds to T_3.

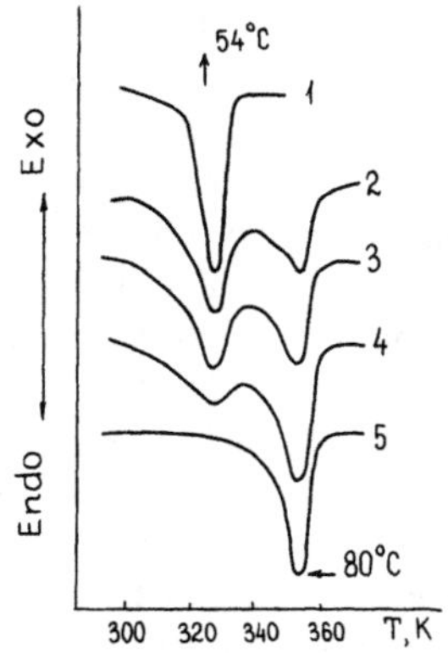

Sample	Tacticity of samples (^{13}C NMR), %	
	Atactic	Isotactic
1	100	-
2	63	37
3	50	50
4	33	67
5	5	95

Fig. 1.10. DSC curves of samples of poly(1-octadecyl ethylene) of different tacticity [108].

The thermal and structural data clearly show that distinct phase separation takes place in such systems; as a result, there are at least two types of structures, which have also been revealed in calorimetric studies (Fig. 1.10, curves 2-4). The results of the studies by Magagnini clearly demonstrate the significant role of the evaluation of the microtacticity of the macromolecules in the determination of the structure and thermal properties of comb-shaped poly(1-alkyl ethylene)s.

The significant difference in the melting points (especially T_1) found for the same polymer by different investigators is apparently due to the above; this clearly follows from a comparison of the melting points of poly(1-alkyl ethylene)s with side chains of the same length (see Table 1.2, Fig. 1.11).

This "difference" in the melting points is due to at least three causes: the use of samples with a different microstructure, the use of different experimental conditions of crystallization, which play a significant role in the preference of formation of one crystalline modification or another as indicated above, and the use of different methods of estimating T_m. The melting points of isotactic poly(1-alkyl ethylene)s as a function of the length of the side chains, taken from different studies, are shown in Fig. 1.11.

The greatest difference in melting points with the same n value is observed for the lower homologs of poly(1-alkyl

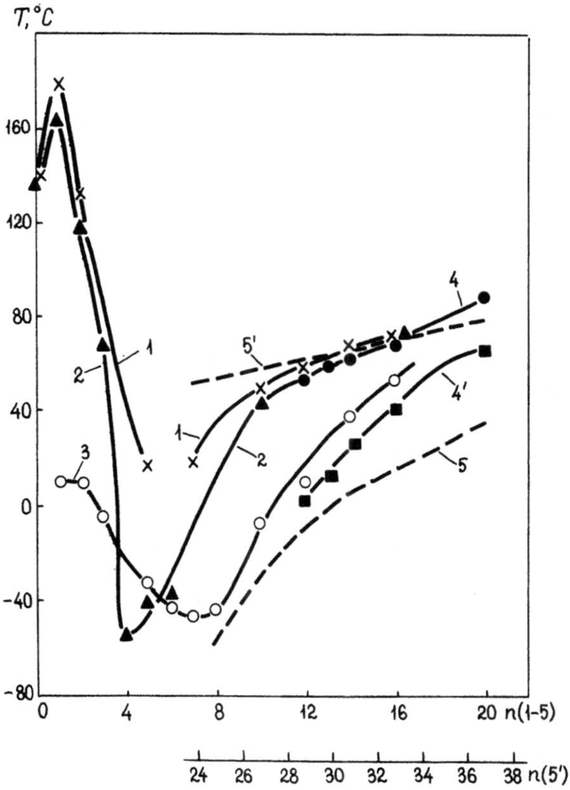

Fig. 1.11. Dependence of melting points T_1 and T_3 and the glass transition temperature T_g of comb-shaped isotactic poly(1-alkyl ethylene)s on the length of the side chains according to different data: 1) T_3 [93]; 2) T_3 (beginning with n > 10) [50]; 3) T_g [93]; 4) T_3 [122]; 4') T_1 [122]; 5, 5') melting points of n-alkanes [126].

ethylene)s which crystallize in modification II, and usually at low temperatures which complicates the precise estimation of their melting points (left branches of curves 1 and 2). The melting points of the higher homologs with side chains of the same length forming crystalline modification II are very similar according to the various data, which can be seen by a comparison of the right-hand branches of curves 1, 2, and 4. According to [93], curve 3 expresses the dependence of the temperature of the so-called α-relaxation process (T_α) on the length of the side chain. The α-relaxation process is usually correlated with the segmental mobility of the macromolecules, and temperature T_α is determined as the glass transition temperature, as in the study by Clark et al. [93] (see Chapter 2 for more detail). In our opinion, the right-hand part of curve 3 corresponds to the change in melting points of the partially ordered hexagonal packing of isotactic poly(1-alkyl ethylene)s, i.e., it reflects the change in T_1 [58] as a function of the length of the side chain.

The results of the study of comb-shaped polymers of the simplest structure, isotactic poly(1-alkyl ethylene)s, thus indicates the relatively complex character of the structural organization of branched macromolecules in the crystalline state. The presence of two types of structural units in each macromolecule, the main chain and long side chains, is responsible for the dual nature of comb-shaped poly(1-alkyl ethylene)s, which reflect the properties of both high-molecular-weight and low-molecular-weight compounds. When the length of the side chains is no greater than 3-4 methylene groups, the tendency of the isotactic main chain to assume a helical conformation predominates and, as a result, both the main and the side chains are helically arranged and are in the crystal lattice. In this case, the crystallization of branched poly(1-alkyl ethylene)s takes place according to the same principles of closest packing observed in the crystallization of linear polymers.

With an increase in the length of the side chain, the three-dimensional packing which could have been constructed from parallel packed helices inevitably becomes increasingly loose. For this reason, at some critical chain length n_{cr}, the polymer may stop crystallizing or its crystallization will be significantly hindered.* Going above the critical length of the side chains will create increasingly favorable conditions for their crystallization due to an increase in the energy of interaction, and comb-shaped polymers begin to crystallize in one of three crystalline modifications (modification I, II, or a partially ordered hexagonal structure). According to Turner-Jones [94], crystal-

*According to Turner-Jones, in the series of isotactic poly(1-alkyl ethylene)s n_{cr} corresponds to 6 carbon atoms (PE-6); however, this polymer totally crystallized when held for 14 days at −20°C [121].

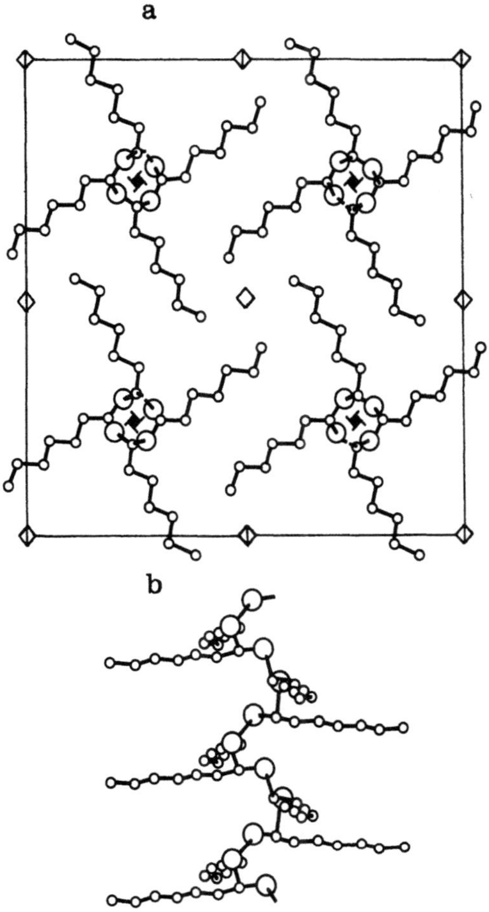

Fig. 1.12. Diagram of the macromolecular packing of isotactic polyheptaldehyde in the unit cell (a) and projection of the macromolecules on a plane parallel to its long axis (b) [129].

line modification II, which also corresponds to the highest melting points, is the most stable modification (curves 1, 2, and 4 in Fig. 1.11).

The theoretical calculations of the possible conformations of isotactic poly(1-alkyl ethylene)s performed using atom–atom potentials led to a similar conclusion [125].

A comparison of the melting points of isotactic poly(1-alkyl ethylene)s with those of n-alkanes (cf. Fig. 1.11,

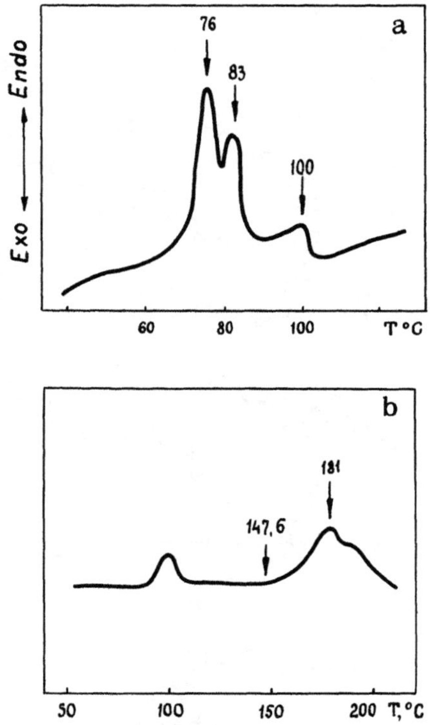

Fig. 1.13. DSC curves of two samples of isotactic polyheptaldehyde obtained with a heating rate of 20°C/min (1) [129] and 80°C/min (b) [127].

curves 5 and 5'), which are analogs of the side chains of comb-shaped poly(1-alkyl ethylene)s, showed that the T_m of paraffins of the same length as the side chains of poly(1-alkyl ethylene)s are significantly lower than the melting points of the latter. At the same time, the melting points of poly(1-alkyl ethylene)s beginning with $n \geq 10$ (right-hand branches of curves 1, 2, and 4) coincide with the melting points of n-alkanes, whose molecular length is twice as long as the length of the side chains (curve 5'). The mechanism found correlates well with the results of the structural studies by Turner-Jones which indicate that the higher homologs of poly(1-alkyl ethylene)s (see Table 1.2) crystallize due to packing of the side chains in two layers of an orthorhombic cell (see Fig. 1.6b), since this is also observed for low-molecular-weight n-alkanes. Consequently, the intensified interaction of the methylene chains with increasing length determines the structure of the higher ho-

TABLE 1.4. Melting Points (T_m), Heats of Fusion (ΔH_m), and Entropy of Fusion (ΔS_m) of Atactic and Isotactic Samples of Polyalkyl Oxiranes [110]

Polymer and its microtacticity	T_m, °C	ΔH_m, kJ/mole	ΔS_m, J/mole·°C
Poly(1-hexadecyl ethylene oxide) (PEO-16)			
atactic	57	25.53	77.35
isotactic	82	25.74	68.65
Poly(1-octadecyl ethylene oxide) (PEO-18)			
atactic	65	32.44	95.98
isotactic	105	33.90	89.70
Poly(1-eicosyl ethylene oxide) (PEO-20)			
atactic	70	35.58	103.72
isotactic	107	39.68	104.44

mologs of poly(1-alkyl ethylene)s to a significant degree, "imposing" a conformation on the main chain which corresponds to the energetically advantageous packing of the methylene chains to the greatest degree. In this case, the role of the main chain is "suppressed" to a significant degree by the interaction of the side chains, and the physical properties of the polymer in the crystalline state differ little from those of low-molecular-weight crystalline paraffins.

Poly(alkyl ethylene oxide)s (so-called polyalkyl oxiranes) have a crystal structure similar to that of poly(1-alkyl ethylene)s [107, 110, 112-113]

$$[-CH_2-\underset{\displaystyle (CH_2)_n-CH_3}{\underset{|}{CH}}-O-]$$

where n = 15, 17, 19.

A comparison of the melting points for fractions of these polymers enriched with isotactic and atactic triads reveals their significant difference (Table 1.4). As a comparison of the data for poly(1-alkyl ethylene)s and poly(alkyl ethylene oxide)s with different side-chain lengths (cf. Fig. 1.11 and Table 1.2) shows, isotactic poly(alkyl oxirane)s are characterized by the highest melting points. The formation of the same type of crystal structure as for isotactic poly(1-alkyl ethylene)s with tilted side chains (~120°) with respect to the axis of the main chain has been proposed for them [107, 110]. At the same time, as Table 1.4 shows, the heats of fusion of isotactic and atactic samples of polyalkyl oxiranes differ only slightly (this will be discussed in more detail below).

In comparing the physicochemical properties of homologs of alkyl ethylenes and alkyl ethylene oxides, Magagnini et al. suggested that the crystal structure of isotactic comb-shaped polymers is primarily determined by the nature of the main chains, while the properties and structure of atactic polymers are a function of the nature and length of the side chains. This conclusion is confirmed by the fact that isotactic samples of two different homologous series of polymers do not mix with each other or with atactic samples. At the same time, atactic samples of poly(1-octadecyl ethylene) and poly(1-octadecyl ethylene oxide) are easily co-crystallized, exhibiting melting points between those of the corresponding homopolymers.

A different type of structural organization is advanced for explaining the crystallization of higher isotactic polyaldehydes. The most detailed, although not totally unambiguous, data were obtained in studying the structure of isotactic polyheptaldehyde [127-219]:

$$\begin{array}{l}—CH—O— \\ \;\;\;| \\ (CH_2)_5—CH_3\end{array}$$

According to [129], this polymer, like the previously studied polyaldehydes with shorter side groups (polypropionaldehyde and polybutyraldehyde), crystallize in a tetragonal cell (space group $14_1/a$; a cell contains 16 monomeric units) (Fig. 1.12). The dual behavior of this polymer during crystallization and melting was expressed in two regions of transitions at 75-105 and 147°C (Fig. 1.13). The thermograms of polyheptaldehyde are characterized by the presence of three sharp endothermic peaks (Fig. 1.13a) whose position and area ratio are a function of the conditions of treatment of the polymer. The existence of these endothermic transitions is due to melting of the side chains of the polymers [128]. Although no unambiguous interpretation of the number of these peaks is given, only the possibility of the formation of "discrete and homogeneously sized crystallites" composed of aliphatic side chains is indicated as one of the causes of their appearance. The second region of melting at 147°C corresponds to melting of the side chains according to the data in [128] (Fig. 1.13b).

The proposed interpretation raises some doubt, however, since the second, barely noticeable melting peak in Fig. 1.13b is near the range of temperatures in which chemical decomposition of the polymer begins (the peak at 181°C corresponds to decomposition of the polymer). The values of the transition temperatures of polyheptaldehyde in the 75-100°C range are far higher than the melting points of n-paraffins and isotactic poly(1-hexyl ethylene) (T_m = 10°C, see Table 1.2), which have the same aliphatic chain length (see Fig. 1.11, curves 5 and 5'), making it impossible to accept the proposed explanation.

An unambiguous conclusion on the structure of polyheptaldehyde could probably be obtained with the x-ray study of the observed transitions and a comparison of the structural data obtained with the thermodynamic parameters.

1.4. THE ROTATIONAL-CRYSTALLINE STATE

Another type of packing, described within the framework of the so-called rotational-crystalline state, is characteristic of such polymers with a comb-shaped structure as PA, PMA, and PV ethers and esters, whose structure has been assumed to be identical to the structure of n-paraffins [131-132]. Kargin, Platé, Shibaev, and others conducted fundamental studies of these compounds [30, 76, 77, 85, 96-104].

1.4.1. Hexagonal Packing

The x-ray studies of PA, PMA, and poly(alkyl vinyl ether)s and esters and most of their copolymers (Table 1.5) [30, 96-104] showed that all of these compounds in the crystalline state produce identical wide-angle x-ray reflections (7-9) regardless of the chemical structure, side-chain length, and microstructure of the polymer. The values of the interplanar distances for these polymers differ from the values for polyethylene, isotactic poly(1-alkyl ethylene)s, and n-

TABLE 1.5. Interplanar Distances (in Å) in Comb-Shaped Polymers [30, 96-104]

Polymer	Config. of main chain*	Number of diffraction maxima (reflections)†								
		small angle						wide angle‡		
		1	2	3	4	5	6	7 (100)	8 (110)	9 (200)
		±1	±1	±0.2	±0.06	±0.05	±0.04	±0.04	±0.02	±0.02
Poly(n-alkyl acrylate)s										
PA-12	At, iso	33 s	—	14.6 m	8.29 w	—	—	4.18 s	2.42 w	2.08 vw
PA-16	At, iso	42 vs	25 vw	14.7 m	8.34 m	6.06 w	—	4.19 vs	2.43 w	2.10 vw
PA-17	At	44.5 vs	26 vw	14.9 m	8.84 m	6.30 w	—	4.17 vs	2.41 w	2.08 vw
PA-18	At	47 vs	27.5 vw	15.8 m	9.50 m	6.60 w	5.40 vs	4.15 vs	2.40 w	2.10 vw
PA-22	At	54 vs	32.0 vw	17.3 m	10.20 m	—	—	4.14 vs	2.40 w	2.08 vw
Polyvinyl Ethers and Esters										
Poly(cetyl vinyl ether)	At	42 vs	—	14.3 m	8.24 m	—	—	4.19 vs	2.43 w	—
Polyvinyl palmitate (PVE-15)	At	40.5 vs	23 vw	13.4 m	8.12 m	—	—	4.19 vs	2.42 w	2.09 vw
Polyvinyl stearate (PVE-17)	At	42 vs	27 vw	14.5 m	8.40 m	6.06 w	—	4.17 vs	2.43 w	2.10 vw
Poly(n-alkyl methacrylate)s										
PMA-16	At	42 vw	29 vs	14.7 s	8.67 m	—	—	4.19 vs	2.41 w	2.08 vw
PMA-16	Iso	42 vw	27 vs	13.5 m	9.16 w	—	—	4.19 vs	2.40 w	—
PMA-18	At	46.5 vw	31 vs	14.9 m	9.90 m	7.27 w	—	4.17 vs	2.42 w	2.08 vw
PMA-22	At	56.5 vw	35 vs	17.6 m	11.20 m	—	—	4.15 vs	2.40 w	—
Poly(n-octadecyl acrylamide) (PAA-18) and Poly(n-octadecyl methacrylamide) (PMMA-18)										
PAA-18	—	—	37 s	16 m	10.5 w	—	—	4.19 vs	2.35 w	—
PMAA-18	—	—	29 s	15.5 m	10.0 w	—	—	4.19 vs	2.35 w	—

*At — atactic, iso — isotactic polymer.
†vs — very strong, s — strong, w — weak, vw — very weak, m — moderate.
‡The indices of the crystallographic planes are indicated in parentheses.

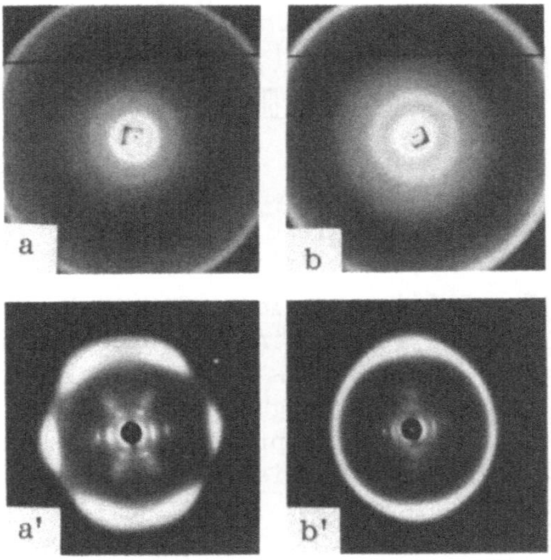

Fig. 1.14. X-ray patterns of unoriented (a, b) and oriented (a', b') samples of PA-18 (a, a') and PMA-18 (b, b') (the x-ray beam is normal to the axis of orientation) [30].

paraffins, which excludes the possibility of their crystallization in a rhombic lattice, as proposed previously [26-29].

The x-ray photographs of these polymers are characterized by a number of features (Fig. 1.14). There are three reflections (cf. Table 1.5) in all wide-angle x rays: d_7 = 4.15-4.10 Å, d_8 = 2.40-2.43 Å, and d_9 = 2.08-2.10 Å, and only one of them, d_7, is sharp and has a high intensity, while the remaining diffraction rings are characterized by such a low intensity that they are only distinguished with difficulty.

The sharp decrease in intensity with an increase in the angle of scattering and a ratio of the interplanar distances d_7 and d_8 equal to $\sqrt{3}$ are characteristic of compounds in the so-called gas-crystalline [131] or rotational-crystalline state [132]. This state, which is one variety of the liquid-crystalline state, is characterized by the fact that each molecule is considered a cylinder of rotation with a more or less uniform distribution of the electron density along the axis of the cylinder [131]. The packing of these cylinders

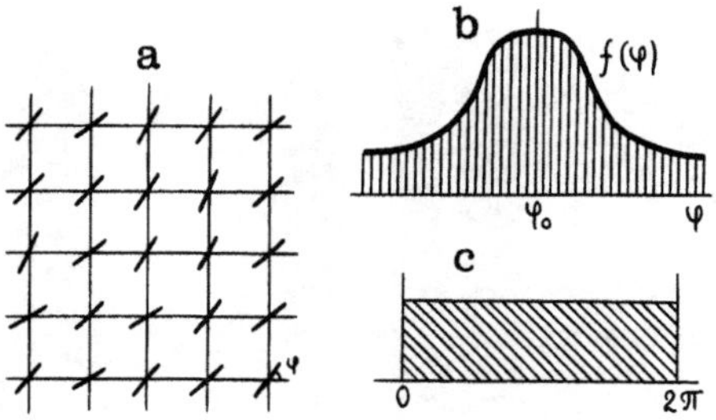

Fig. 1.15. Different azimuthal orientations of the side chains: projection along the long axis of the side chains (a) and their angular distribution function in the presence of rotations (b) and statistical rotation (c).

produces a hexagonal lattice in a plane perpendicular to the axis of the cylinders. Hexagonal packing is characteristic of compounds which contain molecules usually in the anisodiametric shape capable of free or hindered rotation around their long axes. Most long-chain hydrocarbons and their derivatives which usually crystallize in a rhombic or triclinic form undergo a so-called $\beta \to \alpha$ transition near the melting point, acquiring hexagonal packing. In this case, a transition of the "order—disorder" type takes place due to rotations around the long axis of the molecule and the appearance of spatially different orientations: there is order in the arrangement of the centers (or axes) of the molecules and disorder in their azimuthal rotations.

In comb-shaped polymers with PA-type structure and poly(vinyl ether)s and esters, the role of the molecules capable of "rotation" is played by the n-aliphatic side chains whose increased mobility is due to the presence of a C—O bond. Here rotation should be understood as the statistical scattering of the orientations of the side-chain axes around the equilibrium position, where the role of the main chain (or linkage segment) is to fix their position. The angular scattering of the orientations of the side chains, which can be characterized by function $f(\varphi)$ (Fig. 1.15), easily changes into the complete set of all orientations $f(\varphi)$ = const when their section is close to circular, and it is then possible to speak of statistical rotation (i.e., the azimuths of the different side chains of comb-shaped macromolecules are randomly oriented). The side chains approaching contact with each other pack hexagonally. X-ray patterns of oriented samples of PA-16 and PMA-18 corresponding to hexagonal packing are shown in Fig. 1.14.

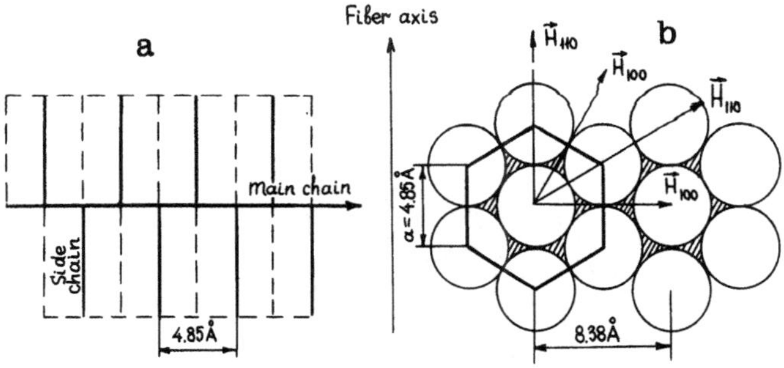

Fig. 1.16. Diagram of the structure of a comb-shaped macromolecule with side chains in the form of cylinders of rotation (a) and diagram of the hexagonal packing of the side chains in a plane perpendicular to their axis (b); the arrows indicate the position of the reciprocal lattice vectors [30].

Diagrams of the structure of a branched macromolecule and hexagonal packing of the side chains with the corresponding reciprocal lattice vectors, plotted using the x-ray vectors of the oriented samples, are shown in Fig. 1.16. The direction of the axis of orientation coincides with crystallographic direction (110). The side chains are positioned normally with respect to the axis of the main chain, forming a hexagonal lattice in a plane perpendicular to their axis. The average diameter of a methylene side chain ($a = 2d_{100}/\sqrt{3}$) can easily be calculated from the value of the interplanar distance of 4.19 Å (for PA-16, PA-17, PMA-16, etc.); the average diameter is equal to 4.85 Å, which is in good agreement with the existing data for the diameter of methylene chains.

The existence of hexagonal packing of aliphatic side chains for comb-shaped polymers is also reflected in their IR spectra. As Chapman showed in detail [133], characteristic absorption bands correspond to the different crystalline modifications of low-molecular-weight long-chain compounds (paraffins, ethers, alcohols, etc.). The hexagonal form is thus characterized by a strong isolated absorption in the region of 720 cm^{-1}, corresponding to vibrations of the C—H groups in the methylene chain. Splitting of this absorption band into two peaks of approximately equal intensity in the region of 720 and 730 cm^{-1}, related to vibration

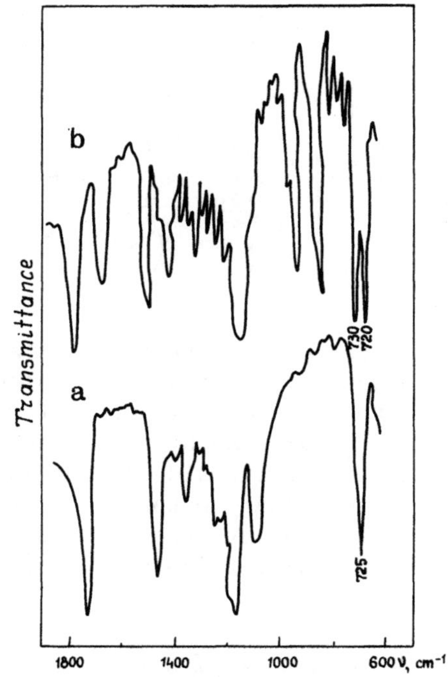

Fig. 1.17. IR spectra of PVE-17 in the hexagonal form (a) and its monomer in the orthorhombic form measured at −196°C (b) [30].

of the C—H groups in phase and out of phase, respectively, is observed for the more stable rhombic form. The IR spectra of all PA, PMA, and polyvinyl ethers and esters which crystallize in the hexagonal form are characterized by a single absorption peak in the 720 cm^{-1} region [30, 134]. As an example, the IR spectra of PVE-17 in the hexagonal form and its monomer which crystallizes in a rhombic cell with a characteristic doublet in the region of 720 and 730 cm^{-1}, are shown in Fig. 1.17 [30].

A similar doublet was observed by Benedetti et al. [109] for comb-shaped atactic and isotactic poly(octadecyl ethylene) (PE-18) and poly(octadecyl ethylene oxide) (PEO-18), which crystallize in an orthorhombic cell, as indicated above (Fig. 1.18). This doublet is most clearly manifest for isotactic samples at low temperatures, while an increase in the temperature, which causes the redistribution of the intensities in the 720-730 cm^{-1} region, results in the appearance of only one absorption band in the 720 cm^{-1} region (Fig.

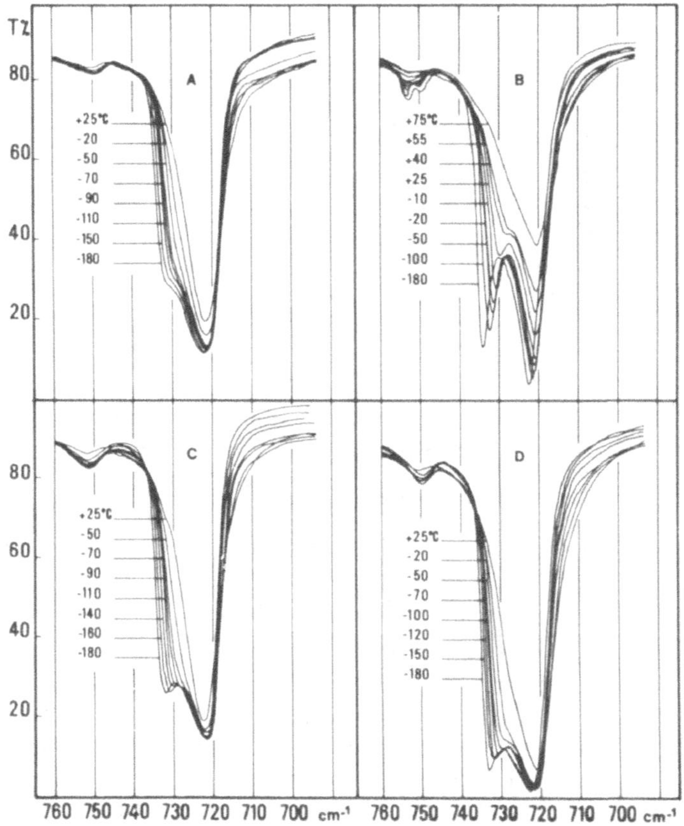

Fig. 1.18. IR spectra of atactic and isotactic comb-shaped polymers at different temperatures: A) atactic PE-18; B) isotactic PE-18; C) atactic PEO-18; D) isotactic PEO-18 [109].

1.18). This indicates the gradual disordering of the packing of the aliphatic side chains, which begins at relatively low temperatures, as Fig. 1.18 shows.

The transition from orthorhombic to hexagonal packing occurs over a wide range of temperatures, probably due to the successive "accumulation" of conformational perturbations, primarily by the main chains and segments of the adjacent side chains.

The study of the polarization IR spectra of comb-shaped polymers clearly reveals the peculiar type of orientation of

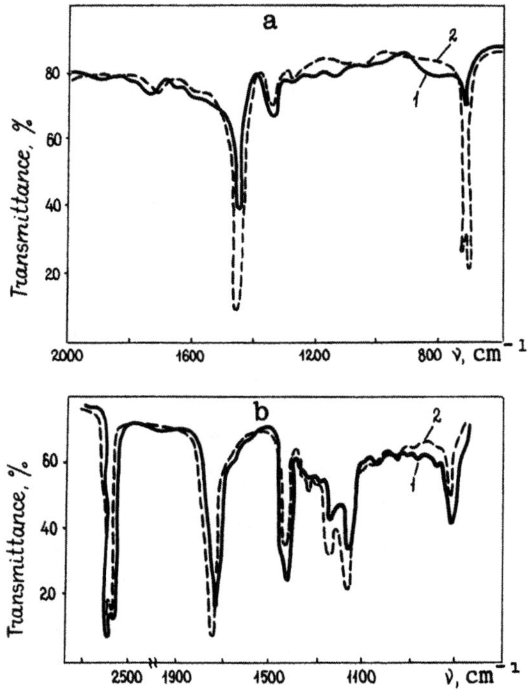

Fig. 1.19. Polarization IR spectra of oriented films of polyethylene (a) and PA-18 (b); the electric vector is parallel (1) and perpendicular (2) to the direction of stretching [134].

the macromolecules (Fig. 1.19 [134]). The calculation of the dichroic ratio for the basic characteristic bands of oriented PA-18, PMA-18, and polyethylene showed that, in contrast to polyethylene, where the methylene chains are primarily positioned along the axis of orientation, the methylene side chains of branched macromolecules have a directly opposite planar orientation in a direction perpendicular to the axis of extension.

The differences in the methods of attachment of the side chains to the main chain of the polymer for higher homologs of isotactic poly(1-alkyl ethylene)s, on the one hand, and PA, PMA, and polyvinyl ethers, on the other hand, results in a significant difference in their crystal structure: crystallization takes place in a rhombic cell in the first case and in a hexagonal cell in all other cases.

We previously mentioned that the hexagonal form is also possible in unfavorable crystallization conditions (for example, in quenching) for comb-shaped isotactic poly(1-alkyl ethylene)s in addition to rhombic packing. In the case of comb-shaped poly(alkyl acrylate)s, poly(alkyl methacrylate)s, and polyvinyl ethers, the hexagonal form is extremely stable up to the temperature of −196°C and occurs under any crystallization conditions. Despite the structural similarity of the side chains of comb-shaped polymers and the molecules of long-chain low-molecular-weight compounds (for example, monomers and their hydrogenated analogs) which usually form a number of different crystalline modifications (rhombic triclinic, monoclinic, etc.) [135], comb-shaped polymers of the type of PA, PMA, and polyvinyl ethers and esters only crystallize in a hexagonal cell.*

The cause of this remarkable stability of the hexagonal form of these polymers should apparently be sought in their structure, in particular in the presence of a linkage containing an ester group which ensures the sufficient mobility of the side chains for their crystallization. Contact among the side chains due to their close packing is always energetically advantageous. However, parallel packing of aliphatic chains inevitably causes strong "conformational perturbation" in the region of the linkages to the main chain. For this reason, the chemical nature, bulk, and degree of flexibility of the main chain and linkage, which determine their capacity for possible conformational changes due to fast crystallization of the side chains, will also define the degree of defects in the crystal lattice. In addition to the statistical scattering of the orientations of the side chains with respect to their equilibrium position, movements of the side chains in the direction of their axes are also possible, as shown in Fig. 1.15a. This will all lead to the radial inhomogeneity of the chain sections, and in the presence of scattering in their axial orientations to "apparent rotation" of the methylene chains. In this sense, the dia-

*The article by Enkelman and Lando [136] on the possible crystallization of polyvinyl stearate (PVE-17) in an orthorhombic lattice with parameters of a = 7.42 Å and b = 4.92 Å is undoubtedly of interest. However, it should be noted that this type of structure is only formed in special conditions: in the polymerization of the monomer in monolayers of the "head-to-head" and "tail-to-tail" type.

gram of the position of the side chains with respect to the main chain (see Fig. 1.16) should be considered somewhat idealized.

Let us first examine how the difference in the radial inhomogeneity of the molecules (in our case, the side chains), which can be expressed as the ratio of the radius of the inhomogeneity of the molecule (r_i) to the average diameter of the molecule ($2r_{av}$) according to the data in [7], should appear in the diffraction pattern:

$$k = \frac{r_i}{2r_{av}} = \frac{r_{max} - r_{min}}{r_{max} + r_{min}} \approx \frac{r_i}{\bar{a}} \tag{1.2}$$

where k is the radial inhomogeneity coefficient, r_{max} and r_{min} are the maximum and minimum radii of the molecules (side chains), and $\bar{a}$ is the average distance between the centers (axes) of the molecules (side chains).

According to Fig. 1.15a, the degree of perturbation of the packing is due to the angles of rotation and the inhomogeneity of the sections of the side chains. The degree of disorder in rotations will be determined by the radial inhomogeneity coefficient k (according to [7]), i.e., not by the absolute value of the radius of inhomogeneity $r_i = r_{max} - r_{min}$, but by the ratio of the radius of inhomogeneity to the average diameter of the molecule $2r_{av} = r_{max} + r_{min}$.

Using the simple geometric model in the theory of x-ray structural analysis, the relation between lattice perturbations Δ_{rad} (usually expressed in radians) and the angular degree of disorder $\Delta\psi$ [7] is

$$\Delta_{rad} \approx 2/\pi r_i \Delta\psi \tag{1.3}$$

Since the ratio Δ_{rad}/a is the criterion of the scattering character, it is possible to correlate this expression with the coefficient of radial inhomogeneity of the side chains. Dividing Eq. (1.3) by $\bar{a}$ and using (1.2), we obtain

$$\Delta_{rad}/\bar{a} \approx \frac{2}{\pi}\frac{r_i}{\bar{a}}\Delta\psi \approx \frac{2}{\pi}k\Delta\psi \tag{1.4}$$

The angular degree of disorder at which $\Delta\psi$ is equal to approximately one radian corresponds to the case of statistical "rotation." Then we obtain the following for the maximum order of diffraction p with consideration of the boundary condition $p\Delta/\bar{a} = 0.25$ and Eq. (1.3):

$$p_{max} = \pi/8k \quad (1.5)$$

If $k \approx 0.3$-0.4, then the appearance of only one reflection is possible for the case of hexagonal packing; when $k = 0.15$-0.20, two successive orders of reflection are possible.

In comb-shaped polymers whose side chains crystallize in a hexagonal cell, $k \approx 0.19$ ($r_{max} = 2.55$ Å and $r_{min} = 1.75$ Å). With this value of k, we find from Eq. (1.4) that $p_{max} = 2$. The hexagonal packing of "rotated" side chains should thus be manifest in the diffraction pattern by the presence of only two orders of diffraction. In addition, in the presence of hexagonal packing between reflections (100) and (200), according to the theory in [7] the appearance of reflection (110) is possible with $d_{110} = d_{100}/\sqrt{3}$. Actually, all crystallizable comb-shaped PA, PMA, polyvinyl ethers and esters, etc., produce typical x-ray patterns with indicated reflections of this type (see Table 1.5).

A similar type of x-ray pattern is also observed for a linear polymer, polyacrylonitrile, whose structure is treated as gas crystalline [131].

With comparatively small values of k, the statistical rotation of the molecules results in lattice perturbations and fast attenuation of the interference function maxima.

Examining the side chains of comb-shaped polymers as cylinders with radius r_0 and distance between their centers $\bar{a}$ and using the results in [137] on the study of the intensity of scattering in cylindrical—symmetrical systems, it is possible to obtain the expression for the intensity of scattering I (in cylindrical coordinates R) of an aggregate of seven cylinders arranged according to the hexagonal packing principle:

$$I(R) = F^2[7 + 24I_0(x) + 6I_0(2x) + 12I_0(\sqrt{3}x)] \quad (1.6)$$

where F is the amplitude of the scattering, $x = 4\gamma\pi r_0 R$, $\gamma = \bar{a}/2r_0$.

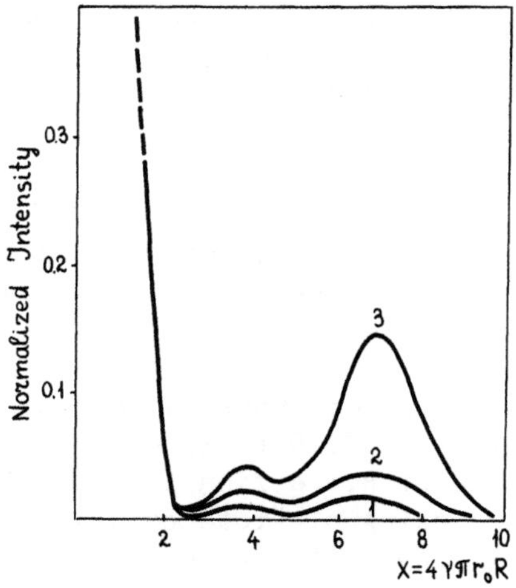

Fig. 1.20. Normalized intensity of scattering per cylinder for independent aggregates of seven cylinders in a hexagonal lattice with different values of γ [137]: 1) γ = 1.00; 2) γ = 1.25; 3) γ = 2.00.

The function expressed by Eq. (1.5) for different values of γ is shown in Fig. 1.20. When the cylinders are in close contact ($\gamma = 1$, $\bar{a} = 2r_0$), the diffraction maxima are very weak, and an increase in γ causes an increase in the maximum corresponding to interplanar distance d_{100}. The pronounced difference in the intensity of the diffraction maxima observed in the x-ray patterns of the comb-shaped polymers (see Fig. 1.14) indicates insufficiently close contact of the side chains (cylinders) during their crystallization.

On the one hand, the presence of order due to the parallel arrangement of the methylene side chains, and, on the other, the appearance of elements of azimuthal "disorder" as a result of slight disorientation of the side chains, causing some sterically different orientations of these chains, are thus a distinctive feature of the structure of comb-shaped polymers of the PA, PME, and PVE types.

This forms the basis for stating that comb-shaped polymers which crystallize due to the hexagonal packing of the side chains form a very unusual type of structural ordering which can be described within the boundaries of the rotational-crystalline state. The appearance of a number of structural features, most clearly seen in the x-ray study of the deformation of polymers of the PA and PMA type, is conceivable within the boundaries of this state as a function of the chemical nature of the main chain of the polymer [85, 103].

Although the x-ray diagrams of a uniaxially oriented sample of PA-16 (even with an insignificant degree of orientation) are characterized by the presence of six sharp reflections corresponding to an interplanar distance of 4.19 Å (see Fig. 1.14), only the appearance of meridional crowding in the Debye ring in the region of 4.19 Å is observed in the x-ray pattern of PMA-16 (with the same degrees of elongation). As an analysis of this pattern with the Ewald sphere and the sphere of normals shows, this distribution of intensity by the azimuth of the Debye rings is due to the special form of the pattern [103], when its axis is located in the section plane of the hexagonal packing of the side chains, and the packing section plane remains perpendicular to the fiber axis. Such a pattern can be considered as a set of axial patterns with axes which change direction in the hexagonal packing section plane. The calculation shows that, in this case, the azimuthal distribution of the intensity can be represented by the following equation:

$$I \approx pd/2\pi\sqrt{1 - \cos^2\theta\cos^2\varphi} \qquad (1.7)$$

where θ is the Bragg angle, and φ is the azimuthal angle of the Debye ring, i.e., the angle between the meridian of the x ray and the radius from the center of the x ray to any point of the Debye ring.

For a given Debye ring, θ = const and the intensity will consequently be a function of azimuthal angle φ alone. It follows from Eq. (1.7) that the intensity attains the maximum value in the meridian of the x ray (when $\varphi = 0$), which is in agreement with the experimental observation for PMA-16 (see Fig. 1.14). Similar results were obtained by Morawetz et al. [138] in the x-ray study of the structure of PA-18 and PMA-18. The appearance of meridional reflections alone has also been observed for oriented samples of poly(octadecyl acrylamide) (PAA-18) [139].

As a consequence, the special form of the observed pattern is determined by the significantly easier disorientation of hexagonal packing segments of the side chains for polymers which have a more rigid main chain (PMA) and greater steric hindrances with respect to the linkages (hydrogen bonds in PAA-18 and PMAA-18) compared to polymers with more flexible main chains (PA). This is indicated by the results of calculating the average dispersion of the orientation of the side chains of branched macromolecules with respect to the axis of extension, obtained with the x-ray diffraction data, and the results of a study of the dichroism of the IR spectra [134]. The role of the main chain in comb-shaped polymers consists of fixing the aximuthal disorder of the methylene chains, which also creates the conditions for the formation of a rotational-crystalline state.

It is interesting to note that the rotational-crystalline state is not only characteristic of comb-shaped polymers but also of polymers with a linear structure. For example, it is known that crystallization of polyethylene usually occurs in the rhombic (less frequently the triclinic) modification. However, γ irradiation of polyethylene in a certain temperature range causes conversion to the hexagonal form whose formation is due to the appearance of inhomogeneities (defects) in the macromolecules of polyethylene according to most investigators [140, 141]. These defects should also include the presence of crosslinks and branches which cause "swelling" of the rhombic lattice of the polyethylene and its transition into the hexagonal form. A similar picture is observed in the perturbation of the regular structure of the chains of the macromolecules of polyethylene, isotactic polypropylene, and trans-1,4-polyisoprene (gutta-percha) due to the introduction of chlorine (or bromine) atoms in their main chain [142-144]. The presence of such halogen-containing "foreign" units (in a certain concentration) perturbs the azimuthal order of the arrangement of the macromolecules of these polymers in the crystal lattice and causes their transition to the rotational-crystalline state, which is the basic form of existence for comb-shaped polymers.

1.4.2. d Spacings. Single- and Two-Layer Packing of the Side Chains

The first indications of the existence of small-angle spacing for some comb-shaped polymers of the acrylic and methacrylic series can be found in [28]. However, in attempts to confirm these investigators, the values of the d

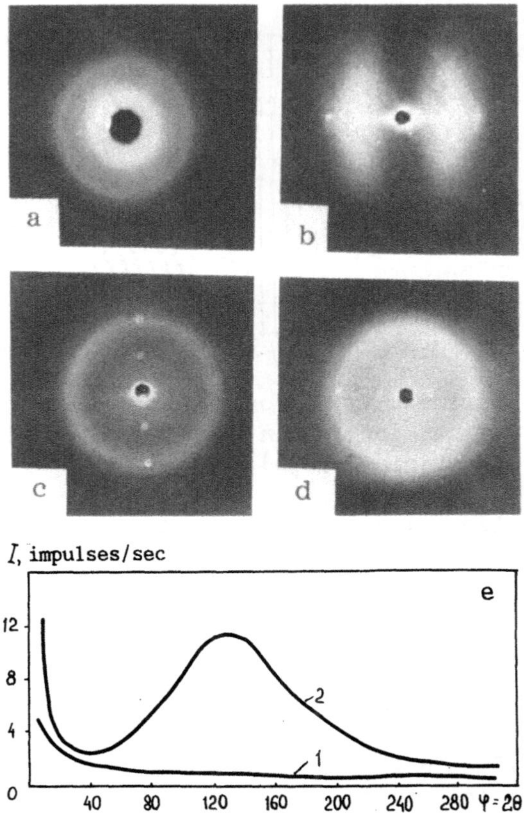

Fig. 1.21. Small-angle x-ray patterns of some comb-shaped polymers [a) unoriented PMA-22; b) oriented PA-18; c) unoriented PMA-18; d) oriented PMA-18] and scattering-intensity distribution (small-angle x-ray patterns) for the oriented sample of PA-16 along the equator (1) and meridian (2) of the x-ray fiber patterns (e).

spacings which they obtained (see p. 11) were virtually unreproducible, and cannot serve as a measure of some ordering of the polymer. Recent studies have produced much more information on the structure of the different comb-shaped polymers using the results of small-angle x-ray scattering.

The x-ray diffraction patterns of a large number of crystallizable comb-shaped polymers such as PA, PMA, polyalkyl amides, etc., are characterized by a number of small-

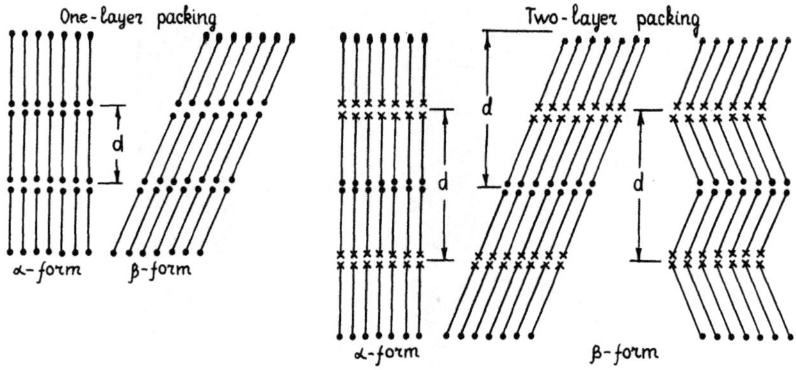

Fig. 1.22. Different types of packing of low-molecular-weight long-chain compounds (•: CH_3 group, ×: functional group).

angle reflections indicating a certain spacing (some of these are shown in Fig. 1.21); the value of the interplanar distances corresponding to the small-angle spacing are reported in Table 1.4, reflections d_1-d_6. It should be noted that the reported values of spacings d_1 and d_2 correspond to the maximum observed values characteristic of oriented and well-annealed samples.

The reflections corresponding to the indicated values of d_1-d_6 are only found at the equator of the x-ray fiber patterns and have an elongated shape along the meridian (see Fig. 1.21). This type of reflection can be considered the result of intersection of the rods in the reciprocal space with the Ewald sphere. The presence of rods parallel to the axis of the pattern in the reciprocal space indicates the existence of layers perpendicular to the axis of the pattern in the space which are formed by the side chains in the given case. The existence of reflections d_1-d_6 is thus due to the periodic distribution of the electron density along the axis of the cylinders, i.e., in a direction parallel to the axes of the side chains packed in a hexagonal cell.

The values of d_1 and d_2, subsequently called d spacings, are similar in a physical sense to the value of the d spacing in the given case (i.e., maximum interplanar distance d_{001}) observed in small angles for low-molecular-weight long-chain paraffins and their derivatives and characterize the length of the molecule. Reflections d_3-d_6 (see Table 1.5) are orders of diffraction from the value of d spacing d_1 or d_2 (within the limits of the precision of the measurements).

In examining the structural similarity of the side chains of comb-shaped polymers and long-chain low-molecular-weight compounds, it is natural to raise the question of the similarity and difference of their packing. It is well known [145-150] that all low-molecular-weight compounds can be divided into two groups as a function of the value of the d spacing: compounds characterized by single-layer packing of the molecules (paraffins, fatty acid ethyl esters, etc.) and compounds with two-layer packing of the molecules (n-aliphatic alcohols, acids, etc.) (Fig. 1.22). Within the boundaries of these two types of packing, low-molecular-weight compounds can form a large number of different crystalline modifications with both a vertical (α form) and a tilted (β form) arrangement of the molecules with respect to the plane of the layers in which the end groups lie. Two-layer packing is stable for compounds whose molecules contain groups which strongly interact (hydrogen bonds in alcohols and acids, dipole–dipole interaction in esters).

For n-aliphatic acid esters and alcohols, the forces of intermolecular interaction (in the absence of OH groups capable of forming hydrogen bonds) are significantly weaker than for alcohols and acids and decrease with an increase in the length of the substituent. For this reason, fatty acid methyl esters are still capable of forming two-layer packing, while ethyl esters only crystallize in single-layer packing [145-150]. In each of the diagrams shown in Fig. 1.22, the spacing corresponds to the distance between planes on which the end groups are located.

As Table 1.5 shows, the presence of two spacings, d_1 and d_2, whose values increase linearly with an increase in the number of carbon atoms in the chain, is characteristic of most of the compounds represented. Note that the intensity of the reflections corresponding to d_1 is significantly higher than the intensity of d_2 for PA and PV ethers and esters, and the inverse relation of the intensities of d_1 and d_2 is observed for PMA. The values of d_2 for all of the polymers nearly correspond to the calculated distance between the parallel main chains if the side chains with a planar zigzag conformation are positioned normal to the axis of the macromolecule. By analogy with the packing of the molecules of low-molecular-weight compounds in a hexagonal cell with

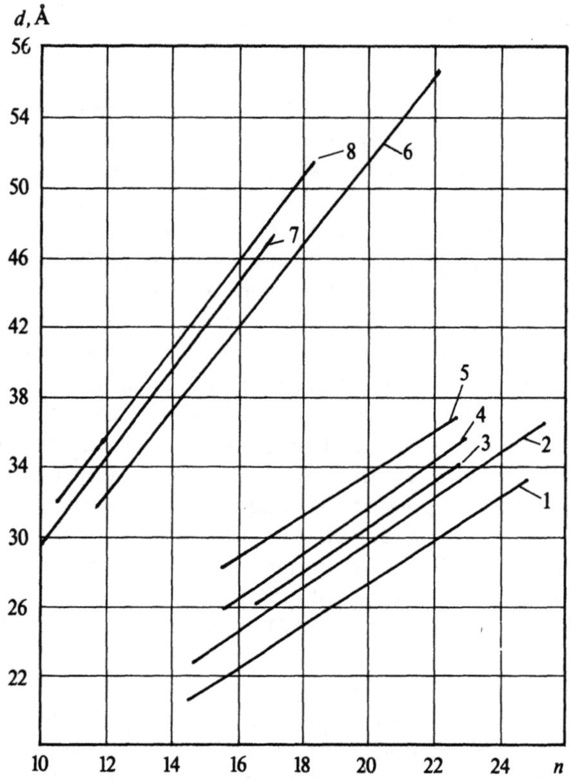

Fig. 1.23. Dependence of the d spacing on the number of carbon atoms in the alkyl group of compounds forming single-layer (1-5) and two-layer (6-8) packing: 1) α form of paraffins [145]; 2) n-aliphatic acid methyl esters (CH_3OCOR_n) [149]; 3) n-aliphatic acid ethyl esters (C_2H_5OCOR) [149]; 4) acetic acid and n-aliphatic alcohol esters (CH_2COOR) [150]; 5) PMA-n [102]; 6) PA-n [102]; 7) isotactic poly(1-alkyl ethylene)s [94]; 8) n-aliphatic acid triglycerides [148].

vertical chains (for example, the α form of n-paraffins, fatty acid ethyl esters, acetic acid esters, and aliphatic alcohols), it should be assumed that the indicated value of d_2 corresponds to single-layer packing of the side chains. This is also indicated by the similar character of the change in the value of d_2 (slope of the line) with an increase in the number of carbon atoms in the PMA series and the indicated low-molecular-weight compounds which crystallize in the α form (Fig. 1.23).

The value of spacing d_1, which has the maximum intensity in the x-ray patterns of homologs of the PA and PVE series as Table 1.5 shows, is almost twice as high as the value of d_2. This value, as Fig. 1.23 shows, is close to the values of d_1 characteristic of two-layer packing of the triglyceride molecules, which are low-molecular-weight analogs of PV esters and crystallize in the α form with vertical chains. For comparison, the values of d_1 of isotactic poly(1-alkyl ethylene)s which also crystallize with the formation of two-layer packing of the side chains (form I) are also shown in Fig. 1.23.

The increase in the value of spacing $d_1^{CH_2}$ per CH_2 group in the case of PA and PV esters corresponds to approximately 2.5 Å, which indicates the two-layer character of the side-chain packing of these polymers. We note, however, that, in this case, the value of d_1 is approximately 8-10 Å smaller than the sum of the lengths of the two side chains with a planar zigzag conformation.* The latter is apparently due to a change in the packing of the end groups of the side chains, especially in the region of their attachment to the main chains; this follows from the lower values of d_1 for PA in comparison to the values of d_1 for isotactic poly(1-alkyl ethylene)s and triglycerides with the same alkyl-chain length and value of the increase in $d_1^{CH_2}$.

Models of some comb-shaped polymers are shown in Fig. 1.24, and diagrams of the packing of comb-shaped macromolecules in the crystal are shown in Fig. 1.25. As Fig. 1.24

*As shown in the footnote on p. 46 in [136], PEV-17 was obtained in the orthorhombic crystalline form; this polymer is characterized by the presence of a d spacing of 55.8 Å, which almost precisely corresponds to twice the length of the side chain of this polymer. Since this polymer was obtained by polymerization in monolayers where the most favorable conditions for the creation of ideal packing of the macromolecules are realized, it can be concluded that the conformational limitations indicated above disappear under these conditions in crystallization of the side chains.

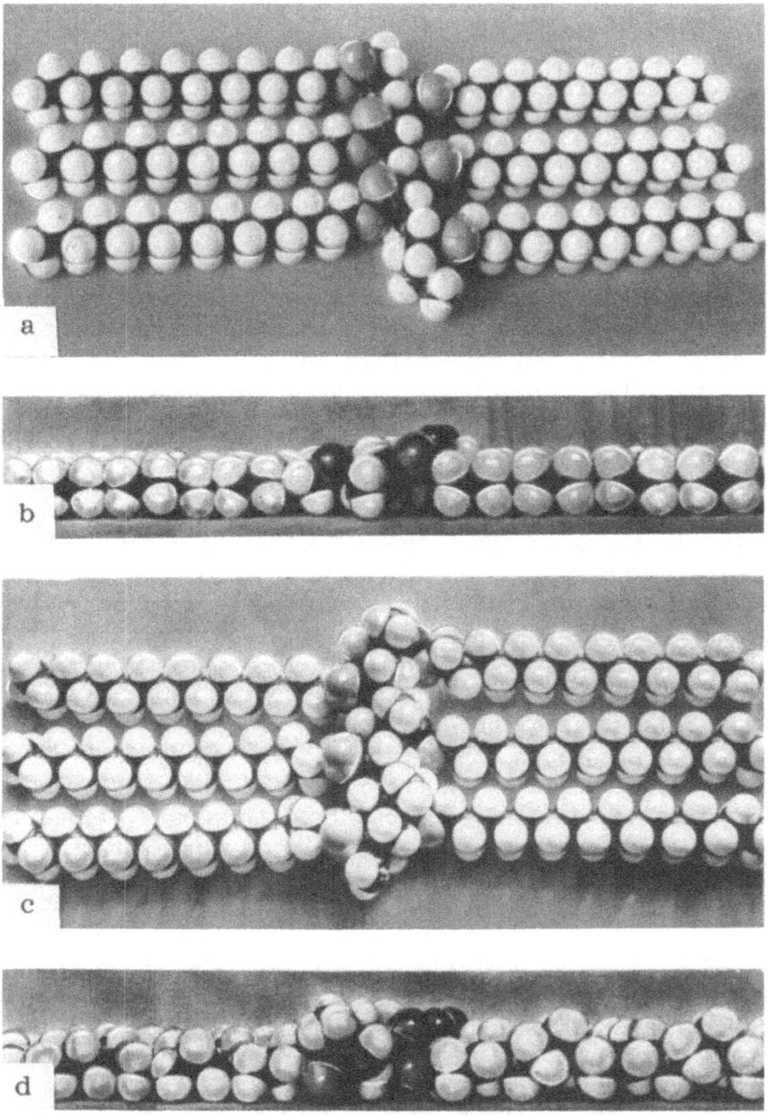

Fig. 1.24. Molecular models of PA-16 (a, b) and PMA-16 (c, d).

shows, each polymer chain is a peculiar layered structure or band consisting of parallel packed methylene chains positioned on both sides of the main chain. The model shown in Fig. 1.24 corresponds to a syndiotactic polymer in which the side chains are in the trans position with respect to the main chain. This position of the side chains is the only

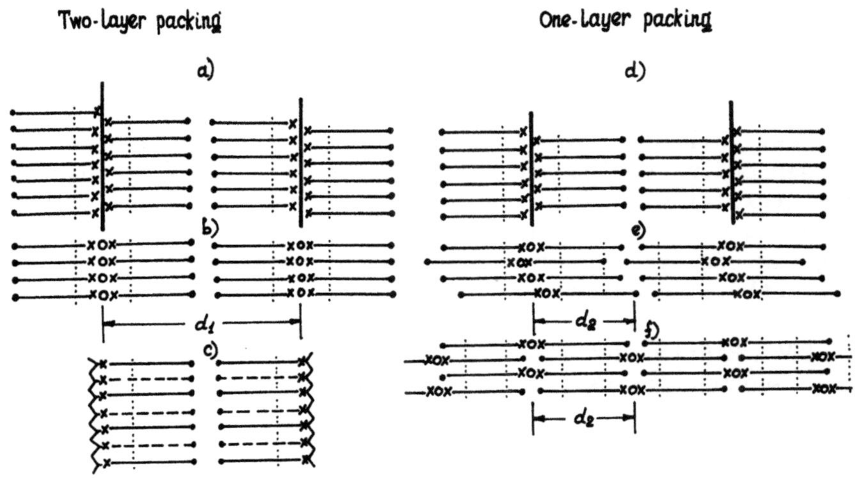

Fig. 1.25. Diagrams of single- and two-layer packing of comb-shaped macromolecules in a crystallite (o: main chain; ×: linkage; •: CH_3 group [85]); a, c, d: main and side chains are located in the plane of the figure; b, e, f: projections along the axis of the main chains of the macromolecules.

one possible for ensuring their close packing in irregular atactic and in isotactic comb-shaped polymers, although in the last two cases realization of the trans position of the side chains is associated with significant distortion of the conformation of the segments of the main and side chains at the sites of their attachment. The degree of flexibility of the main chain and linkages and their capacity for possible conformational changes due to the fast crystallization of the side chains will determine the number of methylene groups (indicated by the broken line in Fig. 1.25) which do and do not participate in crystallization and the formation of crystallites.

In the case of PA and PV ethers and esters with sufficiently flexible main chains, during crystallization the side chains can "impose" the necessary conformation on the main chain which permits relatively close packing of the layered structures, as shown in Fig. 1.25a and b. The value of the spacing corresponds to two-layer packing of the side chains. It should be remembered that the diagram of the structure of a comb-shaped macromolecule shown in Fig. 1.16a, as indicated above, illustrates an ideal variant of the parallel packing of the methylene groups with the end groups of the alkyl chains located in parallel planes. Due to conformational distortions in the region of the main chain and linkages, shifts of the separate layers toward the axes of the main chains (which are not shown in Fig. 1.16) are possible, resulting in some disordering of the end methyl groups in the chains. These shifts are not large enough to cause a change in the type of layered packing for polymers of the PA and PV ether and ester type.

PMA, PAA-18, and PMAA-18 and crystallizable copolymers with a monomer possessing an alkyl side group as one of the components (see Table 1.5) behave somewhat differently during crystallization (cf. p. 87). The increased stiffness of the main chain and the presence of sterically bulky groups in these polymers and copolymers cause significantly greater distortions of the conformation of the main and side chains at the sites of their attachment than in the case of PA and PV esters and ethers. This results in an apparent increase in the bulk of the main chain and perturbation of the uniform packing of the end groups of the side chains, and prevents packing of the layers of comb-shaped macromolecules as shown in Fig. 1.25b. Crystallization of the polymers due to packing of the side chains in a hexagonal lattice is accompanied in this case by shifting of the layers toward the axes of the side chains (see Fig. 1.25d and e). This results in looser packing of their ends, causing the formation of defective regions in the crystallite. This in turn perturbs the coherence of x-ray scattering between its individual segments and results in a decrease in the value of the d spacing. Crystallization occurs with single-layer packing of the side chains.

In some cases, the shifts of the layers of macromolecules can result in an intermediate type of packing, as the data in [138] on a series of poly(n-alkyl acrylamide)s (PAA-m, n) containing amide groups in the side chain, suggest,

$$[-CH_2-\underset{\displaystyle |}{CH}-]_x$$
$$CO-NH-CO-(CH_2)_m-CO-NH-(CH_2)_n-H$$

where m = 2, 5, 10 and n = 4, 10, 16.

The values of the d spacings and the calculated length of the side chains L of polymers of the PAA-m, n series are reported below:

	d, Å	L, Å		d, Å	L, Å
PAA-2, 16	35.6	29	PAA-5, 16	39.1	32
PAA-5, 10	38.1	25	PAA-10, 16	58.5	39

As indicated, d > L, but, at the same time, d < 2L, i.e., significant shifts of the layers of macromolecules are possible for this type of polymer, apparently caused by the formation of intramolecular and intermolecular hydrogen bonds, found for poly(alkyl acrylamide)s, in [151, 152], for example.

The shifts of the individual layers with respect to each other can be even more significant in the case of the packing shown in Fig. 1.25e with an antiparallel position of the side chains. French investigators who studied the structure of poly(octadecyl methacrylate) (PMA-18) prepared by anionic polymerization proposed a similar type of packing for comb-shaped macromolecules [153]. They assigned this polymer a syndiotactic structure, although they unfortunately do not give any information on its microstructure. At the same time, the value of the d spacing, d_2 = 30.6 Å, calculated from the x rays, is in good agreement with the value of d_2 = 31 ± 1 Å obtained in [30] for atactic PMA-18 with predominance of the syndio triad.

The antiparallel position of the aliphatic side chains with their partial "overlapping" was proposed in [114] for poly(10-n-alkylhydroxycarbonyl-n-decyl maleimide)s

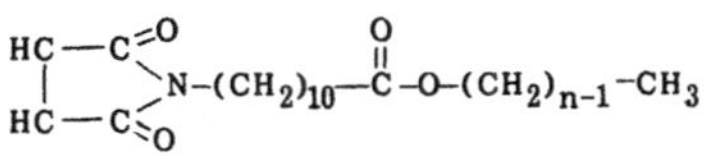

(P-N-MI-10)

where n is an even number within the limits of 2-22.

Of the comb-shaped polymers investigated, these compounds have the longest side chains. The x-ray analysis showed that all polymers, beginning with n ≥ 12, crystallize in a hexagonal cell (reflection d = 4.17 Å), and the value of the d spacing increases linearly with an increase in side-chain length (Fig. 1.26, curve d). The presence of second-order $d^{(2)}$ and third-order $d^{(3)}$ reflections from the value

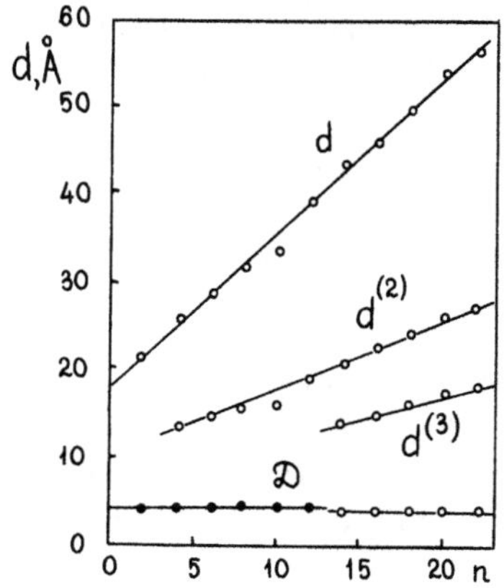

Fig. 1.26. d spacing, second-order $d^{(2)}$ and third-order $d^{(3)}$ reflections, and the distance between side chains D as a function of the number of carbon atoms n in the alkyl side chain of P-N-MI-n polymers [114].

of the d spacing indicates the high degree of perfection in the packing of the side chains. As Fig. 1.27 shows, some intermediate type of packing (similar to PAA-m, n), where the side chains are arranged in layers (similar to the smectic mesophase) so that $L < d < 2L$, is also realized in this case; that is, there is partial "overlapping" of the aliphatic tails.

Crystallization of comb-shaped polymers, like low-molecular-weight compounds, takes place in most cases in either single- or two-layer packing of the side chains, which are the most common structural units of a comb-shaped macromolecule. In addition, as Table 1.5 shows, the coexistence of two types of packing (d_1 and d_2 spacings), with the overwhelming predominance of one type, is possible for some comb-shaped crystallizable polymers. In many cases, packing of the macromolecules with partial overlapping of fragments of the side chains occurs.

The conditions of crystallization of the polymers also play a significant role in the formation of single- or two-layer packing. In conditions of prolonged annealing at a temperature 10-15°C below the melting point or in crystallization from solutions in nonpolar solvents (PA, polyvinyl ethers and esters), two-layer packing usually occurs; a

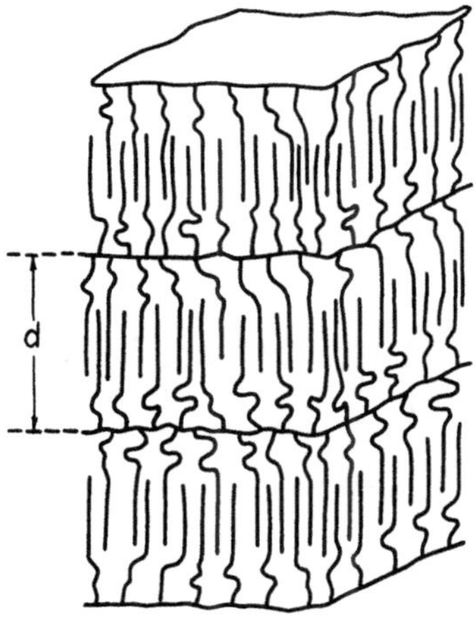

Fig. 1.27. Diagram of the arrangement of the side chains in layers in P-N-MI-n polymers [114].

small-angle reflection corresponding to the value of d_1 can be observed in the x-ray pattern, even for polymers of the methacrylic series crystallized under these conditions. However, quenching of melts of all comb-shaped polymers causes crystallization with single-layer packing of the side chains, and a change to two-layer packing can only occur in some cases with prolonged annealing. The opposite transition can easily take place on heating. In PA, the transition from two-layer to single-layer packing of the side chains takes place when the samples are heated to a temperature 2-5°C below their melting point and is established by dielectric (see Chapter 2) and x-ray methods.

The effect of the temperature on the value of the d spacing for a series of comb-shaped polymers, which illustrates the transition from two-layer to single-layer packing of their side chains, is shown in Fig. 1.28. This transition should be considered as unusual cooperative melting of defective regions. At temperatures 8-10°C below the melting point, a gradual increase in the intensity of reflection d_2 corresponding to single-layer packing is observed in the x-ray patterns of PA-16 and PA-18, which have an intense small-angle reflection corresponding to two-layer packing and a very weak reflection d_2. Reflection d_7 = 4.19 Å, corresponding to the hexagonal packing of the side chains, is preserved, although the proportion of amorphous scattering in the re-

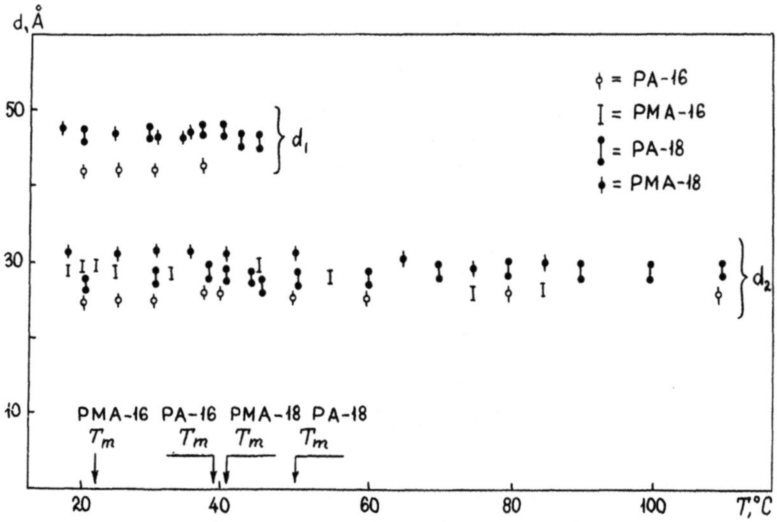

Fig. 1.28. Temperature dependence of spacings d_1 and d_2 for PA-16, PMA-1, PA-18, and PMA-18 [85].

gion of 4.6-4.8 Å increases. In other words, an increase in the temperature, by increasing the amplitude of the vibrations of the parts of the macromolecules in the crystallites, primarily causes the strongest changes in the packing of the end groups of the side chains located at the most defective sites in the crystalline regions. After melting, reflection d_1 totally disappears, but intense reflection d_2, corresponding to single-layer packing of the side chains, appears.

Note the small difference in the values of d_1 for the crystalline and amorphous state of each of the polymers. This indicates a high level of ordering in the position of the methylene side chains in melts of crystalline polymers at temperatures slightly higher than their melting point.

The type of layered packing determines the appearance of orders of reflection of only a certain multiplicity of the value of the d spacing in the x-ray patterns of the polymers. As Table 1.5 shows, polymers which primarily crystallize in single-layer packing (PMA, PAA-18, and PMAA-18) produce both even and uneven orders of diffraction. Polymers which crystallize in two-layer packing only exhibit uneven orders of diffraction of the value of the d spacing [77, 102, 103].

The absence of even orders of diffraction of the value of the d spacing is also characteristic of low-molecular-weight long-chain compounds which crystallize in two-layer packing, such as n-aliphatic alcohols and acids [149, 154]. At the same time, paraffins, which crystallize in single-layer packing, exhibit both even and uneven orders of diffraction of the value of the d spacing. As Fig. 1.20 shows, the presence of two types of reflection planes containing strongly and weakly scattering atomic groups located at different distances from each other results in the disappearance of the even orders of diffraction on the x-ray patterns of compounds which crystallize in two-layer packing.

This periodic distribution of the electron density together with the alternation of amorphous and crystalline regions in the direction of the side-chain axes is also characteristic of comb-shaped macromolecules. The calculation of the intensity of the small-angle scattering I for a different type of packing of the side chains, on consideration of the length of the defective regions formed by the main chains and the ends of the side chains, produced the following equations [103]:

for single-layer packing

$$I \approx \Delta\rho^2 \sin^2 \frac{\pi \ell p}{d} \tag{1.8}$$

for two-layer packing

$$I \approx \Delta\rho^2 \sin^2 \frac{\pi p}{2} \tag{1.9}$$

where $\Delta\rho$ is the difference in the electron densities of the side chain and the defective segment and p is a whole number.

It follows from these equations, in correspondence with the experimental data in Table 1.5, that the value of I is nonzero only with uneven values of p for the case of two-layer packing of the side chains, and with single-layer packing, $I \neq 0$ for both even and uneven values of p.

The subsequent studies of the structure of comb-shaped polymers can, in our opinion, serve as an illustration of the statement in [28] concerning the difficulty of obtaining

TABLE 1.6. Values of Interplanar Distances (in Å) for PVE-17 Obtained by Different Investigators*

Number of diffraction maxima (reflections)							Reference
1	2	3	4	5	6	7	
48±49.6 $(d_1—1)$†	27±0.1 $(d_2—1)$	16.0±0.3 $(d_1—3)$	9.84±0.02 $(d_1—5)$	8.98±0.02 $(d_2—3)$	7.06±0.06 $(d_1—7)$	6.74±0.05 $(d_2—4)$	[144] ‡
—	25.9 $(d_2—1)$	12.95 $(d_2—2)$	8.63 $(d_2—3)$	6.43 $(d_2—4)$	4.37	—	[145]
42±1 $(d_1—1)$	27±1 $(d_2—1)$	14.5±0.2 $(d_1—3)$ $(d_2—2)$	8.40±0.06 $(d_1—5)$ $(d_2—3)$	6.06 $(d_1—7)$	—	—	[30]
—	—	15	9.6	—	—	—	[29]

*Only the values of d > 4.2 Å are reported.
†The orders of reflection of the value of spacing d_1 or d_2 are indicated in parentheses.
‡The polymer contained approximately 20% monomer.

reproducible data with respect to the small-angle spacing. Our analysis of the distribution of the intensity of the small-angle reflections resulted in an interpretation of these results. As an example, the different results of determining the interplanar distances for one comb-shaped polymer, polyvinyl stearate (PVE-17), are reported in Table 1.6.

The apparent "nonreproduciblity" of the values of the d spacing in comparing the results obtained by different investigators is essentially due to the relatively rigorous fluctuation of their values within the boundaries of the results obtained by each investigator. In comparing the value of the d spacing obtained by each of the investigators, it is easy to note that within the limits of the precision of the measurements, these values are p-orders of diffraction (where p = 1, 2, 3, 4, etc.) of the value of the spacing indicated as d_1 and d_2. Corresponding to the reported calculated equations for the intensity of the small-angle scattering, as stated before, the two-layer type of packing (reflection d_1) is characterized by the presence of only uneven orders of reflection, while single-layer packing (reflection d_2) causes the appearance of both even and uneven orders of diffraction. Consequently, the apparent range in the values of the d spacing (and orders of reflection) for the same polymer is not the result of experimental error but is the reflection of certain features of the structure of the polymer apparently caused by the conditions of crystallization. In the case of poly(octadecyl acrylamide) (PAA-18), the deviations in the values of d_2 can reach 25-30% as a function of the conditions of processing the polymer (annealing, quenching, or crystallization from different solvents).

In some x-ray studies [28, 155, 157] of the structure of crystalline comb-shaped poly(alkyl acrylamide)s and poly(alkyl methacrylate)s, small-angle reflections corresponding to spacing of the order of 75-100 Å are also observed. These usually very rare and intense reflections are located on the equator of the x-ray diffraction pattern and have been correlated with the thickness of the crystalline lamellae formed by the side chains.

The attention of many investigators has recently turned to problems of creating and studying comb-shaped polymers with a more complex structure in which the main chains are heterochain polymers. The structural studies of comb-shaped poly(n-alkyl malonamide)s (PAM-n) with the general formula

$$\left[\begin{array}{l}-CO-\underset{\substack{|\\(CH_2)_n\\|\\CH_3}}{CH}-CO-NH-(CH_2)_6-NH-\end{array}\right]_x$$

(where n = 2-17) conducted by Corradini et al. [46, 47] made it possible to propose a single-layer type of packing for the aliphatic chains of these polymers. Beginning with a side chain length of eight carbon atoms, the polymers in this series form a "partially ordered layered structure," schematically shown in Fig. 1.29a.

A linear dependence of the small-angle reflections on the length of the side chain is characteristic of these polymers (Fig. 1.29b). However, in contrast to comb-shaped PA, PMA, and PVE, there is only a halo in the 4.2-4.5 Å region of large angles of scattering on the x-ray patterns of PAM. This circumstance, and the value of the increment of d_1 equal to 1.1 Å with an increase in the length of the side chain by one methylene group, suggested the presence of an irregular conformation of the aliphatic chains, including

a

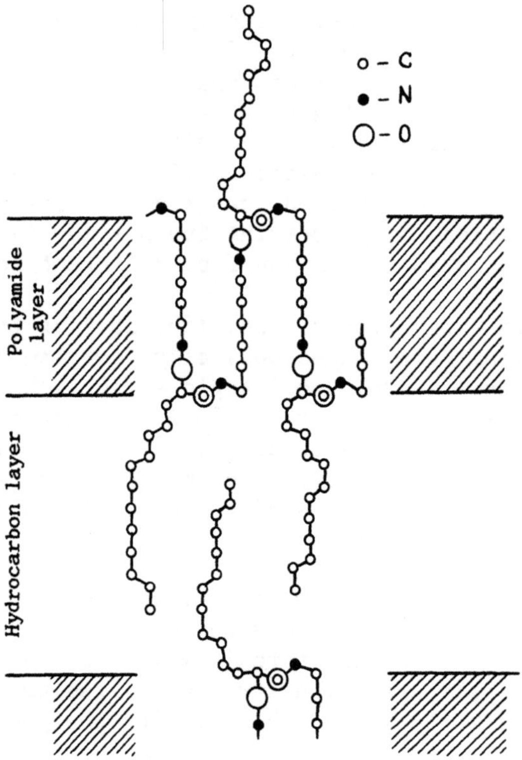

b

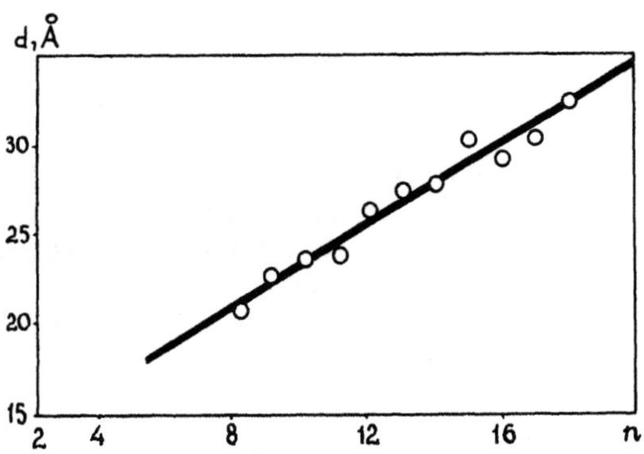

Fig. 1.29. Diagram of the arrangement of the main and side chains in the partially ordered structure of PAM (a) and dependence of small-angle reflection d_1 of the polymers on the length of the side chain (b) [47].

the existence of both trans- and gauche-isomers, as shown in Fig. 1.29a. The suggestion that there is a folded conformation of the polyamide main chain (Fig. 1.29a), the result of comparing the x-ray and IR spectroscopic data for PAM-n and polymers of similar structure of the type

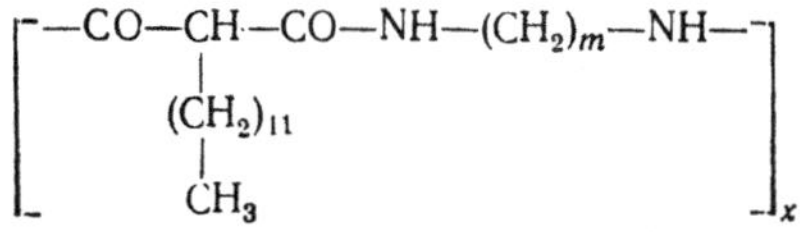

(where m = 4-10) but with a varying distance m between side chains, is of definite interest.

The formation of a folded conformation of the main chain has also been advanced for linear polyethylenimine $[-NH-CH_2-CH_2-]_n$ alkylated by n-hexadecyl units (PEI-16). X-ray and thermographic studies [158] indicate the crystallization of the aliphatic side chains in a lattice of the hexagonal type with insignificant degrees of alkylation (β), where there is an average of one alkyl unit per 14-15 monomeric units in polyethyleneimine. Since the observed value of the d spacing is variable and is dependent on β, and the alkyl groups in the hexagonal lattice are parallel to the direction of the d spacing, a diagram of the structure of the PEI-16 crystallites was hypothesized and is shown in Fig. 1.30. The formation of such a structure with small values of β, as observed in [158], makes it necessary to postulate the formation of a block and not a random distribution of the aliphatic chains in the polyethylenimine macromolecules.

Due to the recent heightened interest in the preparation and study of the properties of rigid-chain liquid-crystalline polymers, a number of studies on the chemical modification of not only aliphatic but also aromatic polyamides by introduction of long aliphatic side chains have been published [159-161].

Synthesis of N-octadecyl poly(n-phenylene) terephthalamide

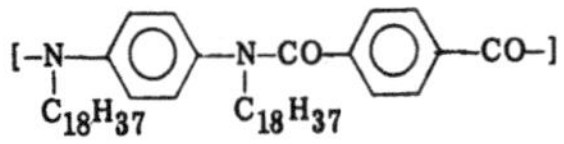

with a degree of substitution by alkyl groups of ~90-97 mole % is described in [159]. The polymer crystallized with the

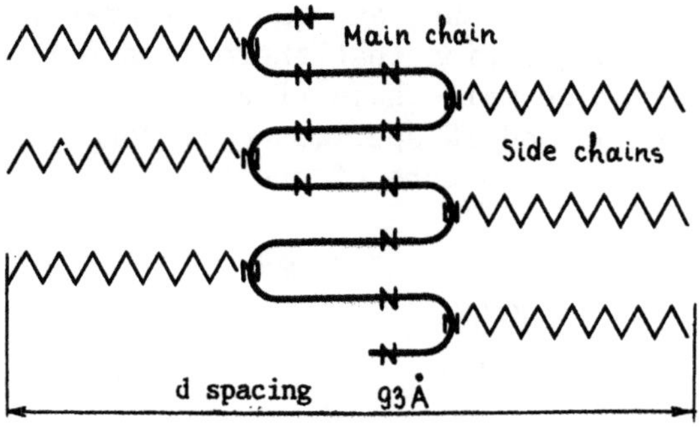

Fig. 1.30. Schematic representation of the structure of PEI-16 crystallites [158].

side chains in a hexagonal lattice with single-layer packing and has a melting point of approximately 41°C. The degree of crystallization is very low, approximately 9%; this is apparently due to the difference in distances between the branching points along the polymer chain, which introduces significant defects in the packing of the aliphatic side chains.

An increase in the distance between the branching points of the side chains, as occurs in N-alkylated copolyamides

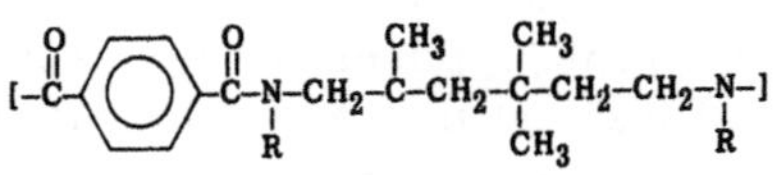

(N-CPA-n)

prepared from terephthalic acid and 2,2,4- and 2,4,4-trimethylhexamethylenediamine [160, 161], where R = $C_{14}H_{29}$, $C_{16}H_{33}$, $C_{18}H_{37}$, and $C_{22}H_{45}$, does not prevent crystallization of the side chain in a hexagonal lattice as before. (The melting points and heats of fusion of these polymers are examined in the next section, and their values are reported in Table 1.7.) However, the absence of small-angle reflections on the x-ray diagrams forced Schulz and Espenschied [160] to draw a conclusion concerning the complete disordering of the main chains, which is not in agreement with the results of studies of other types of comb-shaped polymers.

At the same time, for aromatic rigid-chain copolyesters of the type

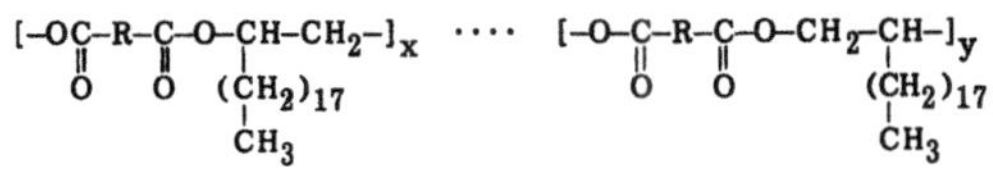

where R corresponds to ortho-, meta-, and para-substituted benzene rings (with random distribution of x and y units), the formation of single-layer packing of the side groups with the main chains located in parallel planes similar to Fig. 1.25e is observed according to [162].

It is interesting to note that ortho-substituted derivatives of copolyesters exhibit maximum values of the melting points and heats of fusion, while para-substituted samples exhibit minimum values of these parameters. This difference is apparently due to the more favorable conformation of the main chain, which promotes crystallization of a large number of methylene units in the ortho-substituted derivatives of the copolymers.

Polyisocyanates with long aliphatic side chains (R)

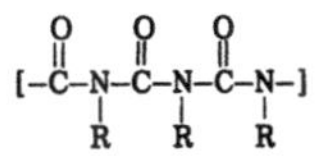

where R = $-(CH_2)_n-CH_3$ (n = 0-17), studied in detail by Aharoni, could be an example of comb-shaped polymers with rigid side chains. Their ability to exist in the crystalline state due to crystallization of the side chains of length $n \geq 13$ was noted, and a hypothesis was proposed concerning the triclinic (or monoclinic) form of the crystal structure of these polymers.

In addition to the synthesis and study of comb-shaped heterochain polymers, the comb-shaped heteroorganic polymers polyorganophosphazines [164],

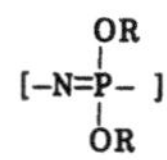

where R varies from CH_3 to $C_{12}H_{25}$, have recently been prepared. The polymers in this series crystallized similarly to PA-n with the formation of two-layer packing of the side chains in the hexagonal form.

The data reported thus form the basis for postulating that the dispersion interaction of the methylene units of the hydrocarbon side chains is the basic factor which determines the structure of a comb-shaped polymer in the solid phase. Regardless of the chemical nature of the main chain, this type of interaction forces the side chains to be arranged in a certain layered order, usually causing the appearance of small-angle spacing in their x-ray diffraction patterns. In some cases, the contribution of this type of interaction can be so strong that it results in a change in the normal conformational state of the macromolecule.

This was most graphically demonstrated on the example of comb-shaped polypeptides based on poly-L-lysine (PL-n) [165-167]:

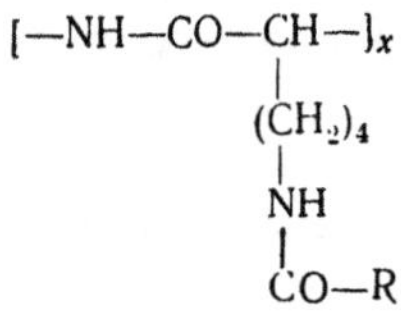

where R = C_5H_{11}, C_7H_{15}, C_9H_{19}, $C_{12}H_{25}$, or $C_{17}H_{35}$.

The possible formation of a certain polypeptide chain conformation (α or β structure) and the tendency of the side chains for intramolecular crosslinking have aroused considerable interest in the study of the conformational state and structure of such compounds.

Macromolecules of poly-L-lysine with a short side chain (PL-5) preserve the conformation of an α-helix characteristic of poly-L-lysine, but starting with PL-9, the side chains begin to become ordered, which initially results in insignificant distortion of the α-helix (PL-9) but then, with its total destruction (PL-12, PL-17), results in the formation of hexagonal packing of the side chains. Crystallization of the side chains does not significantly limit the possibility of the formation of a β-folded conformation of the polypeptide chain. As Fig. 1.27 shows, the main chains, which exist in the conformation of an intramolecular β structure, form layers with the side chains, which crystallize in a hexagonal lattice perpendicular to them. The periodic alternation of the layers formed by polypeptide chains and crystalline regions in the hexagonal lattice of the side chains causes the appearance of small-angle spacing [167].

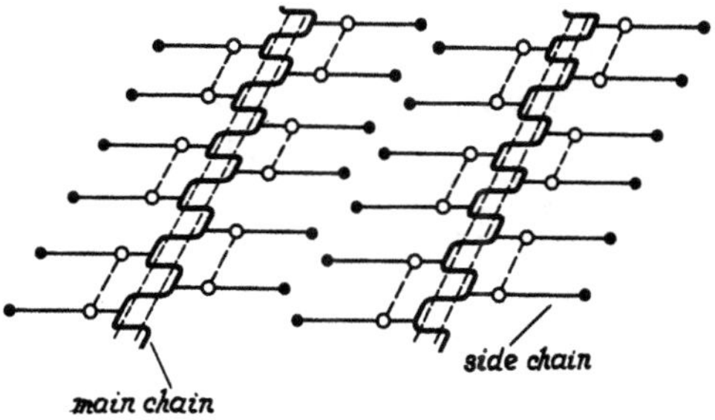

Fig. 1.31. Diagram of the main and side chains of comb-shaped polypeptides with long aliphatic substituents in the crystalline state [167].

Due to the amphiphilic nature of the PL-n macromolecules, it is possible to produce different conformations of the main chains by selective solvation (or precipitation) of the individual structural elements of the macromolecules. In the case of treatment of PL-n with trifluoroacetic acid, which breaks the hydrogen bonds, the order in the main chains is totally destroyed and there is a simultaneous increase in the degree of perfection of the hexagonal packing of the side chains. On the other hand, addition of water to solutions of PL-12 and PL-17 in trifluoroacetic acid results in the appearance of order in the main chains (formation of an intermolecular β structure) with simultaneous preservation of order in the side chains (hexagonal packing). For samples of PL-12 and PL-17, structural order is only possible in the main chains, for example, due to melting of the crystallites formed by the side chains. Reflections corresponding to the β structure of the polypeptide chain are observed in the x-ray patterns of samples of PL-17 prepared at temperatures above the melting point of the crystallites formed by the side chains.

Comb-shaped polypeptides can thus exhibit structural order with respect to both the main and side chains of the macromolecules. Such polymers are essentially the simplest models of protein–lipid complexes in which the protein–lipid interaction is fixed by covalent bonds, and the side chains form layered structures similar to the lipid layers in biomembranes.

The tendency of the aliphatic side chains in comb-shaped polymers toward layered packing was used in [169] to

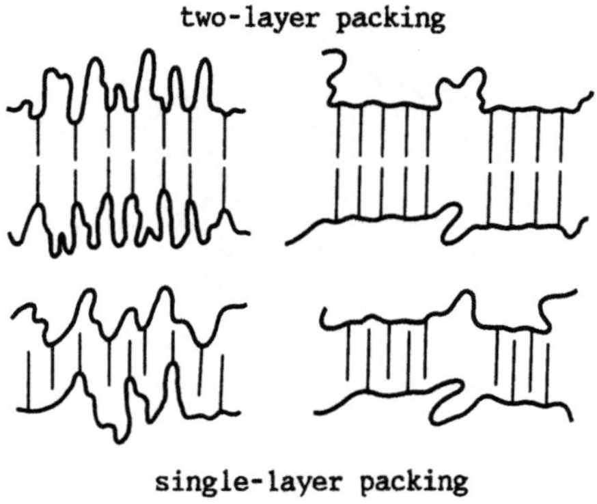

Fig. 1.32. Diagrams of single- and two-layer packing of the side chains of graft copolymers of polyvinyl alcohol (PVA) and paraffins [168].

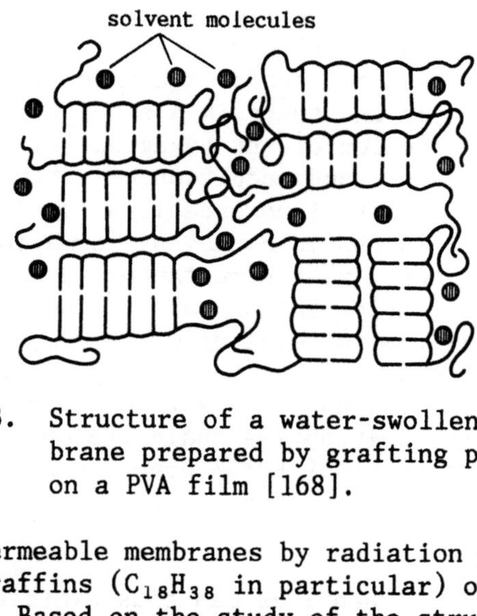

Fig. 1.33. Structure of a water-swollen membrane prepared by grafting paraffin on a PVA film [168].

prepare semipermeable membranes by radiation grafting of long-chain paraffins ($C_{18}H_{38}$ in particular) on a polyvinyl alcohol film. Based on the study of the structure of comb-shaped polymers, diagrams of the single- and two-layer packing of grafted chains of $C_{18}H_{38}$ were proposed for both a random and a regular arrangement in the polyvinyl alcohol macromolecules (Fig. 1.32). The hypothetical structure of a water-swollen membrane prepared from this graft copolymer is shown in Fig. 1.33 [168]. As in the case of comb-shaped polypeptides, the aliphatic side chains play the role of the basic "building material" of the triglycerides of higher fatty acids, which are fundamental in the construction of the lipid bilayers of biological membranes.

1.4.3. Heats of Fusion, Length of Crystallizable Sequences of Units, and Convergence Temperature

As indicated previously, an increase in the length of the side chains in all homologous series of comb-shaped polymers, starting at a certain critical length, is accompanied by an increase in the melting points (see Fig. 1.1, Table 1.2). This mechanism is due to the increasing number of methylene side units involved in crystallization. An increase in the heat of fusion ΔH_f, as in the case of low-molecular-weight paraffins, with an increase in the length of the molecule is observed at the same time as the increase in the melting point.

The hexagonal character of the packing of comb-shaped polymers determines their relatively low heats of fusion. As Table 1.7 shows, the values of ΔH_f vary widely as a function of the chemical nature of the polymer and the length of its side chain [98, 102, 169, 170], but is usually 1.5-1 times smaller than the ΔH_f of paraffins which crystallize in a rhombic lattice with the length of the molecules equal to the length of the side chain of the polymer (see the values of ΔH_f, in kJ/kg, in the last column of Table 1.7).

These values are usually 2-2.5 times smaller than the values of the heats of fusion of paraffins which crystallize in a rhombic (triacontane) and triclinic (eicosane) lattice and have the same length of the molecules as the side chain of the polymer. At the same time, these values of the heats of fusion are close to the ΔH_f of paraffins which crystallize in the hexagonal form [126, 171]. As Table 1.7 shows, good agreement is observed between the estimated T_m and ΔH_f values for the same polymers from studies conducted by different investigators.

Using thermodynamic data to plot the dependence of the heat of fusion ΔH_f on the number of carbon atoms n in the alkyl side chain for comb-shaped polymers and n-alkanes from the expression

$$\Delta H_f = \Delta H_f^{e.g.} + \Delta H_f^{CH_2} n \qquad (1.10)$$

TABLE 1.7. Melting Points and Heats of Fusion of Some Comb-Shaped Polymers

Polymer	T_m, °C	Heat of fusion, ΔH_f J/mole	Heat of fusion, ΔH_f kJ/kg
	Poly(n-alkyl arylate)s (PA-n) [157]		
PA-12 At*	11.9	8,812	36.6
PA-14 At*	31.9	16,735	62.4
PA-16 Iso	36.5	23,520	79.0
PA-16 At	38.0	24,780	83.6
PA-17 At	38.3*	22,650	76.4
	46.0	28,140	90.7
PA-18 At	53.0	31,840	97.9
	50.9*	29,080	89.6
PA-22 At	71.0	43,690	115.0
	66.5*	41,230	108.3
	Poly(n-alkyl methacrylate)s (PMA-n) [157]		
PMA-16 Iso	26.0	14,830	47.9
PMA-16 At	22.0	12,180	39.5
PMA-18 At	38.0	24,570	72.6
PMA-22 At	59.0	35,400	89.9
	Polyvinyl Esters (PVE-n) [157]		
PVE-11	21.0*	5,490	24.3
PVE-15	38.5	19,570	69.3
PVE-17	43.1*	21,630	72.9
	49.5	25,830	83.2
	54.6*	25,850	83.3
	N-Alkylated Copolyamides (n-CPA-n) [160]		
N-CPA-14	−30.0	3,100	9.1
N-CPA-16	17.0	12,800	34.7
N-CPA-18	34.0	19,500	49.1
N-CPA-22	50.0	31,800	70.4
	Poly(N-n-alkyl maleimide)s (N-PMI-n) [172]		
N-PMI-12	−34.8	4,330	0.93
N-PMI-14	−14.6	10,440	2.03
N-PMI-16	7.8	16,760	2.98
N-PMI-18	31.9	23,600	3.86

Table 1.7 (continued)

Polymer	T_m, °C	Heat of fusion, ΔHf J/mole	Heat of fusion, ΔHf kJ/kg
	Polydialkyl itaconates† (PDI-n) [173]		
PDI-11	−32	2,058	4.7
PDI-12	−8	9,878	21.2
PDI-14	19	20,418	39.1
PDI-16	40	32,622	56.4
PDI-18	53	47,490	74.9
PDI-20	58	71,622	103.8
	n-Alkanes		
$C_{19}H_{20}$			
Hexagonal form	21.7	13,827	51.5
Rhombic form	31.8	45,880	170.5
$C_{23}H_{48}$			
Hexagonal form	40.5	21,788	67.0
Rhombic form	47.5	54,051	166.3

*Results obtained in [169, 170].
†The values of ΔHf are expressed in units of 2 kJ/mole.

where $\Delta H_f^{e.g.}$ is the contribution of the end groups of the side chains to the heat of fusion, and $\Delta H_f^{CH_2}$ is the contribution to the heat of fusion per CH_2 group, the number of methylene groups, n_{cryst}, involved in crystallization was calculated. The value of ΔHf for comb-shaped polymers crystallizing in a hexagonal form was assumed to correspond to the value of ΔHf of paraffins in the same hexagonal modification, and in [157], to the value of ΔHf for alkyl acrylates crystallizing in the hexagonal form.

The calculated values of $\Delta H_f^{CH_2}$ for some comb-shaped polymers in comparison with the data for n-alkanes crystallizing in different modifications are reported in Table 1.8.

As Table 1.8 shows, the values of $\Delta H_m^{CH_2}$ for all comb-shaped polymers (except for polyalkylitaconates), alkyl acrylates, and n-alkanes which crystallize in the hexagonal form are similar, a relatively weighty argument in support of the identity of the character of their packing in a hexagonal cell. Polyalkylitaconates are also not an exception, since the presence of two long aliphatic chains in each monomeric unit of their macromolecules

TABLE 1.8. Values of ΔHf^{CH_2} Calculated Using Eq. (1.10) for Compounds Crystallizing in the Hexagonal Modification

Compound	ΔHf^{CH_2}, cal/mole CH_2	Reference
n-Alkanes	950.0*	[174]
	734.9 ± 29.3	[170]
n-Alkyl acrylates	799.4 ± 84.0	[157]
	776.6 ± 47.5	[170]
Poly(n-alkyl acrylate)s	745.4 ± 10.5	[157]
	791.6 ± 27.8	[170]
Poly(n-alkyl methacrylate)s (PMA-n)	792.1 ± 74.0	[157]
N-Alkylated (N-CPA-n) copolyamides	749.4	[160]
Poly(N-n-alkyl maleimide)s (N-PMI-n)	766.5	[173]
Poly(dialkyl itaconate)s (PDI-n)	$\frac{1525}{2}$ = 762.5	[172]

*Orthorhombic modification.

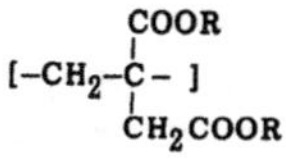

requires a twofold decrease in the value of ΔH_m (J/mole), as shown in Table 1.8. This value, which is twice as small, is in good agreement with the value for other comb-shaped polymers.

The results of calculating the number of methylene units involved in crystallization, n_{crys}, with Eq. (1.10) (method I) for some comb-shaped polymers conducted in [157, 158] are reported in Table 1.9.

The values of the degree of crystallinity calculated in [157, 169] are reported here. A slightly different approach to the calculation of the number of methylene groups involved in crystallization is based on the extrapolation of ΔH_f to 0 and the determination of the number of methylene groups not involved in crystallization, i.e., finding the critical length of the side chain n_{cr} above which the side chains crystallize (method II). It was found that this value is 8 for PA, 9.3 for PVE, and 11.2 CH_2 groups for PMA as a function of the chemical nature of the polymer, i.e., n_{cr} varies within the limits of 8 to 12 carbon atoms, which is in good agreement with the value of n_{cr} determined from the dependence $T_m = f(n)$.

TABLE 1.9. Results of Calculating the Degree of Crystallinity and the Average Number of Crystallized Methylene Groups in the Side Chain (n_{cryst})

Polymer	n_{cryst}				Degree of crystallinity	
	method I		method II			
	[169]	[157]	[169]	[157]	[169]	[157]
Poly(n-alkyl acrylate)s						
PA-12	2.9	—	2.8	—	17	—
PA-14	5.4	—	4.8	—	28	—
PA-16	7.3	7.4	6.8	8.0	35	46
PA-16, iso	—	6.9	—	—	—	—
PA-17	—	8.4	—	9.0	—	49
PA-18	9.4	9.5	8.8	10.0	41	53
PA-22	13—14	13.0	12.8	14.0	49—52	60
Poly(n-alkyl vinyl ether)s						
PVE-11	1.8	—	1.7	—	11	15
PVE-15	7.0	5.8	5.7	5.7	33	39
PVE-17	8.4	7.7	7.7	7.7	38	45
Poly(n-alkyl methacrylate)s						
PMA-16, iso	—	4.4	—		—	28
PMA-16	—	3.6	—	4.8	—	23
PMA-18	—	7.3	—	6.8	—	40
PMA-22	—	10.5	—	10.8	—	48
Poly(N-n-alkyl acrylamide)s						
PAA-12	0.3	—	0.1	—	2	—
PAA-14	2.0	—	2.0	—	10	—
PAA-16	4.5	—	4.0	—	22	—
PAA-18	5.7	—	6.0	—	25	
PAA-22	10.9	—	10.0		40	—

The degree of crystallinity, determined radiographically, is 40-50% for the higher homologs of the acrylic series and poly(alkyl vinyl ether)s (PA-16, PA-18, PVE-17) and 30-40% for polymers of the methacrylic series and alkyl acrylamides (PMA-16, PMA-18, PAA-18), which is in good agreement with the calculated values.

The data in Table 1.9 thus clearly show that not all methylene groups in the side chains participate in crystallization. As indicated above, this is the consequence of conformational distortions in the contiguous segments of the main and side chains, and the "loose" packing of the end groups of the side chains.

However, by selective solvation of the linkages of the main and side chains, which eliminates these distortions, it

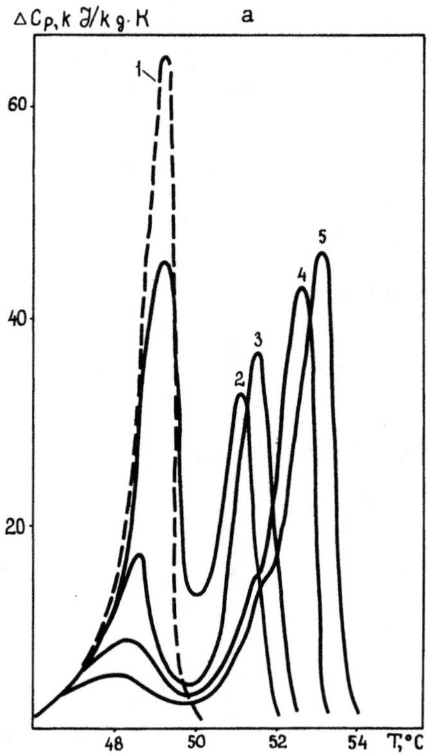

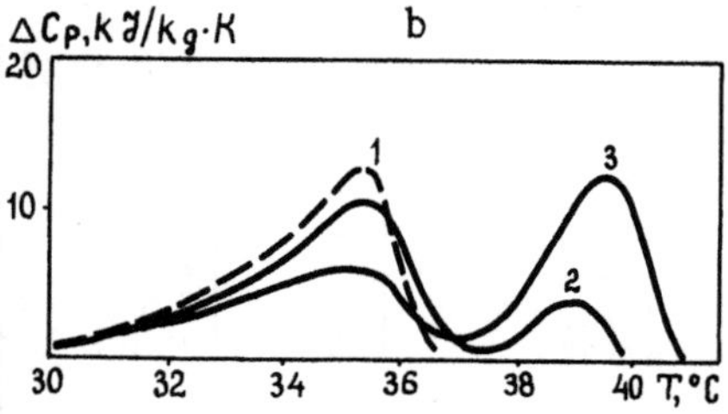

Fig. 1.34. Melting curves of samples of PVE-17 (a) and PMA-18 (b) [175]. a: Quenched sample (1); samples crystallized at 46.3°C for 5 (2), 60 (3), 240 (4), and 1140 min (5) with subsequent quenching at 20°C. b: Quenched sample (1); samples crystallized at 33.5°C for 240 (2) and 400 min (3) with subsequent quenching at 20°C.

is possible to vary the number of crystallized methylene groups within certain limits. This is graphically demonstrated by Jordan et al. [169]. Treatment of poly(n-alkyl acrylamide)s with methanol increases the degree of crystallinity due to the inclusion of a larger number of methylene groups in the crystalline phase and causes the most perfect single-layer packing of the side chains [157].

The detailed study of the thermal behavior of comb-shaped polymers conducted on the example of PA-16, PMA-18, and PVE-17 by Sochava and Sobbotka [175] revealed an interesting feature of the melting and crystallization of these polymers. Using differential scanning calorimetry, they observed the formation

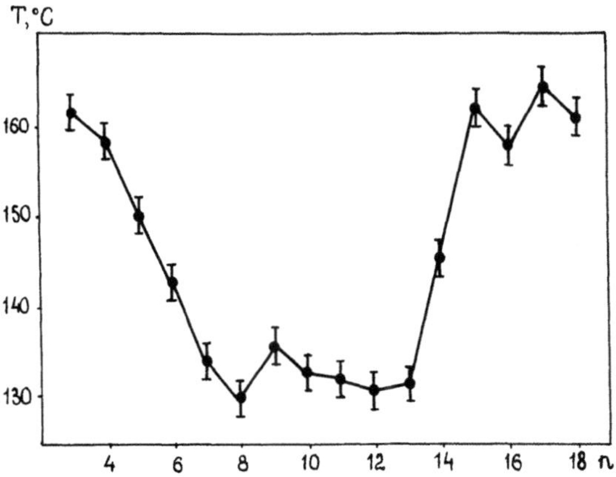

Fig. 1.35. Dependence of the melting points of poly(n-alkyl malonamide)s on the number of carbon atoms in the side chain [47], n.

of two different crystalline modifications for these polymers (Fig. 1.34). The low temperature form (curve 1) arises from quenching of melts of the polymers, and the high-temperature form occurs in isothermal crystallization near the melting range. The difference in the values of the melting points does not exceed 5°C. By varying the duration and temperature of crystallization, two separate endothermic peaks of melting can be seen, as Fig. 1.34 clearly shows. The similar shape of the thermograms of these polymers externally resembles the melting curves of isotactic poly(1-alkyl ethylene)s (see p. 29), although in this case it is difficult to postulate the possibility of separate crystallization and melting of the main and side chains. The observed event of "dual melting" (with a small difference in the melting points) is more probably due to the existence of crystallites based on both single- and two-layer packing of the side chains (cf. preceding section). The doublet nature of the melting thermograms was also observed in [176] in studying comb-shaped poly(fluoroalkyl ether)s

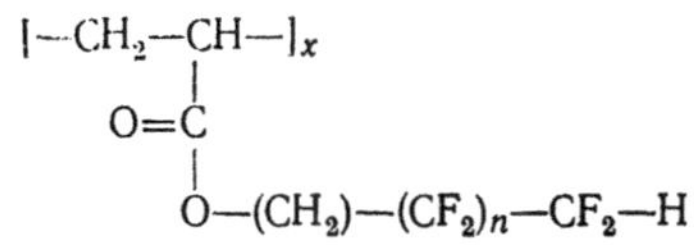

where n = 1, 3, 5, 7, and 9.

In studying the thermal properties of these polymers, sharp endothermic peaks were observed in the thermograms

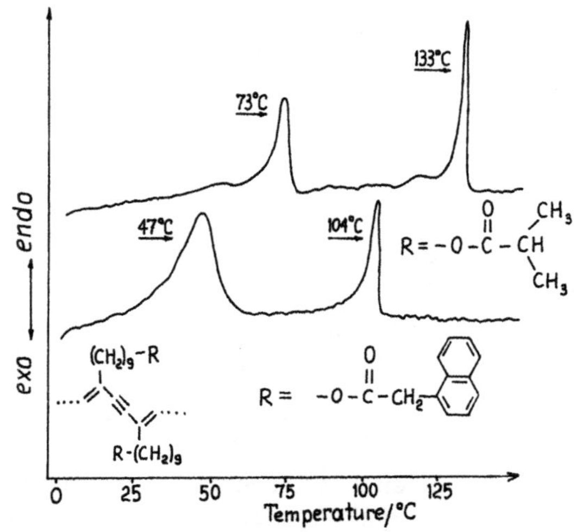

Fig. 1.36. DSC curves of two samples of polymeric diacetylenes [179].

(PPA-7 and PPA-9) indicating the crystallization of these polymers. Two peaks (at 105 and 113°C) were observed for PPA-9, and the ratio of their intensities varied as a function of the crystallization conditions. The possible presence of two crystalline modifications in this polymer was hypothesized, but there are no additional structural data. Nevertheless, the "dual melting" events observed in the last two studies cited and in [158] and [163] with respect to poly(n-alkyl isocyanate)s indicate the relatively complex and not always easily explained structural rearrangements which accompany the process of melting of comb-shaped polymers.

The effects of "dual melting" were observed in studies of the structure and properties of comb-shaped polydiacetylenic derivatives with long alkyl decoupling units (spacers) and different end substituents of the type

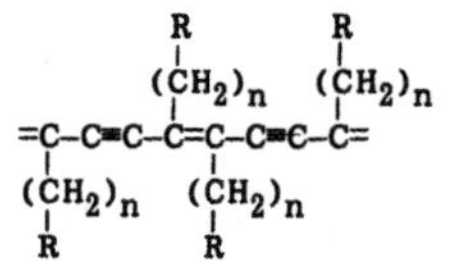

where R are alkyl or aromatic substituents [177-179]. As an example, the thermograms of two polydiacetylenic derivatives are shown in Fig. 1.36, and a diagram of the packing of the polydiacetylene macromolecules is shown in Fig. 1.37. It is

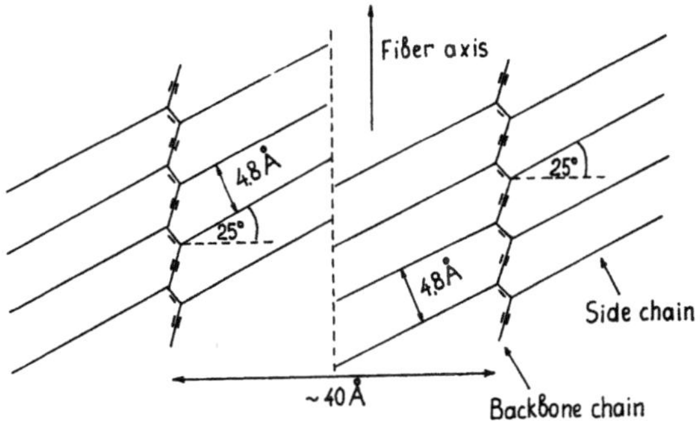

Fig. 1.37. Proposed packing of the long side chains of polydiacetylenes in the oriented fiber [179].

suggested in [179] that the first peak in the thermogram corresponds to melting of the side chains, while the second peak corresponds to melting of the main chains of the polymer.

Systematic thermographic studies of heterochain comb-shaped polymers were conducted by Corradini et al. [47] based on the study of poly(n-alkyl malonamide)s (PAM-n) mentioned previously (cf. p. 67). The dependence, observed in [47], of the melting points of the samples (PAM-n) on the number of carbon atoms in the side chain is shown in Fig. 1.35. The decrease in T_m with an increase in the side-chain length from PAM-3 to PAM-8 is a result of perturbation of the character of the packing of the main chains due to short aliphatic substituents. The further increase in T_m observed for PAM-13 to PAM-18 polymers is attributed in [47] to the specific effect of the long alkyl chains in the formation of the crystal structure. The presence of a minimum on the curve is the result of the effect of these two opposite factors. It should be added that PAM-n, like comb-shaped PMA-n, are characterized by very low values of the heats of fusion which lie in the range of 12 to 29 kJ/kg, caused by the features of the packing of the side chains which can be described within the framework of a rotational-crystalline state.

The dependences of the melting points on the length of the aliphatic side chains n for some of the atactic comb-shaped polymers examined above (and for comparison, isotactic polyoxiranes) are shown in Fig. 1.38; they clearly show that a tendency toward saturation of the values of T_m is observed with an increase in n.

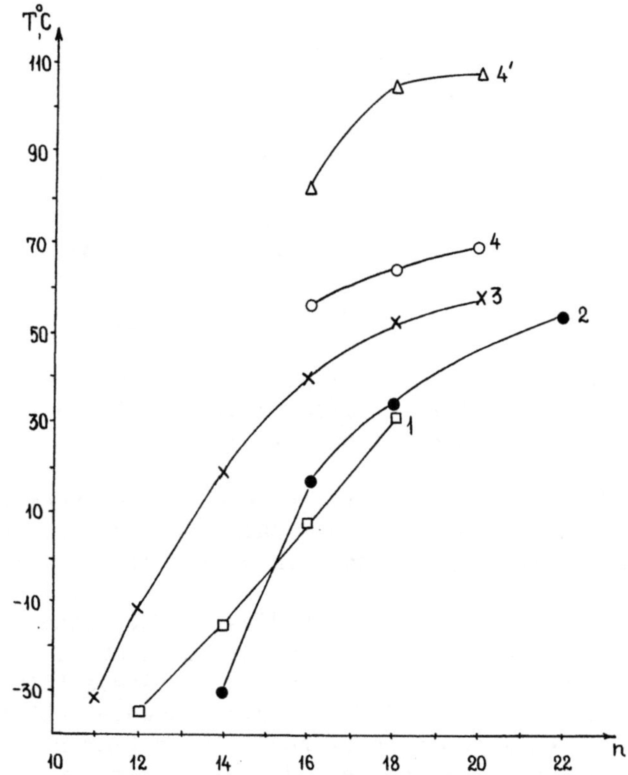

Fig. 1.38. Dependence of the melting point on the number of carbon atoms in the side chain of atactic comb-shaped polymers: 1) poly(N-alkyl maleimide)s (P-N-MI-n) [172]; 2) N-alkylated copolyamides (N-CPA-n) [160]; 3) poly(di-n-alkyl itaconate)s (PDI-n) [173]; 4) atactic poly(alkyl ethylene oxide)s (PEO-n) [112]; 4') isotactic poly(alkyl ethylene oxide)s (PEO-n) [113].

Based on the results of estimating the melting points of PA and PMA comb-shaped polymers in [170, 180], the so-called limit temperature (or convergence temperature) corresponding to the melting point of these polymers with infinitely long side chains was calculated. Using the relation which correlates the T_m of the polymer with the convergence temperature $T_m{}^0$

$$T_m = T_m{}^0[(n + a)/(n + b)]$$

(where a and b are constants, and n is the number of methylene groups in the side chain of the polymer), the values of the convergence temperatures were obtained, equal to 134-135°C for PA [170, 180] and 129°C for PMA [180]. The calculated values of the convergence temperatures should physically correspond to the T_m of the hexagonal lattice of polyethylene, which is found by the x-ray method in special conditions of crystallization of polyethylene with γ irradiation. At the same time, a comparison of the values of the convergence temperatures for these homologous series of polymers and paraffins which crystallize in the hexagonal form exhibited a good agreement (the convergence temperature of paraffins is equal to 135°C).

This method was successfully used by Barrales-Rienda for estimating the convergence temperature T_m^0, the number of crystallized units, and the degree of crystallinity of a series of comb-shaped poly(N-alkyl maleimide)s with T_m^0 of 135°C [172].

The values of T_m and parameters a and b calculated in [170, 172, 180] can thus be used for calculating and predicting the melting points in a number of comb-shaped polymers of the acrylic and methacrylic series.

1.5. CRYSTALLIZATION OF COMB-SHAPED COPOLYMERS

Based on numerous data, it is well known that branching in linear stereoregular polymers, which perturbs the regular structure of the macromolecules, prevents their crystallization [1-6].* Turner-Jones [95] studied the structure of copolymers of 1-ethylene with 1-butyl ethylene and higher 1-alkyl ethylenes and observed a gradual decrease in the capacity to incorporate comonomeric units in the crystal lattice of poly(1-ethyl ethylene) with a high concentration of 1-ethyl ethylene as the length of the alkyl group of 1-alkyl ethylene increased. The 1-butyl ethylene and 1-hexyl ethylene units are already too large for isomorphic substitution,

*A detailed examination of the questions related to crystallization and crosslinking in irregular polymer systems (including statistical, block, and graft copolymers) and a critical analysis of some theoretical statements advanced by Flory and Kilian et al. are found in [21].

but several units can enter the crystal lattice due to swelling of the lattice. The incorporation of 1-heptyl or 1-octyl ethylene units in the crystal lattice is not considered very probable, while the incorporation of the 1-decyl ethylene unit is considered impossible.

The strongest effect of long-chain branches on the capacity of stereoregular polymers to crystallize was found in the example of isotactic copolymers of isopropyl acrylate and hexadecyl acrylate (A-16) [102]. With only 3-4 mole % A-16 in the copolymer, high-melting crystalline polyisopropyl acrylate (T_m = 162°C) passes into the amorphous state.

The observed sharp effect of such a small amount of "destroyer units" on the decrease in the capacity of a stereoregular polymer to crystallize is difficult to explain without postulating some "long-range interaction" of the units of the long-chain comonomer incorporated. The long side chains apparently cause significant conformational distortions of the main chain by strongly shielding the segments and preventing the close packing of the stereoregular sequences of the units. On the other hand, an increase in the concentration of long-chain branches creates favorable conditions for their ordering, and the copolymers can crystallize in this case due to the interaction of the side chains.

The study of the structure of a large group of copolymers of long-chain monomers of the acrylic and methacrylic series with such comonomers as methyl methacrylate, methyl acrylate, methacrylic acid, vinyl chloride, etc. [102, 175, 181] showed that when 50-60 mole % of the comonomer is added, the comb-shaped copolymers retain the crystal structure. For example, as Table 1.10 shows, addition of 70-85 mole % of the comonomer – isopropyl acrylate and methacrylic acid – to PA-16 and PMA-16 does not cause any significant alteration of the crystal structure of the latter, indicated by the preservation of the interplanar distances characteristic of the initial crystalline polymers.

The difference only consists in the fact that the six-point reflection observed in the x-ray patterns of these copolymers (Fig. 1.39) is rotated by 30° with respect to the same reflections characterizing the x-ray patterns of homopolymers (Fig. 1.39a). This is due to the "foreign" units in the copolymer chains which increase the average distance between neighboring aliphatic side chains in the same macro-

TABLE 1.10. Interplanar Distances in Comb-Shaped Copolymers [157]

Comp. of copolymers, mole %	Config. of main chain	T_m of copolymer, °C ± 1°C	Number of diffraction maxima (reflections)							
			1	2	3	4	5	6	7	8
			precision of calculation, Å							
			±1.0	±1.5	±0.3	±0.08	±0.07	±0.1	±0.05	±0.03
			Copolymers of A-16 and Isopropyl Acrylate							
63:33	At	32.0	–	26.5 s	13.6 w	8.69 s	6.34	–	4.19 s	2.43
56:44	At	26.5	–	26.5 s	13.9 w	8.70 w	–	4.6 w. dif	4.19 s	2.43
42:58	At	21.0	–	27.0 s	13.6 w	8.81 w	–	4.7 m. dif	4.19 s	2.40
19:81	At	–	–	28.0 s	14.0 w	–	–	4.8 s. dif	4.19 m	–
39:61	Iso	18.0	–	28.5 s	14.5 w	–	–	4.6 m. dif	4.19 m	–
16:84	Iso	15.0	–	28.0 m	–	–	–	4.8 s. dif	4.19 m	–
4:96	Iso	–	–	–	–	–	–	4.8 s. dif	–	–
PA-16	At	38.0	42.0 vs	25.0 vw	14.7 m	8.34 m	6.0	–	4.19 vs	2.43
			Copolymers of M-16 and Methacrylic Acid							
	Temp. of x-ray photography, °C									
76:24	22	17.5	–	28.0 s	16.5 m	–	–	4.7 s. dif	–	–
36:64	22	17.0	–	28.0 m	17.1 m	–	–	4.7 s. dif	–	–
9:91	22	15.0	–	28.6 w	–	–	–	4.7 s. dif	–	–
76:24	–196	17.5	–	30.0 s	–	–	–	4.6 s. dif	4.15 s	2.46 w
36:64	–196	17.0	–	30.0 s	14.9 m	–	–	4.6 s. dif	4.20 m	2.46 w
9:91	–196	15.0	–	31.0 vw	14.0 m	–	–	4.9 s. dif	4.20 m	–
PMAA	22	–	–	–	–	10.60s.dif	5.3s.dif	–	–	–
PMA-16	–196	22.0	42.0 vw	29.0 vw	14.7 s	8.67 m	–	–	4.19 vs	2.41 w

Notes. 1. The x-ray measurements of the copolymers were made at temperatures below their T_m. 2. The symbols are the same as in Table 1.4 (dif: diffuse halo).

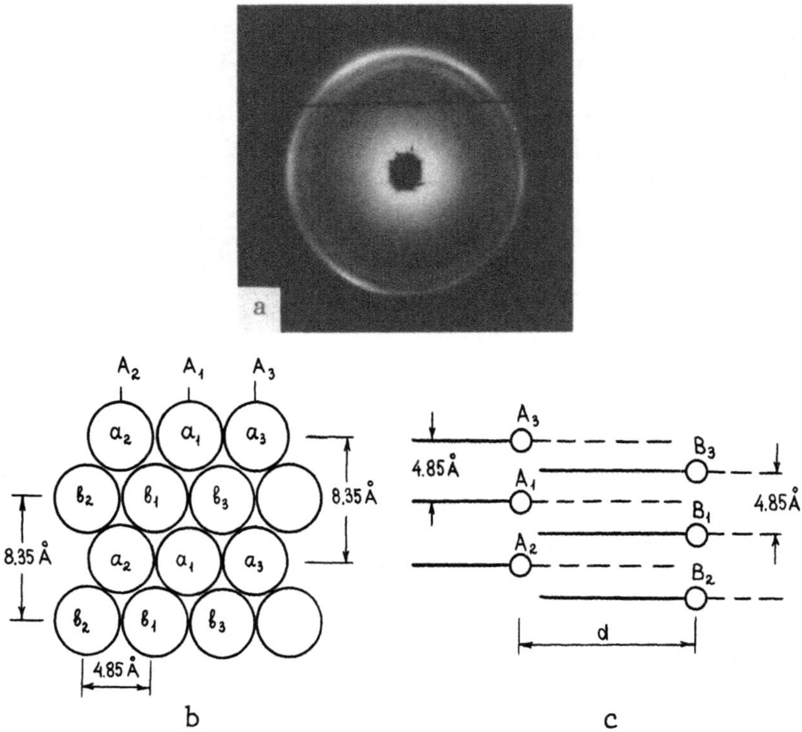

Fig. 1.39. X-ray pattern of a copolymer of A-16 with isopropyl acrylate (a) and diagram of the packing of the main and side chains [projection on the plane of the figure along the axis of the main chain (c) and section of hexagonal packing (b)] (the broken lines correspond to side chains lying in a plane 4.2 Å from the plane in which the other side chains lie, indicated by the solid lines) [139].

molecule, as a comparison of the diagrams of the side chain packing in Figs. 1.24 and 1.39b shows.

Nevertheless, the hexagonal character of the packing of the side chains is also preserved in the case of the copolymers, as seen in the diagram of the packing of aliphatic chains a and b and main chains A and B of comb-shaped macromolecules, according to [139]. The x-ray patterns are characterized by sharp meridional reflections for copolymers of the methacrylic series; in agreement with the data examined above (see p. 51), this indicates the disorientation of segments of the hexagonal packing of the side chains [103]. The 4.19 Å reflection nevertheless remains relatively intense for all of the copolymers, although the proportion of amorphous scattering in the 4.6-4.8 Å region increases with an increase in the concentration of the comonomer. The MA-16—MAA copolymer with a 9 mole % concentration of MA-16 also crystallizes in a hexagonal lattice on cooling. Jordan et al. [170] cited examples of crystallization of copolymers of A-18 with such comonomers as methyl, butyl, and octyl acrylates with a total concentration of A-18 of only 5 mole %. The observed events could be due to the block character of the distribution of the units in the macromolecules.

At the same time, the presence of foreign units in comb-shaped macromolecules sharply affects the method of packing of the layered structures. Actually, as Table 1.10 shows, the reflection corresponding to two-layer packing of the side chains (d_1 = 42 Å) disappears in copolymers of A-16 with isopropyl acrylate. The appearance of significant conformational distortions in comb-shaped macromolecules, induced by "destroyer units," apparently results in shifts in the layered structures which should perturb the two-layer packing of the side chains, as indicated above. These copolymers crystallize in the single-layer type of packing of the side chains, as indicated by the sharp increase in the intensity of reflection d_2 on their x-ray patterns and the disappearance of reflection d_1. Similar mechanisms were also observed in [139] for copolymers of A-18 with methyl acrylate.

The x-ray study of copolymers of MA-16 with MAA (with a 24 and 54 mole % concentration of MAA) at temperatures above their melting point showed the preservation of the layered packing of the side chains (presence of reflection d_2) in the absence of hexagonal packing (replacement of the reflection with d_1 = 4.19 Å by a diffuse amorphous halo in the 4.6-4.9 Å region). However, cooling of the same samples (below the melting point) causes their fast crystallization regardless of the cooling mode, and this is also observed for the initial homopolymers.

A similar picture is also observed for crosslinked copolymers of A-16 with butylene glycol dimethacrylate (DM-4) [157]. The presence of 1-3 mole % of DM-4 has almost no effect on the structure and melting point of PA-16. Crosslinked copolymers with a widely spaced lattice have two-layer hexagonal packing of the side chains like the initial PA-16. An increase in the concentration of comonomeric units to 8 mole % is accompanied by an increase in the proportion of the amorphous phase (appearance of an amorphous halo in the 4.6-4.8 Å region) and causes a change in the type of packing of the layered structures: from two-layer to single-layer (disappearance of reflection d_1 and increase in the intensity of reflection d_2). The melting point decreases slightly, although the hexagonal character of the side chain packing is preserved up to a concentration of the crosslinking agent of 25 mole %.

The preservation of the hexagonal packing of the side n-octadecyl groups was observed in [182] for the alternating copolymer of phthalic anhydride and octadecyloxirane

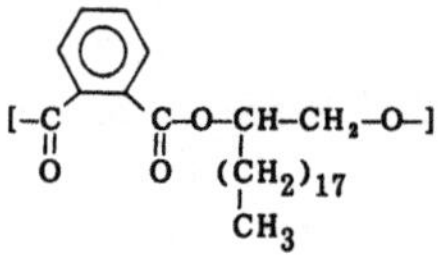

Despite the presence of rigid aromatic nuclei in the main chain, this copolymer is easily crystallized, although it has a low melting point and heat of fusion (T_m = 43°C; ΔH_f = 19.8; compare with Table 1.4).

Despite the small fraction of crystallizable units in the comb-shaped copolymers and the presence of a dense network of chemical bonds between the main chains of the macromolecules in crosslinked copolymers, the side chains are relatively independent in their behavior, which permits them to crystallize in a single-layer packed hexagonal cell.

Examination of the behavior of copolymers containing long-chain comonomers clearly distinguished comb-shaped polymers from linear polymers in which crystallization is usually impossible even with a 20-25 mole % concentration of the uncrystallizable component. The melting points and heats of fusion of comb-shaped copolymers, like linear copolymers, decrease with an increase in the concentration of the uncrystallizable component. However, the experimentally observed decrease in the melting point of the copolymers is slower than that of Flory's theory, developed with respect to crystallization of the main chains of linear polymers without considering the special case of side-chain crystallization.

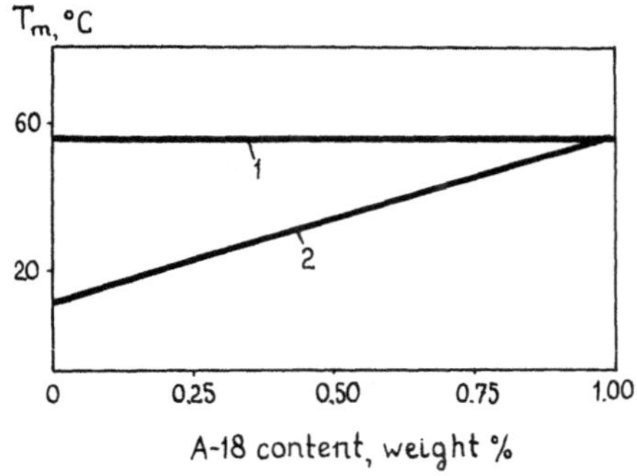

Fig. 1.40. Phase diagram of copolymers of A-18 and VE-17 (1) and A-18 and A-12 (2) [183].

Crystallization due to the packing of the side chains thus takes place easily and in a wide range of compositions for comb-shaped polymers, i.e., introduction of "foreign" units causes comparatively small changes in the packing of the comb-shaped macromolecules. In addition, by using long-chain compounds as the pairs of comonomers, it is possible to observe events of isomorphism. The analysis of the phase diagrams of copolymers of alkyl acrylates and alkyl methacrylates with the same length of the alkyl group – 14, 16, and 18 carbon atoms [28] – and n-octadecyl acrylate (A-18) with n-octadecyl acrylamide (AA-18) [183] shows that these copolymers form true mixed crystals due to the packing of the methylene side chains, and their melting point is almost independent of the molar ratio of components.

Isomorphism is also possible in the case where the copolymer consists of comonomers with different alkyl side-chain lengths. Phase diagrams for copolymers of A-12 with A-18 and A-18 with VE-17 are shown in Fig. 1.40. In both cases, T_m is linearly dependent on the composition of the copolymer in the entire range of ratios of the components. As a consequence, a difference in the length of the side chains is not a factor which perturbs the hexagonal character of the packing of the methylene chains. The capacity for isomorphic substitution of the macromolecules of this copolymer indicates the possibility of significant shifts in the packing of the layers without any perturbation of their crystalline packing. In this sense, the significant difference between such polymeric systems and mixtures of even n-paraffins, which have a different symmetry and do not form continuous solid solutions, should be emphasized [132].

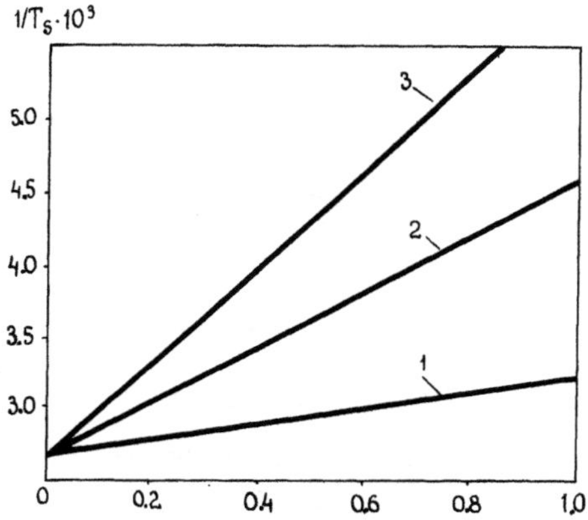

Fig. 1.41. Effect of the mole fraction of some N-alkyl acrylamides (AA) on the softening point of copolymers of acrylonitrile with N-butyl acrylamide (1), N-octyl acrylamide (2), and N-octadecyl acrylamide (3) [183].

The capacity of statistical copolymers which contain a long-chain monomer to crystallize due to the local packing of the side chains is of great interest with respect to the possibility of utilizing this effect for unusual plasticization of the polymers. Introduced in the polymer chain by copolymerization, the long-chain units of the copolymer can crystallize due to the packing of the methylene chains, playing the role of a chemically bound filler in a certain temperature range. Above the melting point of the methylene side chains, the units of the long-chain comonomer can play a different role due to the low glass transition temperature of comb-shaped homopolymers (see Fig. 1.1), i.e., the role of an active plasticizing agent, which significantly decreases the glass transition temperature of the copolymer.

Such systems should be considered unusual physical mixtures containing a polymer, whose chemical structure is defined by the structure of the main chain, as one component, and a melt of a low-molecular-weight compound, a plasticizer, whose chemical structure is defined by the structure of the side chain, as the second component.

The study of copolymers of methyl methacrylate, acrylonitrile, vinyl chloride, etc., with long-chain monomers (such as n-alkyl acrylates, n-alkyl acrylamides, etc.) [182, 183] shows that the plasticizing effect of these compounds is in many respects greater than the effect of low-molecular-weight plasticizers and increases with increasing length of the alkyl side chain in the comonomer. As an example, the dependence of the softening point T_s (this point is only several degrees higher than T_g according to [183]) for copolymers of acrylonitrile with n-alkyl acrylamides with a length of the n-alkyl side group of 4, 8, and 18 carbon atoms is shown in Fig. 1.41. This figure clearly shows that with the same concentration of n-alkyl acrylamide units in the polymer chain, the longer side chains lower T_s and, consequently, T_g, to a greater degree.

These data demonstrate the effect of long side groups on the capacity of the polymers to crystallize and open up prospects for the possible regulation of these processes by the introduction of long-chain monomers in linear polymers by copolymerization. In addition, the possibility of preparing easily crystallized comb-shaped polymers and copolymers even in conditions of radical polymerization is promising for the synthesis of block copolymers containing comb-shaped polymers as the crystallizable units; this would make it possible to prepare materials with properties similar to thermoelastic plastics.

1.6. SUPERMOLECULAR STRUCTURE

The comparative study of the character of supermolecular formations of comb-shaped polymers makes it possible to determine how differences in the molecular structure of polymers and copolymers with long side chains (chemical nature and flexibility of the main chain, method of packing of the side chains, etc.) are made manifest on the supermolecular level.

Amorphous comb-shaped polymers do not form any supermolecular structures which distinguish them from the usual glasses or elastomers. The morphology of crystallizable polymers, which form both globular and fibrillar structures*

*Here and below, "extended aggregates of chains with a developed internal structure" will be called fibrillar formations according to [2].

as a function of the concentration of the solution, provide more interesting possibilities [96, 100, 101]. For samples of PMA-18 and poly(cetylvinyl ether) prepared from very dilute solutions (0.001 wt. %) at room temperature, the formation of globular structures alone was observed [82] (Fig. 1.42a). The minimum size of the observed globules only just exceeds the size of the molecular coils calculated for PMA-18 from its molecular weight and density. Increasing the concentration of the solutions of these polymers to 0.1 wt. % and annealing the prepared samples for a long time at a temperature below their melting points results in the appearance of fibrillar or spherulitic structures (Fig. 1.42b).

All crystallizable comb-shaped PA, PMA, and PV ethers and esters always form fibrillar or spherulitic structures on crystallization from melts regardless of the cooling mode (and even with abrupt cooling of melts to the temperature of liquid nitrogen). Some typical structural formations for comb-shaped PA-16 and PVE-17 prepared by crystallization from solutions and melts are shown in Fig. 1.35b-g. The character of the structures formed is similar to the structure of such well-crystallizable polymers as polyamides, polyesters, etc.

The minimum thickness of the fibrillar formations of these polymers is 40-50 Å and the maximum is several hundred angstroms. Note the total similarity of the supermolecular formations for both isotactic and atactic polymers (Fig. 1.42c-g). Spherulites with the appearance of sheaf-like structures with a very pronounced center of growth as the first stage of formation appear in both atactic and in isotactic polymers. The microdiffraction from the observed fibrillar and spherulitic formations indicates the hexagonal character of the side-chain packing.*

Polymers and copolymers with a comb-shaped structure without hexagonal packing of the side chains do not form fibrillar and spherulitic structures [82, 101]. Copolymers of

*We note that in crystallization of comb-shaped polymers from both melts and solutions, it is not possible to observe the formation of their higher morphological forms, single crystals. However, as the structural studies of gels of these polymers in n-aliphatic alcohols showed, monocrystalline formations are easily formed in crystallization of the polymers in a gel (see Chapter 3).

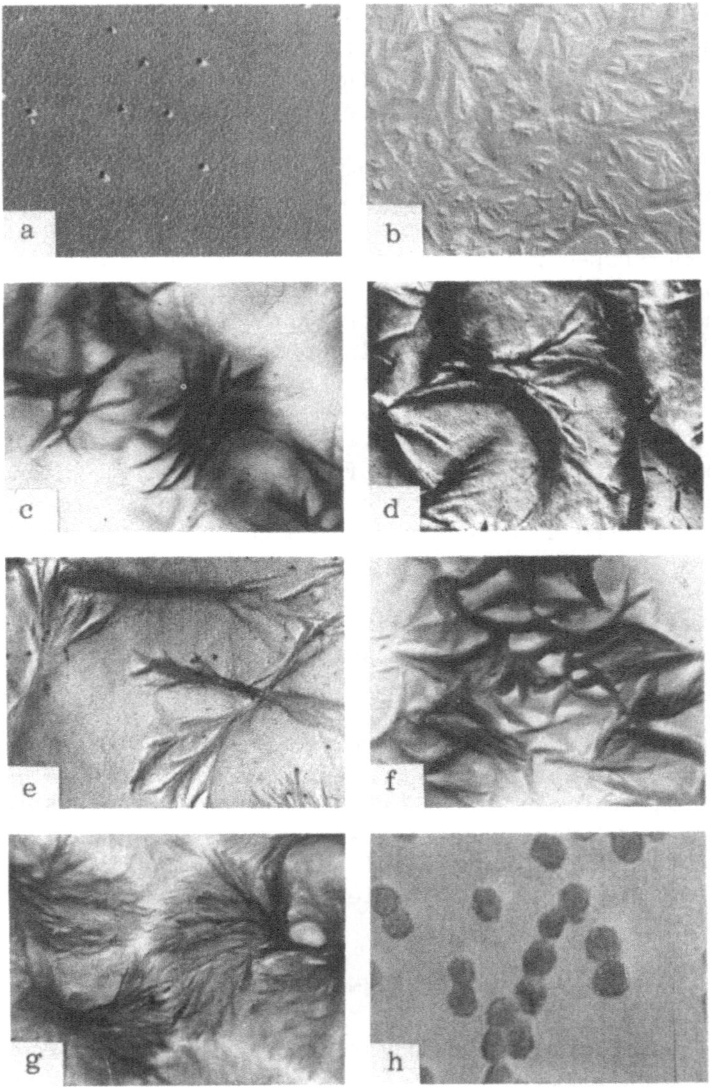

Fig. 1.42. Electron micrographs of comb-shaped polymers and copolymers [96, 100]: a) PMA-18 from 0.001% solution in chloroform; b) PMA-18 from 0.1% solution in chloroform after annealing; c, d) isotactic PA-16 (c) and atactic PA-16 (d) from 0.05% solution in hexane; e) PVE-17 from 0.05% solution in hexane; f) PVE-15 from 0.05% solution in octane; g) PVE-17 film after annealing; h) copolymer of MA-16 and methacrylic acid (MAA) (11 mole % MAA) from 0.1% solution in chloroform.

MA-16 with methacrylic acid (MAA) containing 11 mole % MAA or more only form globular structures of irregular shape (Fig. 1.42g): in the x-ray patterns of such copolymers, the crystal reflection in the 4.19 Å region is replaced by diffuse scattering with an average value of d = 4.7-4.8 Å, characteristic of PMA-16 in the amorphous state. However, the small-angle reflection corresponding to the layered packing of the PMA-16 molecules (approximately 29 Å) is preserved. Other comb-shaped polymers which do not form hexagonal side-chain packing but produce very pronounced layered structures – polyhexylidene and polymethacrylyl hydroxybenzoic acid esters with n-aliphatic alcohols containing 3, 9, and 16 carbon atoms (PPMOB-3, 9, 16) – also have a globular structure similar to the copolymers [82].

Consequently, the length of the side chains, which determines the phase state of the polymer, also determines the character of the supermolecular structures of comb-shaped polymers, and a change in the configuration of the main chain of the macromolecules does not alter the type of supermolecular formations. All crystallizable atactic and isotactic compounds (PA, PMA, PV ethers and esters) are capable of forming different morphological forms (fibrils, dendrites, and spherulites) characteristic of linear stereoregular polymers. At the same time, the formation of globular structures characterized by the hexagonal packing of the side chains [82, 101] can also be observed even for crystallizable comb-shaped polymers under certain preparation conditions; this indicates a high degree of perfection in the intramolecular structure of the globules.

REFERENCES

1. L. Mandelkern, Crystallization of Polymers, McGraw-Hill, New York (1964).
2. P. Geil, Polymer Single Crystals, Wiley, New York (1964).
3. A. Sharples, Introduction to Polymer Crystallization, Arnold Ltd., London (1966).
4. V. A. Kargin and G. L. Slonimskii, Short Essays on the Physical Chemistry of Polymers [in Russian], Khimiya, Moscow (1967).
5. R. Tyudze and T. Kavai, Physical Chemistry of Polymers [in Russian], Khimiya, Moscow (1977).
6. B. Wunderlich, Macromolecular Physics, Vols. 1-3, Academic Press, New York (1973-1980).

7. B. K. Vainshtein, Diffraction of X Rays on Chain Molecules [in Russian], Khimiya, Izd. Akad. Nauk SSSR, Moscow (1963).
8. V. A. Marikhin and L. P. Myasnikova, Supermolecular Structure of Polymers [in Russian], Khimiya, Leningrad (1977).
9. R. Evans, H. Mighten, and P. Flory, J. Am. Chem. Soc., 72, 2018-2028 (1950).
10. P. Flory, Principles of Polymer Chemistry, Cornell University Press, Ithaca (1953).
11. P. Flory, Trans. Faraday Soc., 848-862 (1955).
12. P. Flory, J. Am. Chem. Soc., 84, 2857-2867 (1962).
13. H. Kilian, Kolloid Z. Z. Polym., 176, 149-163 (1963); 189, 23-36 (1963).
14. H. Kilian and R. Muller, Kolloid Z. Z. Polym., 192, 34-45 (1963).
15. E. Helmuth, H. Kilian, and R. Muller, J. Polym. Sci., C6, 101-108 (1964).
16. H. Kilian, Kolloid Z. Z. Polym., 202, 96-105 (1965).
17. L. Mandelkern, Chem. Rev., 56, 903-958 (1956).
18. B. Wunderlich and D. Poland, J. Polym. Sci., A1, 357-368 (1963).
19. B. Wunderlich, Polymer, 5, 125-131 (1964).
20. C. Baker and L. Mandelkern, Polymer, 7, 7-15, 71-82 (1966).
21. N. A. Platé and V. P. Shibaev, Zh. Vses. Khim. Ova. D. I. Mendeleeva, 9, 638-653 (1964).
22. V. A. Kargin, J. Polym. Sci., C4, 1601-1632 (1966).
23. G. Battaerd and D. Tregear, Graft Copolymers, Wiley (1967).
24. W. J. Burlant and A. S. Hoffman, Block and Graft Copolymers, Reinhold (1960).
25. J. McGrath and A. Noshay, Block Copolymers, Academic Press (1977).
26. C. E. Rehberg and C. H. Fisher, J. Am. Chem. Soc., 66, 1203-1206 (1944).
27. H. Kaufman, A. Sasher, T. Alfrey, and I. Fankuchen, J. Am. Chem. Soc., 70, 3147-3150 (1948).
28. S. Greenberg and T. Alfrey, J. Am. Chem. Soc., 76, 6280-6285 (1954).
29. D. Lutz and L. Witnauer, J. Polym. Sci. B2, 31-35 (1964).
30. V. P. Shibaev, B. S. Petrukhin, Yu. A. Zubov, N. A. Platé, and V. A. Kargin, Vysokomol. Soedin., A10, 216-229 (1968).
31. R. Wiley and G. Brauer, J. Polym. Sci., 3, 647-652 (1948).

32. W. Port, E. Jordan, J. Hansen, and D. Swern, J. Polym. Sci., 9, 453-465 (1952).
33. S. Rogers and L. Mandelkern, J. Phys. Chem., 61, 985-992 (1957).
34. P. Graham, J. Polym. Sci., 38, 209-217 (1959).
35. T. Nakahara and K. Motomura, Bull. Chem. Soc. Jpn., 40, 495-501 (1967).
36. W. Burlant, J. Hinsch, and C. Taylor, J. Polym. Sci., A2, 57-64 (1964).
37. C. Overberger, L. H. Arond, R. H. Wiley, and R. R. Garrett, J. Polym. Sci., 7, 431-440 (1951).
38. W. Port, J. Hansen, E. Jordan, T. Dietz, and D. Swern, J. Polym. Sci., 7, 207-215 (1951).
39. H. Teeter, J. Am. Chem. Soc., 40, 143-155 (1963).
40. J. Lal and G. Trick, J. Polym. Sci., A2, 4559-4567 (1964).
41. J. Lal, J. McGrath, and K. Scott, J. Appl. Polym. Sci., 9, 3471-3483 (1965).
42. C. Overberger, G. Fraizier, J. Mandelman, and H. F. Smith, J. Am. Chem. Soc., 75, 3326-3335 (1953).
43. D. Brownawell and I Ming Feng, J. Polym. Sci., 60, 19-21 (1962).
44. A. Klarman, A. Galanti, and L. Sperling, J. Polym. Sci., A7, 1513-1525 (1969).
45. E. Jordan, G. Riser, B. Artymyshyn, W. Parker, J. Pensabene, and A. Wrigley, J. Appl. Polym. Sci., 13, 1777-1794 (1969).
46. G. Maglio, E. Musco, R. Palumbo, and F. Riva, J. Polym. Sci., Polym. Lett. Ed., 12, 129-130 (1974).
47. P. Corradini, N. Lanzetta, G. Maglio, R. Palumbo, and F. Riva, J. Polym. Sci., Polym. Chem. Ed., 13, 1107-1123 (1975).
48. T. Nakahara, K. Motomura, and R. Matuura, J. Polym. Sci., 4A-2, 649-656 (1966).
49. N. Beredjik, "Investigation of monomolecular layers of polymers," in: Newer Methods of Polymer Characterization, B. Ke (ed.), Interscience, New York (1972).
50. E. Reding, J. Polym. Sci., 21, 547-558 (1956).
51. C. Willces and M. Lehr, J. Macromol. Sci. Phys., B7, 225-236 (1973).
52. G. Yeh, Pure Appl. Chem., 31, 65-77 (1972).
53. G. Yeh, J. Macromol. Sci. Phys., B6, 451-465 (1972).
54. P. Smit, Kolloid Z. Z. Polym., 250, 27-45 (1972).
55. H. Balyuzi and R. Burge, Biopolymers, 10, 777-778 (1971).
56. Yu. K. Ovchinnikov, G. S. Markova, and V. A. Kargin, Vysokomol. Soedin., A11, 329-346 (1969).

57. H. Zachman and W. Gölz, Preprints of IUPAC, Vol. 4, Helsinki (1972), p. 359.
58. V. P. Rochupin, T. K. Goncharov, and Z. A. Karapetian, Vysokomol. Soedin., B14, 484-485 (1972).
59. E. B. Bokhyan, Yu. K. Ovchinnikov, G. S. Markova, and N. F. Bakeev, Vysokomol. Soedin., A16, 376-384 (1974).
60. D. Chapman (ed.), Biological Membranes. Physical Fact and Function, Academic Press, New York (1968).
61. S. Tumasheff and G. Fasman (eds.), Structure and Stability of Biological Macromolecules, Marcel Dekker, New York (1969).
62. V. A. Kargin, A. I. Kitaigorodskii, and G. L. Slonimskii, Kolloid. Zh., 19, 131-132 (1957).
63. V. A. Kargin, J. Polym. Sci., 30, 247-261 (1958).
64. S. A. Arzhakov, N. F. Bakeev, and V. A. Kabanov, Vysokomol. Soedin., A15, 1154-1167 (1973).
65. G. Yeh, in: Proceedings of the International Symposium on Macromolecular Chemistry [in Russian], Vol. 1, Tashkent (1978), pp. 74-75.
66. H. Benoit, J. Macromol. Sci., B12, 27-40 (1976).
67. E. Fisher, J. Wendorff, and M. Dettenmaier, J. Macromol. Chem., B12, 41 (1976).
68. M. Meyer, J. Van der Sande, and D. Uhlmann, J. Polym. Sci., Polym. Phys. Ed., 16, 2005-2021 (1978).
69. R. Stein and S. Hong, J. Macromol. Sci., B12, 125-138 (1976).
70. P. Flory, Statistical Mechanics of Chain Molecules, Wiley—Interscience, New York (1969).
71. P. Flory, in: Plenary and Invited Lectures of the 29th International Symposium on Macromolecules, Bucharest, Romania (September 5-9, 1983), Vol. 1, Part 2 (1983), pp. 7-19.
72. P. Hosemann, in: Plenary and Invited Lectures of the 29th International Symposium on Macromolecules, Bucharest, Romania (September 5-9, 1983), Vol. 1, Part 2 (1983), pp. 293-314.
73. J. Wendorff, Polymer, 23, 543-557 (1982).
74. P. Corradini, in: The Stereochemistry of Macromolecules, Vol. 3, A. D. Ketley (ed.), Marcel Dekker, New York (1968), p. 49.
75. A. Turner-Jones, Makromol. Chem., 86, 249 (1965).
76. N. A. Platé, V. P. Shibaev, and B. S. Petrukhin, Vysokomol. Soedin., B13, 757-759 (1971).
77. V. P. Shibaev, B. S. Petrukhin, and Ya. A. Zubov, in: 17th All-Union Conference on High-Molecular-Weight Compounds, Proceedings [in Russian], Moscow (1969), p. 77.

78. R. Miller, R. Boyer, and J. Heijboer, J. Polym. Sci., Polym. Phys. Ed., 22, 2021-2041 (1984).
79. G. Stewart, Phys. Rev., 31, 174-179 (1928).
80. B. Warren, Phys. Rev., 44, 969-978 (1933).
81. A. A. Baturin, Yu. B. Amerik, B. A. Krentsel', and I. I. Constantinov, in: 2nd All-Union Liquid Crystal Conference, Proceedings [in Russian], Ivanovo (1972), p. 35.
82. B. S. Petrukhin, V. P. Shibaev, M. Ren'o, and N. A. Platé, Vysokomol. Soedin., B13, 405-411 (1971).
83. F. Balta-Calleia, J. Ramos, and J. Barrales-Rienda, Kolloid Z. Z. Polym., 250, 474-481 (1972).
84. G. Brandy, E. Wasserman, and H. Wellendoff, J. Chem. Phys., 47, 855-861 (1967).
85. N. A. Platé and V. P. Shibaev, J. Polym. Sci., Macromol. Rev., 8, 117-253 (1974).
86. A. F. Skryshevskii, The Structural Analysis of Liquids [in Russian], Vysshaya Shkola, Moscow (1971).
87. V. I. Danilov, The Structure and Crystallization of Liquids [in Russian], Izd. Akad. Nauk Ukr. SSR, Kiev (1956).
88. L. I. Tatarinova, Electron Diffraction by Amorphous Solids [in Russian], Nauka, Moscow (1972).
89. R. Morrow, Phys. Rev., 31, 10-15 (1928).
90. G. Stewart and R. Morrow, Phys. Rev., 30, 232-244 (1927).
91. Ya. I. Frenkel', Kinetic Theory of Liquids [in Russian], Izd. Akad. Nauk SSSR, Moscow—Leningrad (1959).
92. A. R. Ubbelohde, Melting and Crystal Structure, Clarendon, Oxford (1965).
93. K. Clark, A. Turner-Jones, and D. Sandiford, Chem. Ind., No. 24, 2010-2011 (1962).
94. A. Turner-Jones, Makromol. Chem., 71, 1-25 (1964).
95. A. Turner-Jones, Polymer, 6, 246-268 (1965); 7, 23-37 (1966).
96. V. P. Shibaev, N. A. Platé, and V. A. Kargin, in: 3rd European Regional Conference of Electron Microscopy, Proceedings, Vol. A, Prague (1964), p. 415.
97. V. P. Shibaev, N. A. Platé, B. S. Petrukhin, and V. A. Kargin, in: International Symposium on Macromolecular Chemistry, Tokyo—Kyoto, Japan, Abstracts, Section 7-6 (1966), p. 141.
98. N. A. Platé, V. P. Shibaev, B. S. Petrukhin, and V. A. Kargin, J. Polym. Sci., C, No. 23, 37-49 (1968).

99. V. P. Shibaev, B. S. Petrukhin, N. A. Platé, and V. A. Kargin, International Symposium on Macromolecular Chemistry, Toronto, Canada, Preprints of Scientific Papers, Vol. 6A (1968), p. 27.
100. V. P. Shibaev, B. S. Petrukhin, Ya. A. Zubov, N. A. Platé, and V. A. Kargin, in: Electron Microscopy of Solids and Biological Materials [in Russian], Nauka, Novosibirsk (1969), pp. 150-151.
101. V. P. Shibaev, B. S. Petrukhin, and V. A. Kargin, in: 8th All-Union Conference on Electron Microscopy, Abstracts [in Russian], Section III, Kiev (1969), pp. 47-48.
102. V. P. Shibaev, B. S. Petrukhin, N. A. Platé, and V. A. Kargin, Vysokomol. Soedin., A12, 140-154 (1970).
103. Ya. A. Zubov, B. S. Petrukhin, and V. P. Shibaev, Vysokomol. Soedin., B12, 290-293 (1970).
104. N. A. Platé and V. P. Shibaev, Vysokomol. Soedin., A13, 410-424 (1971).
105. N. A. Platé, V. P. Shibaev, B. S. Petrukhin, Ya. A. Zubov, and V. A. Kargin, J. Polym. Sci., 9A-1, 229-233 (1971).
106. N. A. Platé, V. P. Shibaev, and R. V. Tal'roze, in: Progress in the Chemistry and Physics of Polymers [in Russian], Khimiya, Moscow (1973), pp. 127-144.
107. P. G. Magagnini, F. Andruzzi, and G. Benetti, Macromolecules, 13, 12-15 (1980).
108. A. L. Segre, F. Andruzzi, D. Lupinacci, and P. Magagnini, Macromolecules, 14, 1845-1847 (1981).
109. E. Benedetti, P. Vergamini, F. Andruzzi, and P. Magagnini, Polym. Bull., 2, 241-247 (1980).
110. P. Magagnini, D. Lupinacci, F. Cotrozzi, and F. Andruzzi, Makromol. Chem. Rapid Commun., 1, 557-560 (1980).
111. D. Lupinacci, F. Andruzzi, M. Paci, and P. Magagnini, Polymer, 23, 277-280 (1982).
112. L. Magagnini, Makromol. Chem. Suppl., 4, 223-238 (1981).
113. A. L. Segre, F. Andruzzi, D. Lupinacci, and P. Magagnini, Macromolecules, 16, 1207-1212 (1983).
114. J. I. Gonzalez de la Campa and J. M. Barrales-Rienda, Preprints of Short Communications of the International Symposium on Macromolecules, Mainz, West Germany (September 17-21), Vol. 8 (1979), pp. 1567-1570.
115. J. Barrales-Rienda and C. R. Galicia, Macromolecules, 16, 932-939 (1983).
116. J. Barrales-Rienda, C. R. Galicia, J. Freire, and A. Horta, Macromolecules, 16, 940-945 (1983).

117. W. Philippoff and E. Tornqwist, J. Polym. Sci., C, No. 23, 881-893 (1968).
118. D. Aubrey and A. Barnatt, J. Polym. Sci., 6A-2, 241-247 (1968).
119. K. Holland-Moritz, Colloid Polym. Sci., 253, 922-933 (1975).
120. K. Holland-Moritz, E. Sausen, and D. Hummel, Colloid Polym. Sci., 254, 976-985 (1976).
121. J. Modric, K. Holland-Moritz, and D. Hummel, Colloid Polym. Sci., 254, 342-356 (1976).
122. G. Trafara, R. Koch, K. Blum, and D. Hummel, Makromol. Chem., 177, 1089-1097 (1976).
123. J. Brandrup and E. Immergut (eds.), Polymer Handbook, Wiley, New York (1974), p. III-8.
124. G. V. Fraser, Makromol. Chem., 173, 195-201 (1973).
125. L. M. Beck, P. G. Hägele, M. Junger, and H. P. Grossmann, Preprints of Short Communications of the 28th International Symposium on Macromolecules, Mainz, West Germany (September 17-21, 1979), Vol. 3 (1979), pp. 1213-1216.
126. V. M. Tatevskii (ed.), Physicochemical Properties of Individual Hydrocarbons [in Russian], Gostoptekhizdat, Moscow (1960).
127. O. Vogl, J. Polym. Sci., A2, 4633-4650 (1964).
128. J. Negulescu and O. Vogl., J. Polym. Sci., Polym. Chem. Ed., 14, 2415-2432 (1976).
129. J. Wood, J. Negulescu, and O. Vogl., J. Macromol. Sci., A11, 171-184 (1977).
130. G. Natta, G. Mazzanti, P. Corradini, and H. Bassi, Makromol. Chem., 37, 156-172 (1960).
131. A. I. Kitaigorodskii, Dokl. Akad. Nauk SSSR, 124, 861-865 (1959).
132. A. I. Kitaigorodskii, Molecular Crystals [in Russian], Nauka, Moscow (1971).
133. D. Chapman, The Structure of Liquids by Spectroscopic and X-Ray Techniques, Methuen (1965).
134. T. Poolak, V. P. Shibaev, and N. A. Platé, Vysokomol. Soedin., 14, 127-132 (1972).
135. B. S. Petrukhin, V. P. Shibaev, and N. A. Platé, Vysokomol. Soedin., A12, 687-692 (1970).
136. V. Enkelmann and J. Lando, J. Polym. Sci., Chem. Ed., 15, 1843-1852 (1977).
137. W. Oster and D. Riley, Acta Cryst., 5A, 272-275 (1952).
138. H. Hsieh, B. Post, and H. Morawetz, J. Polym. Sci., Polym. Phys. Ed., 14, 1241-1255 (1976).

139. H. Hsieh, Doctoral Dissertation, Polytechnic Institute of New York, New York (1976).
140. H. Orth and E. Fisher, Makromol. Chem., 88, 188-196 (1965).
141. E. Thomas and S. Sass, Makromol. Chem., 164, 333-348 (1973).
142. N. A. Platé, Tran-Khieu, and V. P. Shibaev, J. Polym. Sci., C, No. 16, 1133-1144 (1967).
143. V. P. Shibaev, N. A. Platé, R. K. Grushina, and V. A. Kargin, Vysokomol. Soedin., 6, 231-239 (1964).
144. N. A. Platé, V. P. Shibaev, and Tran-Khieu, in: Preprints of the International Symposium on Macromolecular Chemistry, Prague (1965), pp. 122-126.
145. A. I. Kitaigorodskii, Organic Crystal Chemistry [in Russian], Izd. Akad. Nauk SSSR, Moscow (1955).
146. T. Malkin, J. Chem. Soc. (London), 2796 (1931).
147. F. Francis, F. Collins, and S. Piper, Proc. R. Soc. London, A158, 691-695 (1937).
148. F. Gunstone, Chem. Ind., No. 1, 84 (1964).
149. T. Malkin, in: Progress in the Chemistry of Fats and Lipids, Vol. 1, Pergamon Press, London (1952), pp. 1-32.
150. S. Aleby, Solid State Behaviour of Long Chain Esters, Acta Universitatis, Gothoburgensis, Abstracts of Gothenburg, Dissertations in Science, No. 13, Gothenburg (1969).
151. N. A. Kuznetsov, V. M. Moiseenko, Z. A. Roganova, A. L. Smolyanskii, and V. P. Shibaev, Vysokomol. Soedin., A19, 399-412 (1977).
152. A. L. Smolyanskii and V. P. Shibaev, Vysokomol. Soedin., A21, 2221-2231 (1979).
153. H. Ailhand, J. Gallot, and A. Skoulious, Compt. Rend. Acad. Sci., C267, 139-144 (1968).
154. G. Shearer, Proc. R. Soc. London, A108, 655-670 (1925).
155. N. Morosoff, H. Moravetz, and B. Post, J. Am. Chem. Soc., 87, 3035-3041 (1965).
156. A. Cemel, T. Fort, and J. Lando, J. Polym. Sci., 10A-1, 2061-2087 (1972).
157. V. P. Shibaev, Dissertation for the Degree of Doctor of Sciences, Moscow State University, Moscow (1974).
158. V. S. Pshezhetskii, V. I. Gerasimov, and A. P. Luk'yanova, Vysokomol. Soedin., B18, 532-534 (1976).
159. M. Takayanagi and T. Katayose, J. Appl. Polym. Sci., 29, 141-151 (1984).
160. B. Espenschied and R. Schulz, Makromol. Chem., Rapid Commun., 4, 633-638 (1983).

161. R. Schulz, Makromol. Chem., Suppl., 6, 195-196 (1984).
162. F. Andruzzi, C. Barone, D. Lupinacci, and P. Magagnini, Makromol. Chem., 5, 603-609 (1984).
163. S. Aharoni, Macromolecules, 12, 94-103 (1979); 14, 222-224 (1981); Polymer, 22, 418-419 (1981); J. Macromol. Sci. Phys., B21, No. 1, 105-129 (1982).
164. N. B. Sokol'skaya, Ya. S. Freidzon, V. V. Kochervinskii, and V. P. Shibaev, Vysokomol. Soedin., A27 (1985).
165. V. P. Shibaev, M. Palumbo, and E. Peggion, Biopolymers, 14, 73-81 (1975).
166. V. P. Shibaev, V. V. Chupov, V. M. Laktionov, and N. A. Platé, Vysokomol. Soedin., B16, 332-333 (1974).
167. V. V. Chupov, V. P. Shibaev, and N. A. Platé, Vysokomol. Soedin., A21, 218-228 (1979).
168. A. Mathis, J. Moigne, and P. Gramain, Eur. Polym. J., 9, 283-289 (1973).
169. E. Jordan, D. Feldeisen, and A. Wrigley, J. Polym. Sci., 9A-1, 1835-1851 (1971).
170. E. Jordan, J. Polym. Sci., 10A-1, 3347-3361 (1972).
171. A. Shaerer, C. Bussa, A. Smith, and L. Skinner, J. Am. Chem. Soc., 77, 2018-2024 (1955).
172. J. Barrales-Rienda, F. Fernandez-Martin, K. Galicia, and M. Chaves, Makromol. Chem., 184, 2643-2659 (1983).
173. J. Veilkovic, Z. Petrovic, D. Petrovic-Djakov, and J. Budinskii, Abstracts of Communications of the International Macromolecular Symposium, Strasbourg, France, Vol. 2 (1981), pp. 927-931.
174. M. Broadhurst, J. Res. Natl. Bur. Stand., 66A, 241-247 (1962).
175. I. Sobbotka and I. V. Sochava, Vysokomol. Soedin., A21, 1322-1326 (1979).
176. A. Pittman and B. Ludwig, J. Polym. Sci., 7A-1, 3053-3060 (1969).
177. H. G. Braun and G. Wegner, Makromol. Chem., 184, 1103-1109 (1983); Mol. Cryst. Liq. Cryst., 96, 121-131 (1983).
178. G. Wegner, Makromol. Chem., Suppl., 6, 347-357 (1984).
179. C. Plachetta, N. Rau, and R. Schulz, Mol. Cryst. Liq. Cryst., 96, 141-151 (1983).
180. V. P. Shibaev and Ya. S. Freidzon, Vysokomol. Soedin., B17, 151-153 (1975).
181. E. Jordan, B. Artymyshyn, A. Speca, and A. Wrigley, J. Polym. Sci., 9A-1, 3349-3365 (1971).
182. F. Andruzzi, D. Lupinacci, P. Magagnini, and A. Segre, Polym. Bull., 11, 241-246 (1984).
183. E. Jordan, G. Riser, W. Parker, and A. Wrigley, J. Polym. Sci., 4A-2, 975-996 (1966).

Chapter 2

Molecular Mobility in Comb-Shaped Polymers

2.1. EFFECT OF THE PHASE STATE ON THE RELAXATION PROPERTIES

The nature of the molecular mobility in polymers is determined by the presence of long-chain molecules and the possibility of internal rotation of the atoms in the main and side chains. At temperatures below the glass transition temperature, movement of the kinetic units located in small volumes (short segment or unit of the main chain, side substituent or part of the side substituent) is possible. This type of motion, related to the mobility of only a few groups of atoms, is usually called group motion, and is indicated by the Greek letters β, γ, δ, etc., as a function of the character of the relaxing groups. At temperatures close to T_g and higher, a second type of motion begins to appear: segmental motion, related to the cooperative movement of the kinetic segments of the main chain, called α-relaxation [1-6]. This division is arbitrary to a significant degree, since the different types of motion which frequently take place simultaneously in a similar region of frequencies and temperatures, thereby impeding their determination and unambiguous interpretation, cannot be distinguished for all polymers.

In examining the nature of the molecular mobility in comb-shaped polymers, it is natural to determine first the degree to which lengthening the side substituent will affect the relaxation behavior of the main and side chains of the branched macromolecules. Unfortunately, there is only an insignificant number of systematic studies of the molecular mobility through a wide range of side-chain lengths in polymeric homologs of varying chemical structure in the extensive literature on the study of molecular motion in polymer systems. Nevertheless, these studies are of significant interest for understanding the nature of the different local forms of motion. In studying comb-shaped polymers with increasing side-chain lengths, it is possible to follow the nature of molecular motion over a wide range of compositions, obtaining information corresponding to the specific number of methylene groups introduced.

On the other hand, a comparison of the data for different homologous series of polymers, each with the same side-chain length, allows the determination of the specific features of the molecular mobility of a given class of polymers. Finally, in going from amorphous to crystallizable compounds (in view of the capacity of higher homologs of comb-shaped polymers to crystallize), it is possible to evaluate the role of the phase state of the polymer in determining molecular mobility.

An examination of the nature of molecular mobility should begin with the structurally simplest comb-shaped polymers such as poly(1-alkyl ethylene)s. Polyethylene, whose structure corresponds to the structure of the side chains of comb-shaped polymers, is the first member of this series.

2.1.1. Polyethylene and Isotactic Poly(1-alkyl ethylene)s

The presence of three major relaxation regions* is characteristic of high-density polyethylene. One of these is the γ-process resulting from the mobility of the $(CH_2)_n$ groups (where $n \geq 3$) in the amorphous regions of the polymer; the temperature corresponding to this relaxation region (from −100 to −110°C at a frequency of 1 Hz) is often called the glass transition temperature T_g of the $(CH_2)_n$ groups [12]. The β-relaxation is due to the motion of segments at the sites of defects in the crystal structure caused by the branches in the main chain; an increase in the number of branching units (ethylene copolymers) significantly increases

*An examination of all observed forms of motion in polyethylene can be found in a number of monographs [1, 5] and surveys [7-11].

the intensity of this process [13] which is usually observed in the temperature range from −20 to −60°C. The α-relaxation has not yet been unambiguously interpreted, although most investigators [1, 11] correlate its appearance with two mechanisms, caused by the motion of folds in the chains and reorientation (including translation) of segments of the macromolecules in the crystalline regions of the polymer. The activation energies of these relaxation processes most frequently encountered in the literature for polyethylene are reported below:

	α		β				γ	
Activation energy	105	[14]	67	[15]	84	[16]	46	[14]
E, kJ/mole	117	[14]	105	[17]	160	[14]	46-63	[14]

The significant differences in the values of E obtained for the β-relaxation even for such a well-studied polymer as polyethylene are first due to the use of methods of varying sensitivity and second to the uncertain identity of the samples (degree and nature of branching of the macromolecules, degree of crystallinity, etc.). The degree of crystallinity, which determines the microstructure of the chain, plays the most important role in the realization of relaxation processes for higher poly(1-alkyl ethylene)s in which the type of molecular motion should be determined by the nature of the packing of the main and side chains.

The first attempt to study the molecular mobility of comb-shaped poly(1-alkyl ethylene)s was made in [18]. In studying the dynamic mechanical properties of these polymers (prepared using Ziegler–Natta catalysts with no evaluation of the microstructure of the macromolecules), three maxima were found for the mechanical losses: one is observed near T_m of the polymers, and the temperatures corresponding to the other two maxima, indicated as T_α and T_β, are reported in Table 2.1.

The low-temperature relaxation process whose temperature (in the range from −110 to −160°C) is indicated by T_β is believed to be similar to the γ-process observed in polyethylene and its copolymers and caused by the motion of the $(CH_2)_n$ groups with $n \geq 3$ [19]. The second loss maximum at temperatures T_α is attributed to the mobility of the main chain, although this has never been proven.

TABLE 2.1. Temperatures of α- and β-Relaxation Processes in Poly(1-Alkyl Ethylene)s Determined by the Dynamic Mechanical Measurements [18]

Poly(1-alkyl ethylene)s	T_α, °C/Hz	T_β, °C/Hz
Polyethylene (PE)	From −20 to −60	−120/300
Poly(1-methyl ethylene) (PE-1)	+15/100	–
Poly(1-ethyl ethylene) (PE-2)	+15/150	−125/430
Poly(1-propyl ethylene) (PE-3)	−3/80	−130/300
Poly(1-butyl ethylene) (PE-4)	−23/170	−130/500
Poly(1-pentyl ethylene) (PE-5)	−31/180	−148/320
Poly(1-hexyl ethylene) (PE-6)	−42/140	−160/300
Poly(1-heptyl ethylene) (PE-7)	−47/170	−160/310
Poly(1-octyl ethylene) (PE-8)	−35/180	−160/590
Poly(1-decyl ethylene) (PE-10)	−6/150	−145/520
Poly(1-dodecyl ethylene) (PE-12)	+10/60	−130/450
Poly(1-tetradecyl ethylene) (PE-14)	+40/30	−125/200
Poly(1-hexadecyl ethylene) (PE-16)	+55/25	−110/170

An increase in the length of the alkyl group to n = 7 shifts the values of T_α to the low-temperature region due to the plasticizing effect of the substituent. It is interesting that an increase in T_α, instead of the expected decrease, is observed with a further increase in n due to intensification of the plasticizing effect of the long substituents. In our opinion, this is due to intensification of the interaction and ordering in the position of the methylene side chains with an increase in their length, which results in limitation of both their own mobility and the mobility of the main chain. As Table 2.1 shows, T_β also increases slightly beginning with poly(1-decyl ethylene), which graphically illustrates the effect of a decrease in the mobility of the segments of the methylene chains.

Slichter and Davis [20] observed similar events in studying the temperature dependence of the spin–lattice relaxation time of a series of poly(1-alkyl ethylene)s with NMR (Fig. 2.1). Two peaks corresponding to $T_{1,min}$, one in the region from −100 to −120°C and the second in the region of 70-100°C, are observed. According to the concepts developed in [21], the first is due to rotation of the end methyl groups of the side chains around the axis of symmetry passing through the C'–C bond. This minimum is observed for all types of comb-shaped polymers whose alkyl side chains have end CH_3 groups. The value of $T_{1,min}$ in this region is proportional to the relative number of protons in the methyl groups and decreases with an increase in the length of the

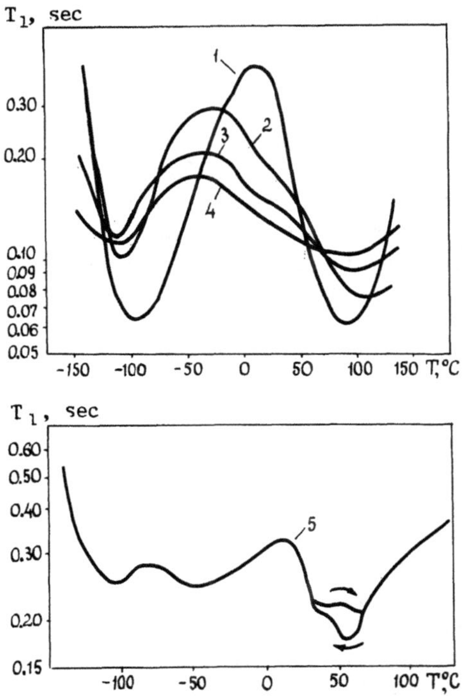

Fig. 2.1. Temperature dependence of spin-lattice relaxation time T_1 for isotactic poly(1-alkyl ethylene)s (at 50 MHz) [20]: 1) poly(1-methyl ethylene); 2) poly(1-propyl ethylene); 3) poly(1-butyl ethylene); 4) poly(1-pentyl ethylene); 5) poly(1-hexadecyl ethylene).

alkyl segment of the side chain and on introduction of a methyl group in the α position.

The second relaxation region $T_{1,min}$ is located in the region of temperatures above T_g of poly(1-alkyl ethylene)s and is attributed to segmental mobility. However, the possibility of the combined motion of segments of the main and side chains is assumed for polymers with short chains (the value of n is unfortunately not reported). As in [18], an extreme dependence of $T_{1,min}$ on the length of the side chain is observed when examining the effect of the length of the alkyl side chain on the temperature of the appearance of a relaxation process (on T_α in [18] and in the given case, on $T_{1,min}$), which confirms our hypothesis concerning the intensification of the side-chain interaction when a certain critical length of the alkyl side chain (in this case, n = 9) is exceeded.

Further intensification of the interaction of the side chains resulting in their crystallization alters the character of the temperature dependence of $T_{1,min}$ [20]. The dependence of $T_{1,min}$ on the temperature for poly(1-hexadecyl ethylene) prepared by heating and cooling the sample is shown in Fig. 2.1. As for the lower homologs of poly(1-alkyl ethylene)s, the low-temperature peak corresponds to rotation of the CH_3 groups. The second minimum in the region of −50°C is assigned to "the onset of side-chain mobility close to the cooperative motion of the macromolecules observed at the glass transition temperature of the normal amorphous polymers" [20]. A slight hint of the appearance of this transition can be seen in Fig. 2.1 for poly(1-propyl ethylene) and poly-(1-butyl ethylene) (in the 0-10°C region). However, the proposed interpretation of the observed values of $T_{1,min}$ in the 0-10°C region for poly(1-propyl ethylene) and poly(1-butyl ethylene) is insufficiently clear, since a similar molecular motion is proposed at these temperatures and at tempertures 100°C higher (appearance of segmental mobility) which is definitely not very probable. The calculated values of the activation energies of segmental mobility were approximately the same for the polymers PE-2, PE-3, and PE-10 (100 kJ/mole) and for PE-4 and PE-5 (50 kJ/mole). Unfortunately, the causes of the observed differences are not discussed.

The appearance of a third minimum for PE-16 with a hysteresis loop caused by the conditions of crystallization is attributed to the motion of the segments of the macromolecules in premelting conditions (42-57°C) and the motion of the main chains in a melt (65°C). The last statement is not sufficiently convincing, since the appearance of a peak of $T_{1,min}$ in this temperature region could simply be the result of the effect of melting of the side chains, as observed in melting of low-molecular-weight long-chain compounds (paraffins, ethers, ketones), for example.

A comparison of [18] and [20] thus shows that the interpretation of the experimental results of the molecular mobility in such apparently structurally simple polymers as poly(1-alkyl ethylene)s is very complicated. In this case, the chemical structure of the main and side chains makes the assignment of the observed forms of motion to actual groups and parts of the macromolecules difficult.

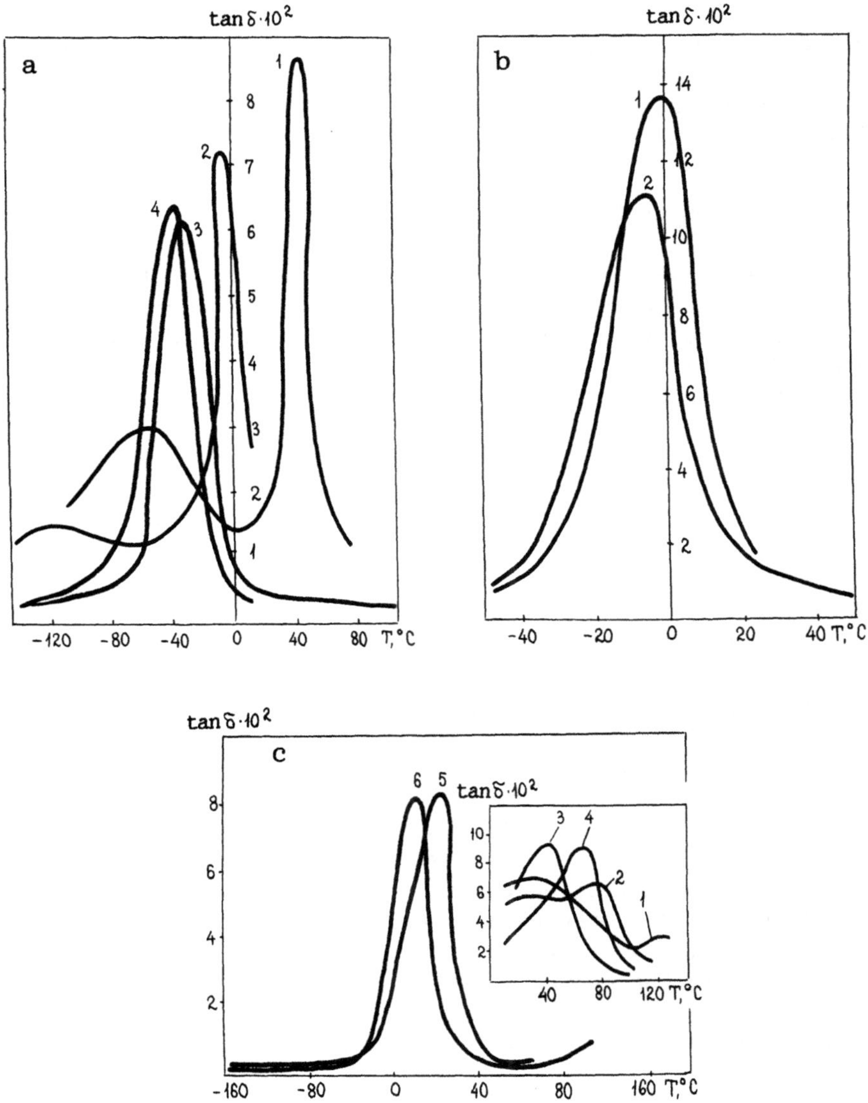

Fig. 2.2. Temperature dependence of tan δ for amorphous polymer homologs of the PA-n (a), PVE-n (b), and PMA-n (c) series. a: 1) PA-1 (2.5 kHz), 2) PA-3, 3) PA-7, 4) PA-10 (2, 3, 4: 1 kHz) [26, 31]; b: 1) PVE-7, 2) PVE-9 (1, 2: 1 kHz) [26]; c: 1) PMA-1, 2) PMA-2, 3) PMA-3, 4) PMA-4 (1, 2, 3, 4: 20 Hz) [27]; 5) PMA-6, 6) PMA-8 (5, 6: 1 kHz) [26].

In this sense, slightly greater possibilities have arisen in the study of molecular mobility in comb-shaped polymers with polar groups; on the one hand, this extends the number of methods of investigation which can be used and, on the other, allows the assignment of the different forms of molecular motion to be significantly more rigorous.

2.1.2. Amorphous Comb-Shaped Polymers

Let us examine the molecular mobility data in the lower homologs of these polymers and the structurally more complex comb-shaped poly(n-alkyl acetal)s [15, 22], poly(N-alkyl maleimide)s (P-N-MI) (see formula on p. 16) [23], and polyphenylene ethers [24] with the general formula

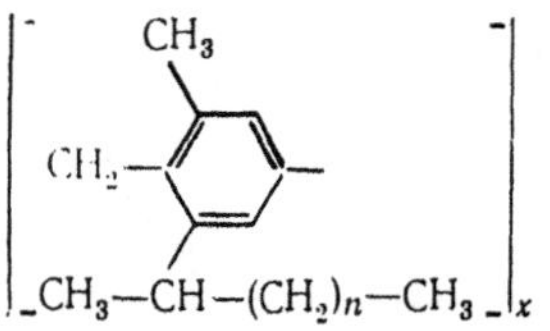

where n = 2-11.

The use of dielectric and mechanical methods for investigating mobility in comb-shaped polymers with n ≥ 3 led to the relatively precise definition of two regions of dielectric and mechanical losses related to the mobility of the main and side chains for PA, PMA, PVE [25-32], and polyvinyl acetals [15] (Fig. 2.2). The maximum of the losses observed at higher temperatures occurs in the region of T_g and, according to [1-3], consequently corresponds to the cooperative segmental mobility of the macromolecules (α-relaxation). The second peak in the curve of $\tan\delta = f(T)$ corresponds to dipole–group losses and is due to relaxation of the dipole–group polarization (β-relaxation) related to the mobility of the side chains.

The existence of two relaxation processes is also found in studying the nuclear magnetic relaxation (together with the γ-process related to rotation of the end methyl groups in the alkyl chains at temperatures from −120 to −140°C) of polymer homologs of the PA, PMA, and PVE series [26] (Fig. 2.3). However, in this case (with the frequencies used for measurement), the high-temperature peak corresponding to the α-relaxation is located several tens of degrees higher than

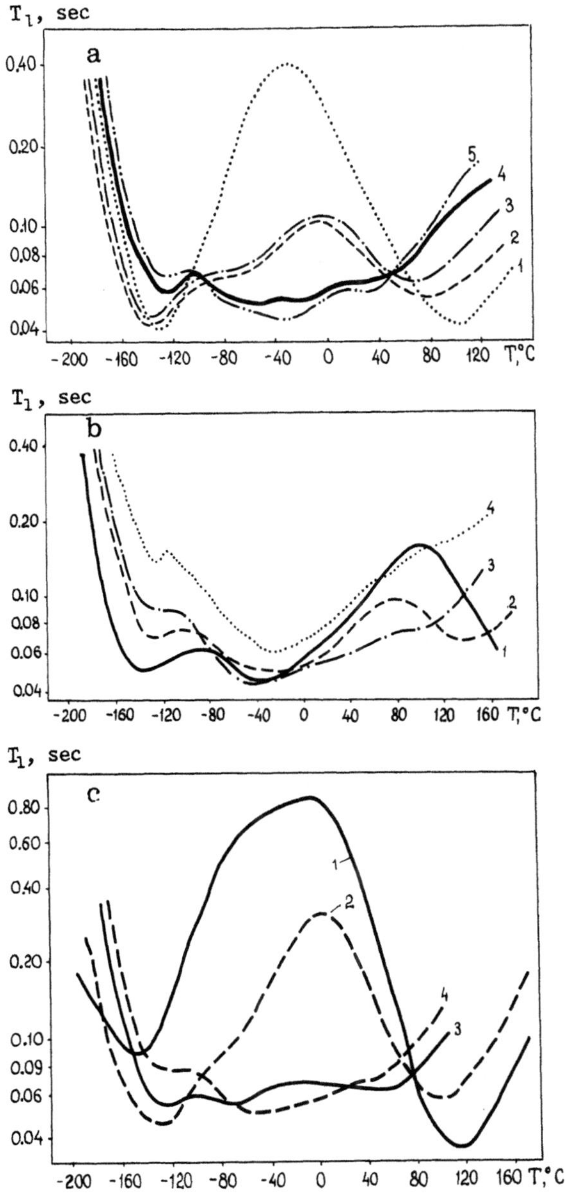

Fig. 2.3. Temperature dependence of the spin-lattice relaxation time for PA-n (a), PMA-n (b), and PV esters (c) at 18.6 MHz. a: 1) PA-2, 2) PA-3, 3) PA-4 [33], 4) PA-7, 5) PA-10 [20, 26]; b: 1) PMA-3 [27], 2) PMA-6, 3) PMA-10, 4) PMA-12 (30 MHz) [20, 26]; c: 1) PVE-1, 2) PVE-3 [33], 3) PVE-7, 4) PVE-9 [19, 20, 26].

TABLE 2.2. Activation Energies of α-, β-, and γ-Relaxation Processes in Comb-Shaped Polymers and their Lower Homologs*

No. of carbon atoms in side chain, n	Eα, kJ/mole	Eβ, kJ/mole
	Poly(n-alkyl acrylate)s	
1	176, D [29], 239, D [35]	31, D [38], 40, D [37]
		63, D [35], 59, D [36]
2	164, D [29]	30, D [38], 34, D [29]
		38, D [37]
3	139, D [29]	24, D [29]
7	97, D [26]	
10	88, D [26]	
	Poly(n-alkyl methacrylate)s (PMA)	
1	580, D [48], 462, M [45]	71, M [42], 76, M [40, 41]
	420, D [27]	80, D [43, 45], 88 D [27]
2	210, D [43], 336, M [40]	78, D [43], 88, D [27]
	386, D [27]	100, D [36]
3†	206, D [27]	88, D [27]
4‡	157, D [39], 122, D [43]	100, M [44], 117, D [46]
	100, M [44]	
6	88, D [44], 105, M [26]	55, M [44]
	113, M [44]	
8	76, M [44], 80, D [26]	109, D [46]
	88, M [44], 134, M [46]	
9	97, D [43]	
	Poly(n-alkyl vinyl ester)s (PVE)	
1	206, D [29], 251, D [49], 252, D [50]	36 [52], 42 [50, 51]
		44, D [29]
2	185, D [29]	37, D [29]
3	130, D [29]	20, D [29]
7	92, D [26]	
9	88, D [26]	
	Poly(n-alkyl acetal)s	
2	231, D [22]	
3	205, D [23]	
4	193, D [22]	
6	168, D [22]	
	260, D [53]	
8	147, D [22]	

Poly(n-alkyl maleimide)s, D [23]

n	Eα, kJ/mole	Eβ, kJ/mole	Eγ, kJ/mole
1	–	150	55
2	710	90	56
5	529	80	53
7	399	80	70
12	399	110	73

TABLE 2.2 (continued)

Total No. of carbon atoms in alkyl group	T_c, °C	T_γ (1 Hz), °C	E_γ, kJ/mole	T_δ (1 Hz), °C	E_δ ± 8, kJ/mole
		Poly(phenyl ether)s, M [24]			
5	100	−102	38	−192	25
6	85	−90	42	−185	30
7	60	−78	50	−178	25
8	54	−67	59	−180	25
10	20	−49	76	−179	30
44	−10	–	–	–	–

*Method of determination of E: D – dielectric, M – mechanical.
†E_γ = 16.8 kJ/mole [47].
‡E_γ = 14.3 kJ/mole [47].

T_g. The second maximum observed in the region from −30 to −80°C (for polymers with n ≥ 3) corresponds to the motion of the alkyl side chains.

An increase in the length of the side chain accompanied by an increase in the free volume [2] and a decrease in the packing density [34] decreases the steric hindrance for manifestation of segmental mobility, and a shift in the high-temperature maximum on the curves of $\tan\delta_{max} = f(T)$ and $T_{1,min} = f(T)$ to the region of lower temperatures is the result. This unusual internal plasticization decreases T_g and the activation energy of segmental motion E_α, as Table 2.2 indicates.* Although the values of E_α differ by almost three times (two times for P-N-MI) for lower homologs of PA and PMA, a further increase in the length of the alkyl side chain gradually levels the difference in the flexibility of the main chains. The values of E_α for PA and PMA containing 6-8 carbon atoms in the side chain almost coincide. We note that

*The large difference in the activation energies of the same relaxation processes for the same polymer homologs in Table 2.2 clearly shows how carefully the interpretation of their absolute values should be approached. These data are of much greater value for the comparative examination of molecular mobility in homologs of the same or different classes of chemical compounds.

the activation energy of segmental motion is also 100 kJ/mole for poly(1-decyl ethylene) (PE-10). As a consequence, the kinetic characteristics of the main chain, with which the activation energy and dipole-segmental polarization relaxation time are correlated, is weakly dependent on the length of the side chains in vinyl polymers with sufficiently long methylene side chains ($n \geq 6$-8).

The plasticizing effect of the side substituents is most clearly manifest in the examination of the PMA homolog series. Since the segmental motion is strongly hindered by the presence of a methyl group for PMA-1, an inverse ratio of the heights of the maxima of $\tan\delta$ corresponding to α- and β-processes in comparison to PA and PV esters is observed. Progression from the lower to higher homologs of these series of polymers is accompanied by an increase in the values of $\tan\delta_{max}{}^{dip\,seq}$ for PMA and a slight decrease in the values of $\tan\delta_{max}{}^{dip\,seq}$ for PA and PVE. This means that internal plasticization is significantly more effective for polymers with more rigid chains (PMA) in which lengthening of the side chains increases the segmental mobility to a greater degree than in polymers with flexible chains (PA, PVE) in which interactions between the side groups, which decrease the segmental mobility, begin to play an important role. This actually determines the possibility of using long-chain monomers to increase the segmental mobility of rigid-chain polymers (i.e., decreasing their T_g) by copolymerization of the corresponding monomers with long-chain vinylic derivatives.

As an illustration, the curves plotted with the results in [54] are shown in Fig. 2.4; they show how a long-chain monomer, octadecyl acrylate (A-18), affects the decrease in T_g of two different types of polymers: rigid-chain [poly(methyl methacrylate) (PMA-1), polyacrylonitrile (PAN)] and flexible-chain [polymethacrylate (PA-1) and poly(ethyl acrylate) (PA-2)]. The most pronounced decrease in T_g is observed in the case of copolymers of A-18 with MA-1 and acrylonitrile, while the change in T_g with an increase in the concentration of A-18 is not as significant for PA-1 and PA-2.

With respect to the second relaxation process observed for polymers with side chain groups (β-relaxation), as indicated above, this process is only interpreted more or less unambiguously for lower homologs of polymers of the acrylic and methacrylic series [1, 2, 4]. It is believed that the

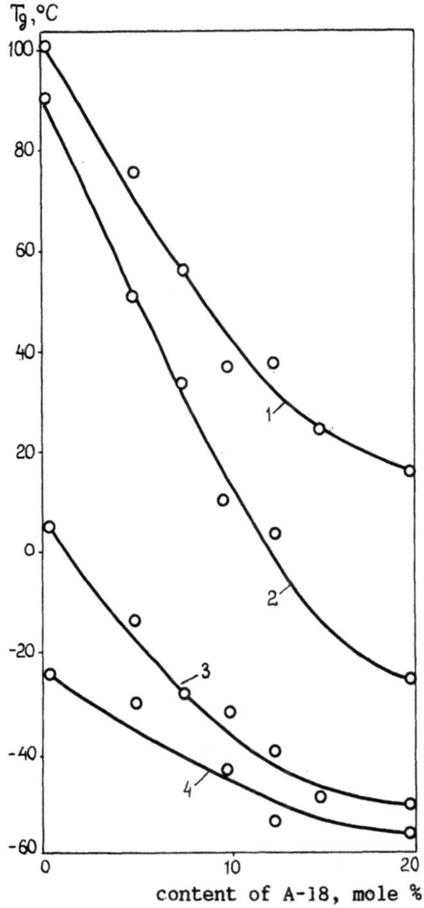

Fig. 2.4. Dependence of T_g of copolymers of octadecyl acrylate (A-18) with methyl methacrylate (1), acrylonitrile (2), methyl acrylate (3), and ethyl acrylate (4) on the concentration of A-18 [54].

β-relaxation corresponds to rotation of ester groups with respect to the C—C bond binding the methoxycarbonyl group to the main chain for the first members of the homologs: PMA-1 and PA-1. This process, in the form of dipole—group polarization, as the dependence of tan δ on T shows (Fig. 2.2), takes place at lower temperatures than the α-process, and the activation energy of this process is significantly lower than the activation energy of the α-process (see Table 2.2). The maximum values of E_β are characteristic of rigid-chain polymers of the PMA and P-N-MI series.

Lengthening of the side chain shifts the maximum of $\tan\delta$ to the region of low temperatures; as Fig. 2.2 shows, some congruence of the α- and β-relaxation processes is observed for polymers containing 3-4 carbon atoms, and the β-relaxation cannot be observed up to temperatures from −170 to −160°C with a further increase in the length of the side chain.

As Table 2.2 shows, the values of E_β for PMA and P-N-MI is weakly dependent on the length of the main chain, while in the case of PA and PVE, E_β decreases by 2-3 times in going from the first to the third homolog. In flexible-chain polymers, the possibility of rotation of the side ester group is apparently significantly facilitated by the lability of the main chain (this also causes a decrease in the activation energy of the β-process), while the mobility of the ester and carbonyl groups is strongly hindered due to the elevated rigidity of the main chain in such polymers as PMA, P-N-MI, and polyacetals.

The analysis of the data on the temperature dependence of the spin-lattice relaxation time for the polymers shown in Fig. 2.3 and for poly(1-alkyl ethylene)s (see Fig. 2.1) provides additional information concerning the β-process. As these figures show, the appearance of $T_{1,min}$ in the region from 0 to −80°C is observed for all polymers with $n \geq 2$. The position of this minimum on the temperature scale is dependent on the length of the alkyl side chain and is slightly shifted toward higher temperatures with an increase in n. The intensity of this peak is determined by the relative number of protons contained in the side groups. This relaxation process is least pronounced in poly(1-alkyl ethylene) with the same values of n (see Fig. 2.1) for which it is manifest at the highest temperatures (0-10°C); the region of the minimum is located at −40°C in PMA and at even lower temperatures in PA and PVE.

All of the above indicates the onset of mobility of the alkyl side groups in this region either independently of the motion of the ester group or together with it. The difference in the temperatures of the β-process observed by dielectric and NMR measurements and the change in the temperature corresponding to the maximum of $T_{1,min}$ and $\tan\delta$ with increasing side-chain length support the first hypothesis. On the other hand, a comparison of the temperatures corresponding to $T_{1,min}$ for such classes of comb-shaped polymers as poly-

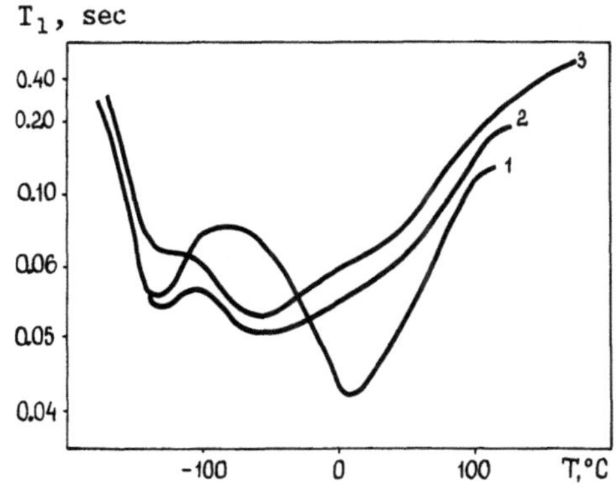

Fig. 2.5. Temperature dependence of the spin-lattice relaxation time for polydecane (1), PA-7 (2), and PVE-9 (3).

(1-alkyl ethylene)s, PA, PMA, and PVE (with the same n) indicates the highest inhibition of motion of the alkyl groups in poly(1-alkyl ethylene)s. As Fig. 2.5, which shows the dependences of $1/T_1 = f(T)$ for one of the poly(1-alkyl ethylene)s, stereoregular 1-polydecene, PA-7, and PVE-9, indicates, the regions of appearing mobility of the alkyl groups in poly(1-alkyl ethylene)s, PA, and PVE with approximately equal n differ by 50-60°C. This is undoubtedly due to the presence of ester groups which significantly facilitate the motion of the alkyl groups and possibly also participate in this motion in the side chains of PA, PMA, and PVE.

In addition to the β-relaxation, many of the amorphous lower homologs of the series of poly(1-alkyl ethylene)s, PA, PMA, and PV ethers and esters exhibit small additional relaxation maxima designated as γ, δ, etc. at even lower temperatures (from −190 to −200°C) or higher frequencies, according to NMR, dielectric, and mechanical measurements [1, 5, 55-57]. Unfortunately, different investigators in the literature frequently assign different values to the motion of the same groups, which significantly impedes the classification of the observed relaxation processes. For example, the appearance of a γ-transition (see Table 2.2) can be correlated with both the hindered rotation of the side-chain segments containing 2-4 carbon atoms (of the crankshaft type according to Schatzki) [58] and with reorientation or rotation of the CH_3 groups at the ends of the side chains; the last type of motion is usually designated as a δ-process (see Table 2.2) [55].

The number of studies of the γ-processes in comb-shaped polymers with n ≥ 4 is extremely limited [24, 47]; some of the results of determining E_γ are reported in Table 2.2. The values of E_γ for amorphous PMA significantly differ from the values for polyethylene, but, at the same time, the values of E_γ for P-N-MI and polyphenylene ethers do not differ significantly when n changes from 1 to 12 (although they uniformly increase with an increase in n) and are close to the values for E_γ for polyethylene.

The appearance of two glass transition temperatures at relatively low temperatures (below 0°C) was observed by Cowie et al. [59, 61] for comb-shaped homopolymers and copolymers of itaconic acid, where each monomeric unit contained two aliphatic chains. The values of these temperatures for some n-alkyl polymer homologs with different lengths of the side substituents R are indicated below:

R	T_1, °C	T_2, °C
C_7H_{15}	−25	−101
C_8H_{17}	−33	−95
C_9H_{19}	−28	−86

The lower glass transition temperature T_1 was attributed to the independent cooperative relaxation of side chains frequently positioned along the chain and temperature T_2 was attributed to the glass transition process determined by the presence of main chains and the cooperative motion of the entire molecule.

The attention of investigators has recently been focused on the study of relaxation processes in increasingly structurally complex objects such as organometallic comb-shaped polymers including the comb-shaped dialkoxy-substituted polyphosphazones mentioned in Chapter 1 (Table 2.3) [62].

Table 2.3, where the values of the temperatures and activation energies E corresponding to relaxation transitions are compared, shows that lower values of E in comparison to the polymers of different homologous series examined above are characteristic of polydiphosphazenes. In addition, the appearance of the so-called α,β-relaxation processes, believed to be due to the segmental motion of the main chain for lower homologs (with n = 1-2), and reorientation of the side chain around the P—O bond for higher homologs [62] is characteristic of polydiphosphazenes.

TABLE 2.3. Characteristics of the Relaxation Transition in Poly(di-n-alkyl phosphazene)s (PAP-n) $[-N{=}P(OR_n)(OR_n)-]$

Polymer	γ-Process		β-Process		α-Process		αβ-Process	
	T_γ, °C	E_γ, kJ/mole	T_β, °C	E_β, kJ/mole	T_α, °C	E_α, kJ/mole	$T_{\alpha\beta}$, °C	$E_{\alpha\beta}$, kJ/mole
PAP-1	−203	2.1	−187	32	78	206	−14	22
PAP-2	−113	8.8	−186	29	63	172	−40 to −15	10-16
PAP-4	−133	4.6	−162	35	97	109	−62	9
PAP-6	−138	6.7	−158	33	95	130	−50	11
PAP-8	−138	7-9	−152	31	87	193	−45	12
PAP-12	−143	5-11	−120	54	−	−	+12	12

We note the qualitatively different shape of the dependence of the temperatures of the α- and β-transitions on the length of the side chain in comparison to organic polymers with similar side chains. This is apparently due to the predominant effect of crosslinking of the side chains on the characteristics of the relaxation processes and to the separation of the motions of the side and main chains with small values of n, which is due to the elevated flexibility of the main polyphosphazene chain.

A detailed study of the relaxation processes and a comparison of their characteristics to similar processes in low-molecular-weight compounds, analogs of the side chains which have been intensely studied recently, will allow definitive conclusions to be drawn concerning their molecular mechanism.

2.1.3. Crystalline Comb-Shaped Polymers

In examining the nature of the molecular mobility in crystallizable comb-shaped polymers, it is necessary to note that studies particularly involved the wide use of a series of monomers to investigate the kinetics and mechanism of solid-phase polymerization [63-64]. Since the nature of the packing and mobility of the molecules in the solid phase has a predominant effect on the rate of polymerization and the molecular structure of the polymer chains formed, it is important to establish the correlation between the structure and

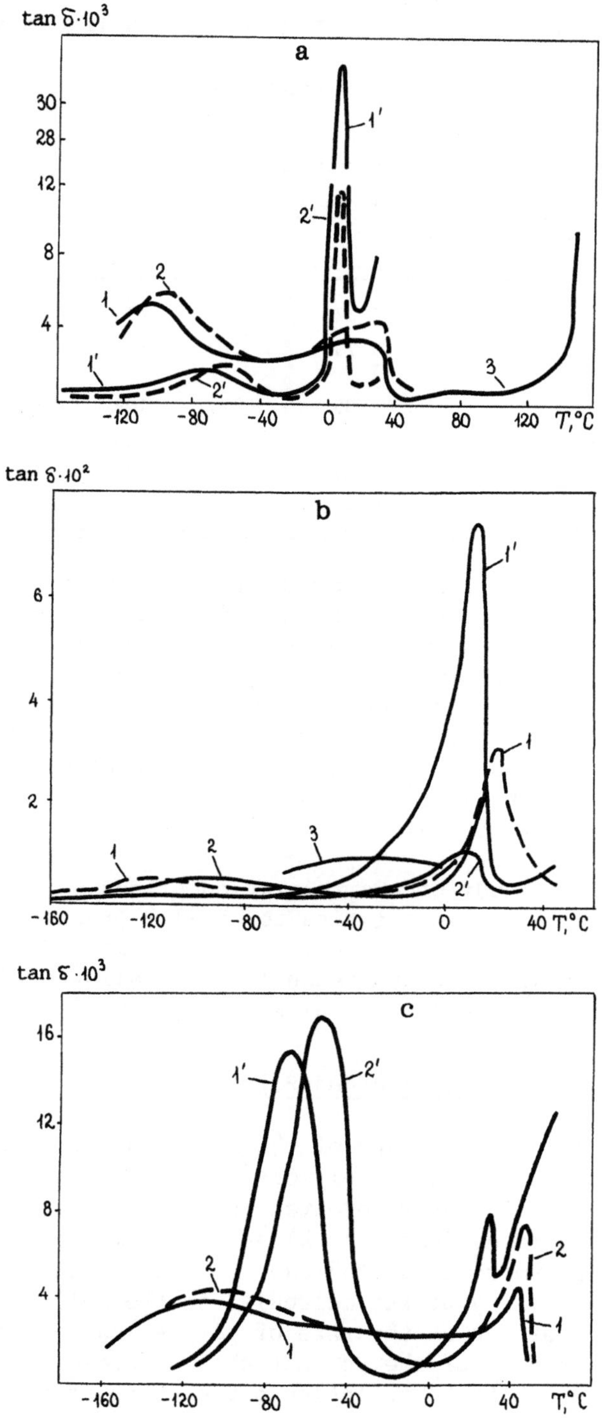
tan δ·10³
a
30
28
12
8
4
1′
2′
1
2
3
-120 -80 -40 0 40 80 120 T,°C
tan δ·10²
b
6
4
2
1′
1
2′
3
2
1
-160 -120 -80 -40 0 40 T,°C
tan δ·10³
c
16
12
8
4
1′
2′
2
1
-160 -120 -80 -40 0 40 T,°C

relaxation behavior of the polymeric and the corresponding monomeric compounds. For this reason, in the subsequent discussion, the relaxation behavior of higher homologs of PA, PMA, and PV esters is compared with the corresponding mechanisms observed in the crystalline monomers whose structure was examined in [65].

The temperature dependences of tan δ of PA-16, PMA-16, and PVE-17 polymers and their corresponding monomeric compounds are shown in Fig. 2.6 [25]. The presence of regions of $\tan\delta_{max}$ at temperatures from −110 to −90°C and near the melting point is characteristic for all of the polymers. Monomers A-16 and VE-17 also have two regions of $\tan\delta_{max}$: from −70 to −50°C and in the melting region. Monomer MA-16 only has one region of $\tan\delta_{max}$ near T_m. The processes at low temperatures are relaxation processes, as indicated by the shift of the region of $\tan\delta_{max}$ to higher temperatures with an increase in the frequency of the measurements. The values of the activation energies of the dipole polarization U are reported below:

	PA-16	PMA-16	PMA-18
U, kJ/mole	46.2 [25]	37.8 [25]	22.3 [47]
	PVE-17	A-16	VE-17
U, kJ/mole	46.2 [25]	75.6 [25]	63 [25], 53.3 [66]

In crystalline polymers and monomers, the polar COO group retains its mobility at low temperatures, causing relaxation of the dipole polarization which probably takes place according to the same mechanism. This is indicated by the similar values of the relaxation time τ and the activation energy of polarization of the COO groups and approximately similar order of magnitude of $\tan\delta_{max}$.

As indicated above, in comparing the amorphous homologs of the PA and PMA series, the presence of an α-methyl group

Fig. 2.6. Temperature dependences of tan δ for crystallizable comb-shaped polymers [25] PA (a), PMA (b), PVE (c), and their monomers: a) PA-16 (1-3) and A-16 (1', 2') at 1 kHz (1, 1', 3) and 10 kHz (2, 2'); b) PMA-16 (1-3) and MA-16 (1', 2') at 1 kHz (1, 1'), 10 kHz (2, 2'), and 30 MHz (3); c) PVE-17 (1, 2) and VE-17 (1', 2') at 1 kHz (1, 1') and 10 kHz (2, 2').

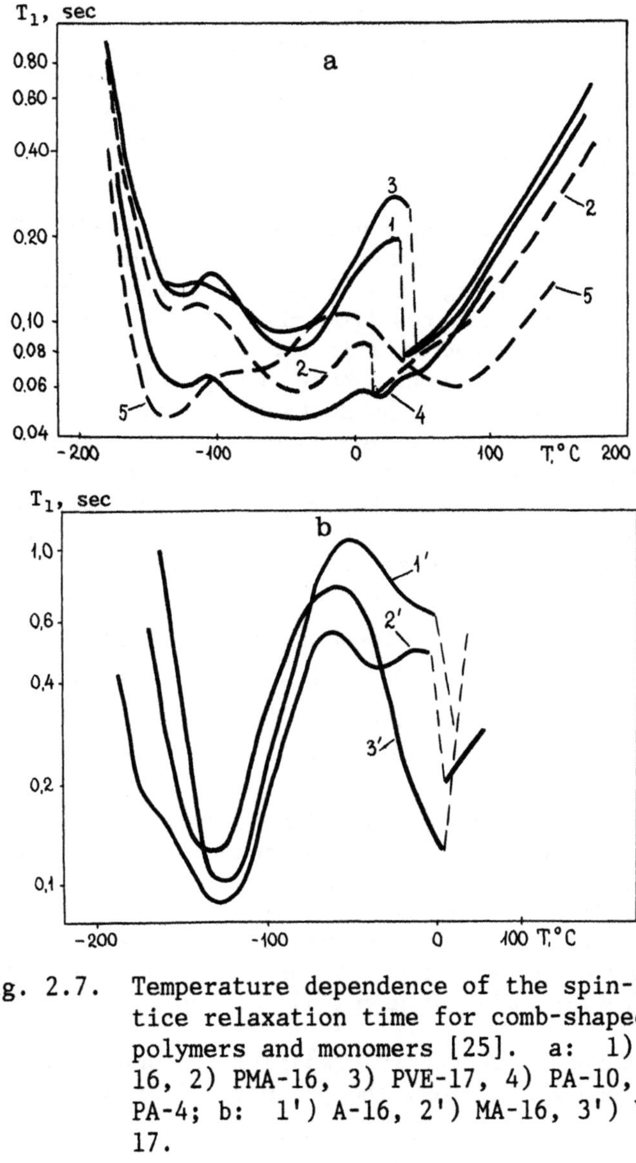

Fig. 2.7. Temperature dependence of the spin-lattice relaxation time for comb-shaped polymers and monomers [25]. a: 1) PA-16, 2) PMA-16, 3) PVE-17, 4) PA-10, 5) PA-4; b: 1') A-16, 2') MA-16, 3') VE-17.

results in hindrance of the main chain, and this in turn causes elevated values of τ and U in PMA in comparison to PA (see Table 2.2). In the crystalline higher homologs PA-16 and PMA-16, the ratio of τ and U is inverse: higher mobility, i.e., smaller τ and U, is characteristic of PMA-16.

This is due to differences in the packing of the PA-16 and PMA-16 macromolecules: the more defective packing of PMA causes the increased mobility of their ester groups. For polymers with two-layer packing of the side chains (PA-16 and PVE-17) but a different method of addition of the alkyl groups to the main chain, the values of $\tan \delta_{max}$ and U are relatively similar.

The study of the molecular motion in monomers revealed the presence of dipole polarization relaxation for A-16 and VE-17 in the region of −70 to −50°C and the total absence of a low-temperature relaxation process for MA-16. The difference in the values of $\tan \delta_{max}$ and U for A-16 and VE-17 is determined by the concentration of different crystalline forms (rhombic, triclinic, and hexagonal) in these monomers, and the ratio between them is a function of the preparation conditions of the samples [65]. The absence of dipole polarization relaxations in MA-16 at low temperatures is apparently due to the difficulty of obtaining hexagonal packing which permits rotation of the individual segments of the molecules, including the COO groups.

A comparison of the molecular mobility parameters (U, τ, and $\tan \delta_{max}$) for PA-16 and PVE-17 and their monomers shows that the polar group has greater mobility in the polymer than in the monomer [25]. The temperatures of $\tan \delta_{max}$ of the low-temperature polarization process are 40-50°C lower for PA-16 and PVE-17 than the temperatures for the corresponding monomers. This is due to the hexagonal packing of the side chains of the macromolecules, which is stable up to a temperature of −196°C (see p. 44) and permits a certain freedom of motion of the segments of the macromolecules, including the COO group. In the low-temperature region, the monomers exist in triclinic (A-16), rhombic (VE-17), or mixed crystalline modifications which significantly limit the freedom of rotation of the polar groups [65].

The analysis of the temperature dependences of $\tan \delta$ of the crystalline polymers PA-16, PMA-16, and PVE-17 (cf. Fig. 2.6) shows that the motion of the kinetic units which intensifies the mobility of the polar groups, causing the appearance of dipole polarization relaxation near the melting point, is characteristic of the premelting state. For polymers with two-layer packing of the side chains (PA-16 and PVE-17), the premelting region is slightly broader than for PMA-16.

For PA-16 (see Fig. 2.6a), another tan δ peak is observed in the premelting region whose temperature coincides with the melting of all crystallizable comb-shaped polymers. The appearance of this peak can be correlated both with the transition from two-layer packing of the side chains to single-layer packing, examined previously (cf. p. 61) for poly(n-alkyl acrylate)s and PA-16 in particular, and with the presence of the two crystalline modifications found in [67].

We note that some mobility of macromolecules in the premelting region was also observed for PE-16 [20] in a study of the temperature dependence of the spin-lattice relaxation time (see Fig. 2.1), but this was not clearly explained. The presence of two minima in Fig. 2.1 in the melting region could correspond to the transition between two crystalline modifications of this polymer (see p. 63); in our opinion, this could be due to the transition from two-layer to single-layer packing of the side chains.

Above the melting point, the constancy of the values of tan δ in the 40-130°C temperature range for PA-16 (see Fig. 2.6) indicates the invariant nature of the molecular motion of the kinetic units containing a COO group, i.e., in this range of temperatures, no segmental mobility is observed.

The study of the temperature dependence of the spin-lattice relaxation time of protons (T_1) for crystallizable polymers provides additional information on the nature of the molecular motion (Fig. 2.7) [25]. Two regions of $T_{1,min}$ are clearly distinguished in crystallizable PA-16, PMA-16, and PVE-17 at temperatures below the melting point, as in amorphous homologs with $n \geq 3$: at −130 and −40°C. The region of $T_{1,min}$ at −130°C, as in the lower homologs of these series, is due to rotation of the CH_3 group at the end of the side chain.

The second relaxation region at −40°C is especially interesting. By analogy with the amorphous homologs of PA, PMA, and PV esters, this region can be correlated with the mobility of the side chain as a whole or with the motion of kinetically independent segments of the alkyl side chains. In comparison to PA-4, for example (see Fig. 2.7a, curve 5), the examined region of $T_{1,min}$ in amorphous PA-10 (Fig. 2.7a, curve 4) and in crystallizable PA-16, PMA-16, and PVE-17 is shifted toward higher temperatures, which indicates an increase in the side-chain interaction with an increase in their length [25, 26]. The coincidence of the temperature regions of $T_{1,min}$ for PA-10 and crystalline polymers arises from the partial crystallization of PA-10 during the experiment, since this polymer also crystallizes in the hexagonal form on prolonged cooling, although to a lesser degree.

In crystallizable PA-16, PMA-16, and PVE-17, the intensity of $T_{1,min}$ is significantly lower than in PA-10, although the proportion of protons in the side chains increases. In the crystallization of PA-16, PMA-16, and PVE-17, parts of the side chains, in particular their alkyl segments, are probably more hindered due to the higher degree of crystallinity of these polymers in comparison to PA-10. This is equivalent to a decrease in the relative number of protons in the mobile alkyl chains. The lower value of $T_{1,min}$ in PMA-16 in comparison to PA-16 and PVE-17 indicates the greater defectiveness of its crystal structures.

At temperatures above T_m, a poorly expressed relaxation region (60-80°C) appears in the course of $T_1 = f(T)$ for PMA-16; it is apparently due to the segmental motion of the macromolecules. For PA-16 and PVE-17, which have no α-methyl groups to significantly hinder the motion of the main chain, segmental mobility should be at this resonant frequency at lower temperatures when its appearance is added to the effects caused by melting. The process of dipole–segmental relaxation was not found in dielectric loss measurements. If we consider the mechanisms of the temperature change corresponding to $\tan\delta_{max}$ (at a constant frequency) or T_g in the homologous series of PA or PMA (see Fig. 1.1), the appearance of a region of $\tan\delta_{max}$ of dipole–segmental losses should be expected at the temperatures at which the polymer is in the crystalline state. However, the dipole polarization relaxation at temperatures of −110 to −90°C cannot be considered as the dipole–segmental relaxation since parameters τ and U of the low-temperature process of dipole polarization in polymers do not exceed those of the corresponding crystalline monomers. This indicates the localization of the kinetic unit in the bulk of no more than one unit. It can thus be concluded that segmental motion does not occur when $T \leq T_m$, and the main chains of these polymers are kinetically rigid.

In the crystalline monomers A-16, VE-17, and MA-16 and in the corresponding polymers, $T_{1,min}$ is located at −130°C, arising from rotation of the CH_3 groups at the ends of the alkyl chains. A similar type of mobility is also found in long-chain paraffins, for example in $C_{32}H_{66}$ [8].

The spin-lattice relaxation region which appears in the polymers at −40°C is absent in monomers due to the significant hindrance of the motion of the alkyl chains which crystallize in the triclinic or rhombic form. The protons of the alkyl segment only acquire mobility in the premelting region, as indicated by the decrease in T_1 at a temperature above −50°C up to melting, similar to what takes place in the melting of paraffins [5, 8].

Mobility of the polar side groups adjacent to the main chain has thus been found in the crystalline state in polymers with crystallizable side chains by the dielectric method. According to the spin-lattice relaxation data, not only rotation of the methyl groups, but also motion of the alkyl side chains, takes place in these crystalline polymers. If the mechanisms of dielectric and spin-lattice relaxation are the same, i.e., the polar group and alkyl chain form one kinetic unit, then the most probable dielectric relaxation time and correlation time at the temperature of the minimum of T_1 should be close to each other [25].

A comparison of these values suggests that the motion of the COO group and the alkyl segment (possibly part of this segment) is independent and involves different kinetic units. Based on the conclusions drawn above on the dipole polarization relaxation in PA, PMA, and their copolymers [31], it is possible to postulate that the COO groups in PA-16 and PMA-16 move together with several carbon atoms in the main chain.

2.1.4. Concluding Comments

In summarizing the data reported above, the relation between the structure of comb-shaped polymers and their molecular mobility can be represented as follows. An increase in the length of the alkyl side chain in a series of comb-shaped polymers (up to n = 6-8) sharply decreases the temperature of the appearance of segmental mobility of the molecules and the glass transition temperature of the polymers. This is most clearly demonstrated by the dependences of T_g and the activation energies of dipole–segmental polarization

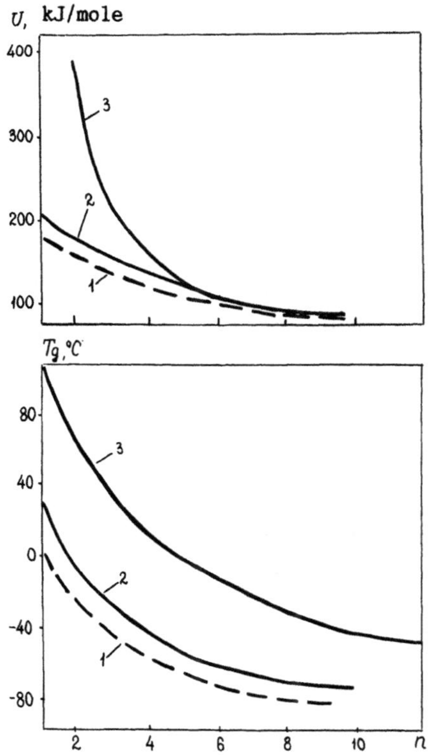

Fig. 2.8. Dependence of the glass transition temperature and activation energy of dipole–segmental polarization on the length of the side chain in PA-n (1), PV esters (2), and PMA-n (3).

on the length of the alkyl chain shown in Fig. 2.8, plotted with the data from different studies [27, 29, 69-70]. For the lower homologs, i.e., for polymers of methyl methacrylate, methyl acrylate, and vinyl acetate, the differences in U and T_g (and consequently the difference in the dipole–segmental polarization relaxation times at the same temperature) are very significant. A progression to higher homologs is accompanied by a decrease in these differences so that the values of U become very close together, beginning with n ≈ 6, and the glass transition temperatures are close together when n = 10-12. It thus follows that in comb-shaped polymers with sufficiently long linear side groups (n > 10), the kinetic characteristics of the main chain, with which the activation energy and dipole–segmental polarization relaxation time are correlated, almost become independent of their length above a critical side-chain length ($n_{cr} \geq 6$-8). In addition, as the analysis of the mechanisms of the change in U and T_g shows (see Fig. 2.8), lengthening of the side chain levels the differences in the kinetic flexibility of polymer homologs of different chemical structure.

In our opinion, lessening the increase in the kinetic flexibility of the main chain with an increase in the number of carbon atoms in the side chains is due to the intensified (with an increase in n) interaction of the side chains, which results in a slight increase in the rigidity of the main chain on one hand (see Chapter 3 for more detail) and decreases the mobility of the alkyl chains, which is inhibited to the greatest degree in crystallizable polymer homologs, on the other hand.

Crystallization in the side chains of comb-shaped polymers results in the total loss of segmental mobility. The character of molecular motion in this case is determined by the degree of perturbations contributed by the main chains to the crystalline packing of the side chains, i.e., the type of layered packing in the side chains. The onset of mobility within the limits of only a few kinetic units — motion of linkages and motions of segments of the alkyl side chains, which either occur separately [poly(alkyl acrylate)s and methacrylates] or together [poly(1-alkyl ethylene)s] as a function of the chemical nature of the groups and their immediate environment — is possible within the boundaries of the rotational-crystalline state.

The parameters of the observed forms of local mobility (activation energy and relaxation times) in comb-shaped macromolecules in the crystalline state are nearly the same as those of their low-molecular-weight analogs (for example, monomers), and the observed differences are determined by the type of the corresponding crystalline modifications and the degree of their defects.

A special type of structural ordering of amorphous homologs of comb-shaped polymers which consists in the layered ordering of the side chains and the features of their relaxation behavior makes it possible to use them as "active" matrices in conducting some photochemical processes [71]. The study of the photochromic properties of indoline spiropyrans dissolved in different homologs of poly(n-alkyl methacrylate)s revealed in some of them a number of anomalies.

The features of the photochromism of indoline spiropyrans consisting of a strong hypsochromic shift in their photoinduction spectrum and slowing of photochemical and thermal processes are most pronounced in poly(n-hexyl methacrylate) (PMA-6) and partially in poly(n-octyl methacrylate) (PMA-8). The relatively successful combination of the layered ordering of the aliphatic side groups, the value of the free space required for cis–trans isomerization of the spiropyran molecules, and the high mobility of the main chains which permits the necessary rearrangements in the added impurity compounds is apparently realized in these polymers.

2.2. RHEOLOGICAL PROPERTIES OF COMB-SHAPED POLY(ALKYL ACRYLATE)S AND POLY(1-ALKYL ETHYLENE)S

As indicated previously, the structure of comb-shaped polymers is basically determined by the tendency of the n-aliphatic side chains to form close packed structures, which results in their crystallization when the critical side-chain length is exceeded. In the conditions of the crystalline state, in the absence of segmental mobility of the main chains the appearance of some forms of localized molecular motion involving small kinetic units of the macromolecules is possible. No manifestation of the role of the high-molecular-weight component of comb-shaped macromolecules, the main chain, is actually observed in the crystalline state. Realization of segmental mobility is either possible in the amorphous state of comb-shaped polymers (above the melting point for crystallizable polymers) or in dilute solutions of the polymers.

The study of the dynamic behavior of polymers with chain branches possessing segmental mobility in the highly elastic [72-74] and viscous flow [75-77] states upon application of a mechanical field makes it possible to obtain information on the behavior of the main chains of the macromolecules (which is very difficult when studying the properties of these polymers in static conditions) and to evaluate the contribution of each component of comb-shaped macromolecules to its physical and mechanical properties. In addition, the high degree of structural organization of comb-shaped polymers even in the region of the amorphous state (see p. 17) arouses interest in their rheological properties, examined in this section using poly(alkyl acrylate)s and poly(1-alkyl ethylene)s as examples.

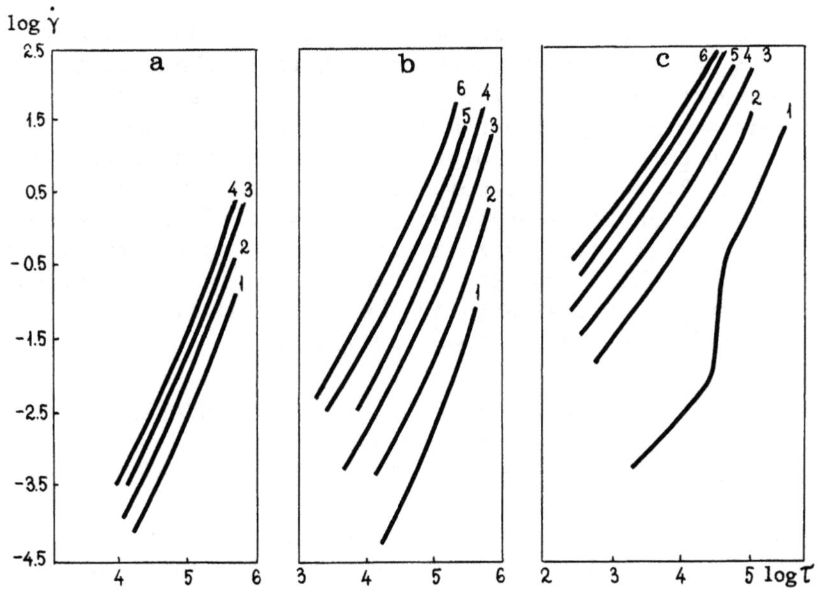

Fig. 2.9. Flow curves of poly(n-alkyl acrylate)s [75, 76]: a) PA-4 ($\bar{M}_\eta = 8.3 \cdot 10^5$); b) PA-10 ($\bar{M}_\eta = 10^6$); c) PA-16 ($\bar{M}_\eta = 6.3 \cdot 10^5$). Temperature, °C: a – 1) 120, 2) 140, 3) 160, 4) 180; b – 1) 20, 2) 40, 3) 70, 4) 100, 5) 120, 6) 140; c – 1) 34-36, 2) 45, 3) 70, 4) 100, 5) 120, and 6) 140.

The flow curves for samples of the PA and PMA series plotted in coordinates of $\log \dot{\gamma}$ vs. $\log \tau$, where $\dot{\gamma}$ is the shear rate and τ is the shear stress, are shown in Figs. 2.9 and 2.10 [75-76]. A very pronounced anomaly in the viscosity is characteristic of the polymers PA-5, PMA-4, PA-10, and PMA-10 in the entire range of variables studied, with no tendency toward a transition into the region of Newtonian flow with low shear rates. The flow curves are described by the Oswald power law; in the region of low shear rates, the effective viscosity attains 10^6-10^7 Pa·sec even at the maximum temperatures studied. For polymers with longer side chains – PA-16, PMA-16, PA-18, and PMA-18 – flow curves with a relatively weak viscosity anomaly and a relatively easily attainable region of the highest Newtonian viscosity are observed in the range of temperatures above the melting point. Consequently, an increase in the length of the side chains in the PA and PMA series (with the same temperatures and shear stresses) causes a decrease in the viscosity of melts of these polymers on the one hand, and broadens the region of flow with the highest Newtonian viscosity, on the other hand.

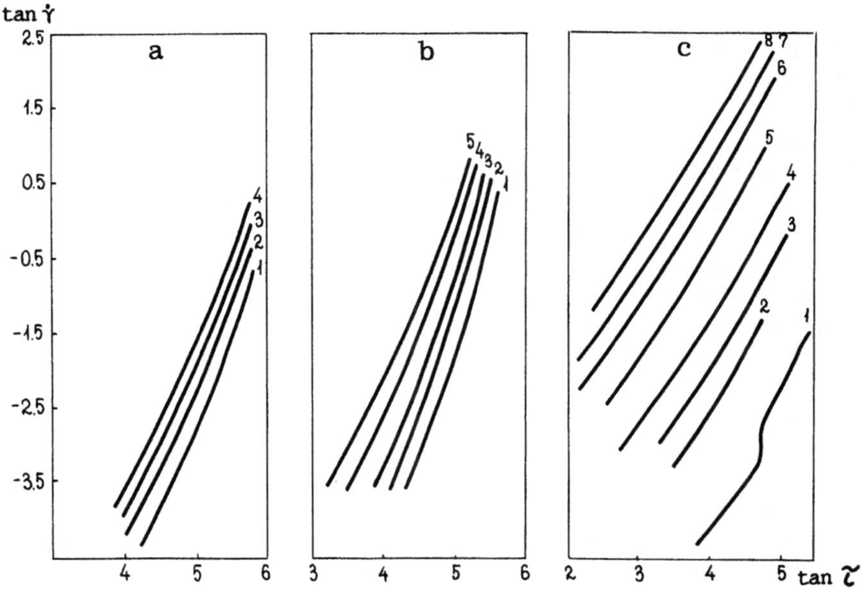

Fig. 2.10. Flow curves of poly(n-alkyl methacrylate)s [75, 76]: a) PMA-4 ($\bar{M}_\eta = 9.8\cdot10^5$), b) PMA-10 ($[\eta] = 1.04$ in heptane at 21°C), c) PMA-16 ($\bar{M}_\eta = 8.1\cdot10^5$). Temperature, °C: A — 1) 120, 2) 140, 3) 160, 4) 180; b — 1) 45, 2) 70, 3) 100, 4) 120, 5) 140; c — 1) 15-20, 2) 25, 3) 30, 4) 50, 5) 70, 6) 100, 7) 120, 8) 140.

The difference in the shape of the flow curves in the region of approximately the same shear rates for samples with short and long side chains graphically demonstrates the plasticizing effect of the n-alkyl side chains, which decrease the glass transition temperature of these polymers* as indicated previously (see Chapter 1). For this reason, the polymers compared are in significantly different physical states in the similar temperature ranges of the measurements.

The flow of polymers with side chains with 16-18 carbon atoms can be identified with the flow of low-molecular-weight hydrocarbons with a common effective viscosity, where the effect of the main chain is reduced to only a slight increase

*The slightly different values of the degree of polymerization of these polymers can also have some effect [75].

in the viscosity of such "filled" systems. This results in the fact that the viscosity in isothermal conditions in polymers with long side chains (PA-16, PA-18, PMA-16, PMA-18) is significantly lower and the viscosity anomaly is significantly less pronounced than in polymers with short side chains. The effective viscosity of polymers in the PMA series is higher than in polymers of the PA series with the same alkyl side-chain length. This indicates the increased kinetic rigidity of the main chains of PMA in comparison to PA. The values for the viscosity of PA and PMA determined with the same shear stress (10^3 Pa) at 100°C are reported below [78]:

	PA-4	PA-10	PA-16	PMA-10	PMA-16
$\log \eta$ (Pa·sec)	7.3	3.8	1.3	6.7	3.5

On the other hand, the differences in the values of the effective viscosity of PA and PMA with side chains of the same length can be attributed to differences in the values of the friction coefficient of the monomeric unit ε_0 determined by Ferry [78] (Table 2.4). According to Ferry, the friction coefficient is a measure of the work relative to the monomeric unit required to stretch a segment of the chain at a constant rate through its local environment. It should be a function of the free volume, the intermolecular interaction, and the steric hindrance for rotation around univalent bonds.

A comparison of the values of $\log \varepsilon_0$ for PMA-1 and for PA-1 indicates the marked steric hindrance caused by the methyl groups in PMA-1. A comparison of the values of ε_0 for the PMA series shows that long side chains decrease the friction, and the effective viscosity is thus decreased in both the PMA series and in the PA series.

The most interesting feature of the flow curves of crystalline samples of PA-16 and PMA-16 (and PA-18 and PMA-18) is the preservation of the possibility of flow at temperatures several degrees below T_m and the appearance of a discontinuity in the flow curves. It should be noted that such an effect has previously not been observed for melts of linear polymers. This event should not be identified with the disruption described in [79, 80] for monodisperse polymers related to the transition from the flowing to the highly elastic state. The difference between these values is not only due to the significantly different stress levels at which the discontinuity is observed, but also to the fact

TABLE 2.4. Friction Coefficients of the Monomeric Units of Some Comb-Shaped Polymers [78]

Polymer	$\log \varepsilon_0$ (N·sec/m)		
	at 25°C	at 100°C	at 125°C
PA-1	−2.68	−8.99	−9.83
PMA-1	–	–	−2.34
PMA-2	–	−3.21	−5.35
PMA-4	–	−6.44	−7.66
PMA-6	−3.75	−8.25	−9.16
PMA-8	−5.29	−8.93	−9.65
PMA-12	−7.69	−9.57	–
PMA-22	–	−10.20 (62°C)	–

that flow stops with the disruption and, in this case, the usual stable flow characteristic of polymeric liquids occurs under stresses which exceed the stresses at the discontinuity.

It can be assumed that the discontinuity in the flow curves of filled polymer systems at the level of some critical stress which is due to perturbation of the quasibrittle structure formed by the particles of solid filler is the closest analogy for this effect [81, 82]. In the case examined, small crystallites formed by the side chains bound by transfer chains can play the role of such a filler. With low stresses, flow of the polymer is not accompanied by any perturbation of its crystal structure, since the stresses in the interstructural bonds can be relaxed. However, exceeding the critical stress results in the appearance and propagation of defective regions which causes the transition from two-layer packing of the side chains to single-layer packing, described above. There is a sharp change in the flow of the polymers due to this structural rearrangement induced by the effect of mechanical stresses (similar in nature to a thermal effect).

In this sense, the behavior of comb-shaped polymers resembles the behavior of so-called plastic crystals in which similar transitions, induced by the effect of mechanical stresses, can occur at temperatures below the melting point due to migration of vacancies and dislocations [83-85].

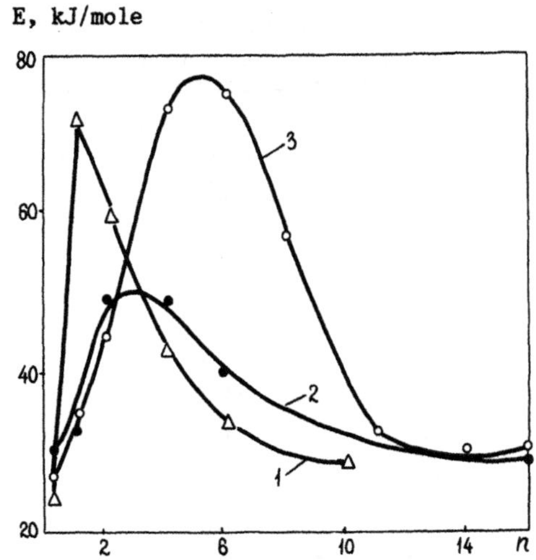

Fig. 2.11. Dependence of the activation energy of viscous flow of poly(1-alkyl ethylene)s on the length of the side chain according to [85] (1), [86] (2), and [87] (3).

To understand the mechanism of the viscous flow of polymers, it is worthwhile to compare the activation energies of this process. According to the existing concepts on the flow of polymers, considered an elementary transfer event in viscous flow, the segmental motion and activation energy of viscous flow (E) are determined by the chemical nature of the polymer, the rigidity of the molecular chain, and the degree of branching to a significant extent. However, there are currently several points of view concerning the effect of the last factor on E. In [84], it was shown that E increases with an increase in the molar volume of side groups in vinylic polymers (such as polyethylene, polystyrene, polyvinyl acetate, etc.), which is clearly confirmed by the data in this study.

An attempt was made in [84] to establish a direct correlation between the value of E and the rigidity parameter of the polymer chains $\sigma^2 = \bar{r}_0{}^2/\bar{r}_{fv}{}^2$ (where $\bar{r}_0{}^2$ is the unperturbed size of the macromolecules in θ conditions and $\bar{r}_{fv}{}^2$ is the root-mean-square distance between the ends of the macromolecules with fixed valence angles in the condition of free rotation of the units [85]). For many linear polymers (polyamides, polyethers, etc.), E actually increases linear-

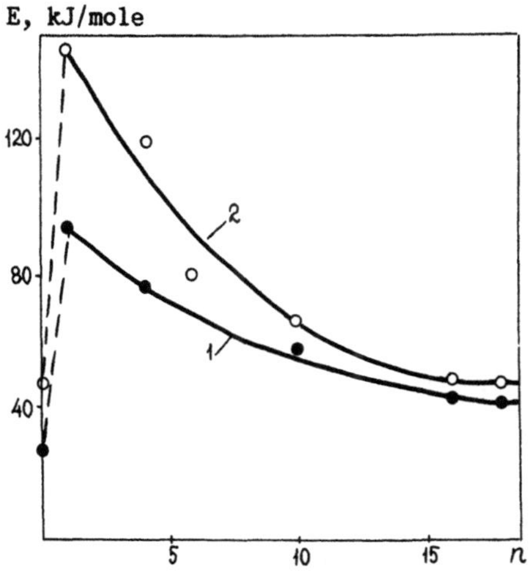

Fig. 2.12. Dependence of the viscous flow activation energy of PA-n (1) and PMA-n (2) on the number of carbon atoms in the alkyl side chain [76].

ly with an increase in σ. However, the results of calculating the viscous flow activation energies for a series of comb-shaped poly(1-alkyl ethylene)s, PA, and PMA demonstrate the opposite effect. As discussed in this section and subsequently examined in detail, an increase in the length of the side chain in comb-shaped macromolecules is accompanied by a slight increase in the equilibrium rigidity of the main chain due to intensification of the interaction in the side groups, which should result in an increase in E in the series of comb-shaped polymers as n increases according to the concepts developed in [84].

The dependence of E on the side-chain length for comb-shaped samples of poly(1-alkyl ethylene)s, plotted with the data from different studies [85-87] and the data in [76] for PA and PMA, is shown in Figs. 2.11 and 2.12. The character of the dependence of E on n is far from unchanging, and the absolute values of E for poly(1-alkyl ethylene)s obtained in the different studies do not always coincide (especially in the range of n = 4-8). The causes of these differences in the values of E could be due to both the specific features of the conditions of synthesis of the polymers studied, which sometimes result in the uncontrolled appearance of random long-chain branches, and to the different microstructure of the samples studied. The differences in the experimental conditions in the estimation of E could also cause differences in its value.

Nevertheless, these dependences exhibit a common characteristic: E remains constant with an increase in n (and σ^2), as stipulated in [84], or E decreases with increase in n, approaching the limit when $n \geq 10$. The limit value of E for comb-shaped poly(1-alkyl ethylene)s is close to the values of E for linear polyethylene (25-30 kJ/mole). This coincidence is in all probability not accidental and is apparently due to a change in the molecular mechanism of the flow of comb-shaped polymers with an increase in n. It is known [83] that the activation energy of the viscous flow of n-paraffins attains the limit value (25-30 kJ/mole) when the molecule is 20-30 carbon atoms long, which is approximately two times higher than the n value corresponding to the limiting value of E in comb-shaped poly(1-alkyl ethylene)s.

The limit of the values of the activation energy of viscous flow in melts of comb-shaped polymers with E equal to that of polyethylene is thus attained faster (i.e., with smaller n) than in melts of hydrocarbons. In our opinion, this can be understood if it is assumed that two methylene side chains in the trans position with respect to the main chain and chemically bound to it are included in the elementary transfer event. The main chain in comb-shaped macromolecules is responsible for the greater cooperation of translation of the methylene chains joined in layered structures in comparison to hydrocarbons. (The possible decrease in the number of linkages with an increase in n also cannot be excluded.)

A comparison of the values of the activation energy of viscous flow of a series of poly(1-alkyl ethylene)s with PA and PMA with the same side-chain length indicates (see Figs. 2.11 and 2.12) that the ester groups in the macromolecules of the latter bring about a sharp increase in E for the lower members of the PA and PMA series. This is clearly seen from a comparison of the values of E for polypropylene and poly(methyl methacrylate) (PMA-1) and linear polyethylene and poly(methyl acrylate) (PA-1). The presence of a polar carbonyl group in the macromolecules of PA and PMA, which intensifies the intra- and intermolecular interactions, is apparently the cause of the higher E values of these polymers in comparison to poly(1-alkyl ethylene)s.

The increase in the length of the alkyl side chain in the PA and PMA series results in a decrease in the values of E, and although the values of E differ significantly for the first members of the series, an increase in n to 10-12 levels out this difference for PA and PMA, as in the series of poly(1-alkyl ethylene)s. In the limiting case of long alkyl chains, the activation energy of viscous flow is equal to 42 kJ/mole (for PA) and 50 kJ/mole (for PMA), which in both cases is higher than the E of linear polyethylene and close to the values of E for branched polyethylene.

For comb-shaped polymers, there is thus a change in the basic elementary transfer event which causes irreversible chain translation as the length of the alkyl radical increases. Actually, more advantageous transfer of the long and flexible alkyl side chains takes place by activation instead of transfer of a segment of the main chain; in conditions of viscous flow, the entire chain is translated, apparently due to the motion of the layered structures. The method of addition of alkyl side chains to the main chain affects the value of the activation barrier of the motion of the side chains while not changing the molecular nature of the flow of comb-shaped polymers of the PA, PMA, and poly(1-alkyl ethylene) series, which probably takes place according to one mechanism.

REFERENCES

1. N. McCrum, B. Read, and G. Williams, Inelastic and Dielectric Effects in Polymeric Solids, Wiley, London (1967).
2. J. D. Ferry, Viscoelastic Properties of Polymers, Wiley, New York (1981).
3. P. Hedvig, Dielectric Spectroscopy of Polymers, Wiley, New York (1977).
4. G. M. Bartenev and Yu. V. Zelenev (eds.), Relaxation Phenomena in Polymers [in Russian], Khimiya, Leningrad (1972).
5. R. Boyer (ed.), Transitions and Relaxation in Polymers, Wiley—Interscience, New York (1966).
6. R. T. Bailey, A. M. North, and R. Pethrick, Molecular Motion in High Polymers, Clarendon Press, Oxford (1981).
7. B. Crist and A. Peterlin, J. Polym. Sci., A-2, 7, 1165-1173 (1969).
8. H. Olf and A. Peterlin, J. Polym. Sci., A-2, 8, 753-808 (1970).

9. W. Rechold, J. Polym. Sci., C, No. 23, 123-140 (1971).
10. R. H. Boyd, Polymer, 26, 323-348 (1985).
11. G. Hoffman, G. Williams, and E. Passaglia, in: Transitions and Relaxation in Polymers, R. Boyer (ed.), Wiley-Interscience, New York (1966).
12. A. Willbourn, Trans. Faraday Soc., 54, 717-728 (1958).
13. D. Kline, J. Sauer, and A. Woodward, J. Polym. Sci., 22, 455-470 (1956).
14. D. Sandiford and A. Willbourn, in: Polythene, A. Renfrew and P. Morgan (eds.), Hiffe, London (1960), Chapter 8.
15. S. P. Kabin, Zh. Tekh. Fiz., 26, 2628-2636 (1956).
16. S. Okamoto and K. Takeuchi, J. Phys. Soc. Jpn., 14, 378-389 (1959).
17. S. Saito and T. Nakaijama, J. Polym. Sci., 36, 533-547 (1959).
18. K. Clark, A. Turner-Jones, and D. Sandiford, Chem. Ind., No. 24, 2010-2012 (1962).
19. R. Gray and N. McCrum, J. Polym. Sci., A-2, 7, 1329-1338 (1969).
20. W. Slichter and D. Davis, J. Appl. Phys., 35, 10-17 (1964).
21. I. Ya. Slonim and A. N. Lyubimov, Nuclear Magnetic Resonance in Polymers [In Russian], Khimiya, Moscow (1966).
22. N. Erlich and P. Scherbak, Zh. Tekh. Fiz., 25, 1575-1585 (1955).
23. H. Block, R. Groves, and S. Walker, Polymer, 13, 527-561 (1972).
24. B. Caylor, A. Eizenberg, J. Harrod, and P. Rocaniere, Macromolecules, 5, 676-670 (1972).
25. T. I. Borisova, L. L. Burshtein, V. A. Shevelev, N. A. Platé, and V. P. Shibaev, Vysokomol. Soedin., A13, 2332-2340 (1971).
26. T. I. Borisova, L. L. Burshtein, V. A. Shevelev, N. A. Platé, and V. P. Shibaev, Vysokomol. Soedin., A15, 674-679 (1973).
27. G. P. Mikhailov and T. I. Borisova, Zh. Tekh. Fiz., 28, 137-148 (1958).
28. T. I. Borisova, G. P. Mikhailov, and M. M. Koton, Vysokomol. Soedin., A11, 1140-1144 (1969).
29. G. P. Mikhailov and L. V. Krasner, Vysokomol. Soedin., 4, 1071-1075 (1962).
30. L. V. Krasner and G. P. Mikhailov, Vysokomol. Soedin., 1, 542-544 (1959).

31. G. P. Mikhailov, T. I. Borisova, N. N. Ivanov, and A. S. Nigmankhodzhaev, Vysokomol. Soedin., A8, 2181-2190 (1966); A9, 773-778 (1967).
32. H. Thurn and K. Wolf, Kolloid Z. Z. Polym., 148, 16-28 (1956).
33. G. P. Mikhailov and V. A. Shevelev, Vysokomol. Soedin., A9, 2442-2449 (1967).
34. G. L. Slonimskii, A. A. Askadskii, and A. I. Kitaigorodskii, Vysokomol. Soedin., A12, 494-512 (1970).
35. Y. Ishida, Kolloid Z. Z. Polym., 174, 124-136 (1961).
36. L. Brouckere and G. Offergeld, J. Polym. Sci., 30, 105-112 (1958).
37. G. P. Mikhailov, Makromol. Chem., 35, 26-38 (1960).
38. D. Scheiber, J. Res. Natl. Bur. Stand., Ser. C, 65, 23-38 (1961).
39. T. I. Borisova, L. L. Burshtein, and G. P. Mikhailov, Vysokomol. Soedin., 4, 1479-1485 (1962).
40. K. Deutsch, E. Hoff, and W. Reddish, J. Polym. Sci., 13, 565-579 (1954).
41. J. Heijboer, Kolloid Z. Z. Polym., 148, 36-48 (1956).
42. N. McCrum and E. Moveis, J. Appl. Phys. Jpn., 4, 542-556 (1965).
43. I. Ishida and N. Yamafuji, Kolloid Z. Z. Polym., 177, 97-108 (1961).
44. S. Kurath, T. Yin, J. Berger, and J. Ferry, J. Colloid Sci., 14, 147-161 (1959).
45. S. Saito and T. Nakajiama, J. Appl. Polym. Sci., 2, 93-108 (1959).
46. H. Sasabe and S. Saito, J. Polym. Sci., A-2, 6, 140-152 (1968).
47. E. Hoff, D. Robinson, and A. Willbourn, J. Polym. Sci., 18, 161-176 (1955).
48. G. P. Mikhailov and T. I. Borisova, Usp. Khim., 30, 895-906 (1961).
49. D. Mead and R. Fuoss, J. Am. Chem. Soc., 63, 2832-2845 (1941).
50. I. Ishida, M. Matsuo, and K. Yamafuji, Kolloid Z. Z. Polym., 180, 108-118 (1962).
51. H. Hikichi and T. Furuichi, Rep. Progr. Polym. Phys. Jpn., 4, 69-80 (1961).
52. P. F. Veselovskii and A. I. Slutsker, Zh. Tekh. Fiz., 25, 939-946, 1204-1210 (1955).
53. T. Sutherland and B. Funt, J. Polym. Sci., 11, 177-188 (1953).
54. E. Jordan, J. Polym. Sci., A-1, 9, 3367-3374 (1971).
55. J. Sauer, J. Polym. Sci., C, No. 32, 69-78 (1971).

56. Y. Kawamura, J. Polym. Sci., A-2, 7, 1559-1568 (1969). (1969).
57. A. Woodword, in: Transitions and Relaxation in Polymers, R. Boyer (ed.), Wiley—Interscience, New York (1966).
58. I. Schatzki, J. Polym. Sci., 57, 496-508 (1962); Polym. Preprints, 6, 646-648 (1965).
59. J. M. G. Cowie, Z. Haq, I. McEwen, and J. Velickovič, Polymer, 22, 327-331 (1981).
60. J. M. G. Cowie, I. McEwen, and M. Y. Pedram, Macromolecules, 16, 1151-1155 (1983).
61. J. M. G. Cowie, R. Ferguson, I. McEwen, and M. Y. Pedram, Macromolecules, 16, 1155-1158 (1983).
62. V. V. Kochervinskii, I. B. Sokol'skaya, V. V. Kireev, and Ya. V. Zelenov, Vysokomol. Soedin., A24, 1275-1282 (1982).
63. V. A. Kargin and V. A. Kabanov, Zh. Vses. Khim. Ova. D. I. Mendeleeva, 9, 602-612 (1964).
64. A. D. Abkin, P. M. Khomikovskii, E. V. Volkova, and V. I. Lukhovitskii, Zh. Vses. Khim. Ova. D. I. Mendeleeva, 18, 246-253 (1973).
65. B. S. Petrukhin, V. P. Shibaev, and N. A. Platé, Vysokomol. Soedin., B12, 687-690 (1970).
66. M. Broadhierst and E. Fitzgerald, J. Chem. Phys., 33, 210-217 (1960).
67. I. Sobbotka and I. V. Sochava, Vysokomol. Soedin., A21, 1322-1326 (1979).
68. S. Rogers and L. Mandelkern, J. Phys. Chem., 61, 985-995 (1957).
69. N. A. Platé and V. P. Shibaev, Vysokomol. Soedin., A13, 410-423 (1971).
70. J. Lal and G. Trick, J. Polym. Sci., A, 2, 4559-4567 (1964).
71. V. D. Arsenov, S. D. Maltsev, V. S. Marevtsev, M. I. Cherkashin, Ya. S. Freidzon, V. P. Shibaev, and N. A. Platé, Vysokomol. Soedin., A24, 2298-2303 (1982).
72. V. A. Kargin, D. Kh. Khalikov, V. P. Shibaev, N. A. Platé, and A. F. Lemenovskaya, Dokl. Akad. Nauk SSSR, 190, 376-379 (1970).
73. D. Kh. Khalikov, V. P. Shibaev, N. A. Platé, and V. A. Kargin, Vysokomol. Soedin., A14, 218-225 (1972).
74. N. A. Platé, D. Kh. Khalikov, V. P. Shibaev, and V. A. Kargin, Vysokomol. Soedin., A14, 226-237 (1972).
75. T. Kh. Poolak, A. Ya. Malkin, V. P. Shibaev, N. A. Platé, and G. V. Vinogradov, Izv. Akad. Nauk Eston. SSR, 21, 229-238 (1972).

76. N. A. Platé, A. Ya. Malkin, V. P. Shibaev, T. Kh. Poolak, and G. V. Vinogradov, Vysokomol. Soedin., A16, 437-448 (1974).
77. T. Kh. Poolak and V. P. Shibaev, in: 18th All-Union Conference on High-Molecular-Weight Compounds, Abstracts [in Russian], Nauka, Moscow (1973), p. 151.
78. J. Ferry, Viscoelastic Properties of Polymers, Wiley, New York (1961).
79. G. V. Vinogradov, Vysokomol. Soedin., A13, 294-302 (1971).
80. G. V. Vinogradov, A. Ya. Malkin, B. G. Yanovskii, E. G. Borisenkova, B. V. Yarlikov, G. V. Berezhnaya, V. P. Shatanov, V. G. Shonganova, and V. P. Yudin, Vysokomol. Soedin., A14, 2425-2433 (1972).
81. G. V. Vinogradov, A. Ya. Malkin, E. P. Plotnikova, Yu. Yu. Sabsai, and N. E. Nikolaeva, Problems of Heat and Mass Transfer [in Russian], Energiya, Moscow (1970).
82. P. A. Rebinder, Supplement, in: Rheology: Theory and Applications, F. Eirich (ed.), Vol. 1, Academic Press (1956).
83. G. Burdjee and V. Burdjee, in: Rheology: Theory and Applications, F. Eirich (ed.), Vol. 1, Academic Press (1956).
84. R. S. Porter and J. F. Johnson, J. Polym. Sci., C, 15, 365-385 (1966).
85. Wang Jeon-Shyong, R. Porter, and J. Knox, J. Polym. Sci., 138, 671-682 (1970).
86. R. L. Combs, D. Slonaker, and H. W. Coover, J. Appl. Polym. Sci., 13, 519-527 (1969).
87. K. Shirayama, T. Matsuda, and S. Kita, Makromol. Chem., 151, 97-108 (1972).

Chapter 3

Comb-Shaped Macromolecules in Solutions and Intramolecular Interactions

The dual character of comb-shaped polymers, which consists of the independent behavior of the main and side chains of the macromolecules in the condensed state, is manifested to the greatest degree in dilute solutions where the intermolecular interaction of the chains is reduced to a minimum. The unusual structure of comb-shaped polymers which contain long and relatively independently behaving segments of methylene groups of varying length in association with the polar groups of the macromolecules is the basis for considering them as amphiphilic systems with a given number and length of the branches.

The study of the conformational structural features of such uniformly constructed branched chains is significant for clarifying problems of conformational statistics of macromolecules. The possibility of intramolecular contacts between the side chains, responsible for stabilization of some structural ordering, determines the capacity of these polymers to crystallize due to the packing of the side chains in the condensed state, as shown in Chapter 1. The interaction of the side groups may be similar to the short-range interaction forces of the hydrogen-bond type which stabilize the α- and β-structure in polypeptides, not only in the condensed state, but also in dilute solutions. In the study of solutions of comb-shaped polymers, questions naturally arise concerning the structural organization of comb-shaped macromolecules in solutions (arising from their amphiphilicity), the possibility and conditions of the occurrence of conformational transformations, the role of long-range interactions, etc. One of the current problems is the study of the crosslinking of comb-shaped polymers in solutions under conditions close to the precipitation point, since the primary ordering which arises in solution determines the structure and physicochemical properties of the polymer material to a great extent.

3.1. OPTICAL ANISOTROPY

As indicated in [1], the size and structure of the side groups of macromolecules have an important effect on their conformation and optical properties, particularly on the anisotropy of the polymer chains. An increase in the length of the side chains is accompanied by a decrease in the positive anisotropy of the macromolecules and can even result in a change in the sign of the anisotropy and a subsequent increase in its negative value. At the same time, it can be predicted that the monotonic increase in the negative anisotropy should slow as the side chain lengthens due to its flexibility. This mechanism was deduced theoretically by Tsvetkov [2, 3], who calculated the anisotropy of the side chain ($\gamma_{II}-\gamma_{I}$) in the axes of its first element and used this to estimate the anisotropy of the side chains of poly(1-alkyl ethylene)s [4], poly(n-alkyl acrylate)s [5], and poly(n-alkyl methacrylate)s [6]. Let us examine the basic relations which correlate the molecular parameters of branched polymers with the optical anisotropy of their macromolecules.

The optical anisotropy of a comb-shaped macromolecule is the tensor sum of the anisotropies of the main and side chains. According to the theory of persistent anisotropy [2, 7], the contribution of Δb_L of the side chain to the anisotropy of the monomeric unit of a comb-shaped macromolecule is expressed as

$$\Delta b_L = -\tfrac{1}{2}(\gamma_{\parallel} - \gamma_{\perp}) = \frac{s\Delta a}{12}[1 - \exp(-6m/s)] \qquad (3.1)$$

where Δa is the anisotropy of the side chain per CH_2 group, and m and s are the valence bonds in the side chain and its segment.

As a consequence, both the flexibility of the side chains and the anisotropy Δa per valence bond in the side chain can be estimated based on the contribution of the side chain methylene groups of different length to the anisotropy

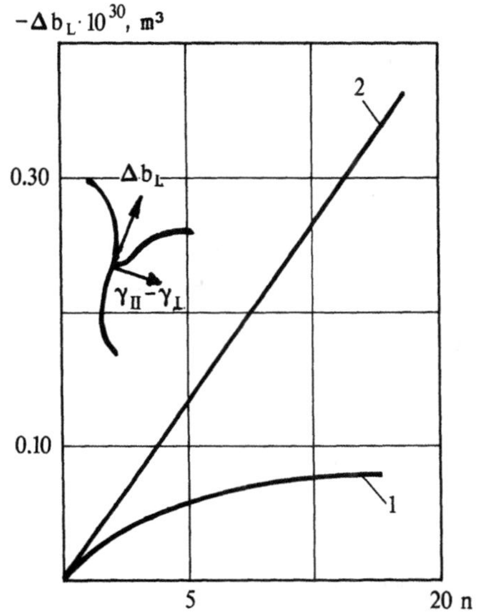

Fig. 3.1. Dependence of the anisotropy of the monomeric unit Δb_L of poly(1-alkyl ethylene)s on the number of carbon atoms in the side chain: 1) theoretical curve corresponding to noninteracting chains with a flexibility equal to the flexibility of the polyethylene chain [8] [Eq. (3.2)]; 2) experimental data from [4].

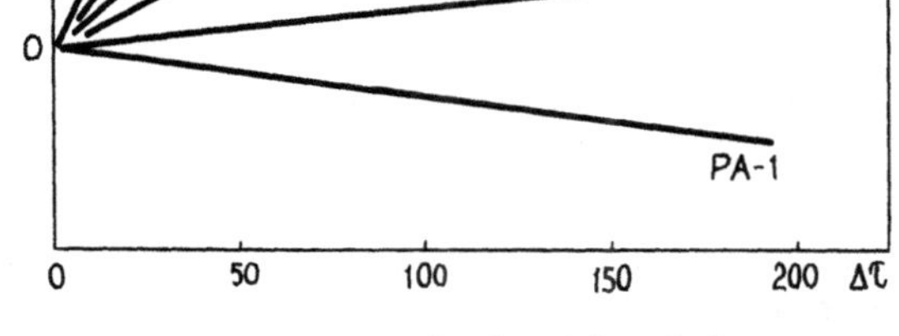

Fig. 3.2. Dependence of the birefringence Δn on the shear stress $\Delta\tau = g(\eta - \eta_0)$ for PA-n polymers [5].

TABLE 3.1. Number of Monomeric Units ν in the Kuhn Segment, Segmental Anisotropy $(\alpha_1 - \alpha_2)$, and Anisotropy of the Monomeric Unit $(a_{\parallel} - a_{\perp})$ of Some Comb-Shaped Polymers

Polymer	Solvent	$(\alpha_1 - \alpha_2)\cdot 10^{30}$, m³	ν	$(a_{\parallel} - a_{\perp})\cdot 10^{30}$, m³	Reference
		Polymethacrylic acid esters			
PMA-1	Benzene	0.2	7	0.02	[9]
PMA-4	Benzene	−1.4	6.7	−0.2	[10]
PMA-6	Benzene	−4.0	8.6	−0.5	[11]
PMA-8	Benzene	−4.7	7.9	−0.6	[11]
PMA-12	Benzene	—	12.2	—	[12–13]
PMA-16	Benzene	−17.0	19.0	−0.9	[13]
PPMOB-16	Benzene	−250	26.0	−10	[14, 15]
		Polyvinyl ethers and esters			
PVE-1 (ester)	Acetone	—	6.9	—	[16]
PVE-12 (ester)	n-Heptane	—	12	—	[17]
PVE-16 (ester)	n-Octane	−13	28±2	−0.47	[18]
PVE-16 (ether)	n-Heptane	—	16–20	—	[19, 20]
		Poly(n-alkyl acrylate)s*			
PA-1	Toluene	+1.6	8.6	0.2	[11, 21]
PA-4	Toluene	−1.0	9.2	−0.1	[21]
PA-8	Toluene	−5.7	13.4	−0.4	[22]
PA-10	Toluene	−9.5	20.0	−0.5	[23]
PA-16	Toluene	−16.4	22.0	−0.7	[24]
PA-18	Toluene	−23.2	28.8	−0.8	[22]

*The values of ν for PA were determined in n-heptane.

of the monomeric unit, i.e., by obtaining the dependence in coordinates of Δb_L vs. length of the side chain. Using the data in [4] on the optical anisotropy of poly(1-alkyl ethylene)s, the value of Δb_L was calculated in [8] with Eq. (3.1) under the condition that the flexibility of the hydrocarbon side chain of poly(1-alkyl ethylene)s is equal to the flexibility of the polyethylene chain (i.e., assuming that s = 16):

$$\Delta b_L = 0.73\cdot 10^{-30}[1 - \exp(-3m/8)] \tag{3.2}$$

The theoretical curve corresponding to Eq. (3.2) is shown in Fig. 3.1. As should be expected, the anisotropy Δb_L attains a constant value determined by the degree of flexibility of the side group (in the given case, with n = 8-10 carbon atoms) with an increase in m due to the flexibility of the side chain. Curve 2 was plotted with the ex-

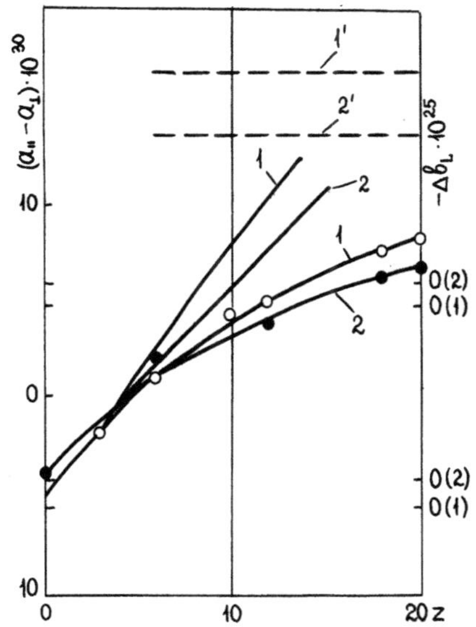

Fig. 3.3. Dependence of the anisotropy of the monomeric unit $(\alpha_\parallel - \alpha_\perp)$ (1, 2) [1) toluene, 2) decalin] on the number of valence bonds z in the side chain [5]. Solid curves: theoretical dependences of $\Delta b_L = f(z)$; 1') $\Delta\alpha = 0.26\cdot10^{-30}$ m^3, s = 80; 2) $\Delta\alpha = 0.21\cdot10^{-30}$ m^3, s = 80 [5].

perimental data in [4], obtained for a series of poly(1-alkyl ethylene)s. Note that the experimental points fit on a line whose slope is equal to the initial slope of curve 1, but the experimental curve does not become parallel to the abscissa, in contrast with the theoretical curve. This difference means that the flexibility and orientational ordering (which determines the value of the observed anisotropy) of the side chains of poly(1-alkyl ethylene)s significantly exceeds the values which can be obtained in the experimental study of the conformational and optical properties of "isolated" hydrocarbon chains in solution.

Using the results of the determination of the flow birefringence and hydrodynamic properties of a series of comb-shaped PA and PMA from Eq. (3.2), the flexibilities of the side chains of PA and PMA were calculated in [5, 6]. As an example, the results of the dynamooptical studies of a series of PA-n in toluene are shown in Fig. 3.2. Based on the experimentally obtained values

$$[n]/[\eta] = \lim_{\substack{g\to 0\\ c\to 0}} \Delta n/g(\eta - \eta_0) \quad (3.3)$$

where g is the flow rate gradient and η and η_0 are the viscosities of the solution and the solvent. The segmental anisotropy $(\alpha_1 - \alpha_2)$ was calculated from these data, and the anisotropy of the monomeric unit was determined using the values of ν (number of monomeric units in the statistical Kuhn segment) (Table 3.1).

By substituting exponent m in Eq. (3.2) with z = m + 2 (to account for the two valence bonds in the ester group of the PA), we obtain the theoretical dependence for the PA series:

$$\Delta b_L = -\frac{s\Delta a}{12}(1 - \exp(-6z/s)] \quad (3.4)$$

The theoretical curves which best satisfy the experimental data are shown in Fig. 3.3. The asymptotic limits are indicated by horizontal dashed lines 1' and 2'. The values of Δa and s were determined by comparing the theoretical curve calculated using Eq. (3.3) with the experimental points of the dependence $(a_\parallel - a_\perp) = f(z)$; $\Delta a = 0.21\cdot10^{-30}$ m^3 in decalin, $\Delta a = 0.26\cdot10^{-30}$ m^3 in toluene, and s = 80 in decalin and toluene. The curves intersect segment $(a_\parallel - a_\perp)_{z=0}$, equal to $0.42\cdot10^{-30}$ m^3 (decalin) and $0.54\cdot10^{-30}$ m^3 (toluene) on the ordinate. These points are the origin of a system of coordinates $(\Delta b_L, z)$ where the theoretical curve of $\Delta b_L = f(z)$ corresponds to Eq. (3.1); z = 0 corresponds to the monomeric unit of polyethylene for which the anisotropy of the chain per methylene group Δa_{PE} in decalin is equal to $0.21\cdot10^{-30}$ m^3 [25], which corresponds satisfactorily to the values obtained for the side groups of PA: $\Delta a = 0.27\cdot10^{-30}$ m^3 (in toluene) and $\Delta a = 0.21\cdot10^{-30}$ m^3 (in decalin).

The value of s = 80 corresponds to the rigidity of the molecular chain, which exceeds that of vinylic monomers of similar chemical structure by 5-6 times but has no long-chain branches. A similar result (s = 60) was obtained when studying the conformational properties of polymethacrylic acid esters [6].

As Table 3.1 indicates, an increase in the length of the side chain in a series of comb-shaped polymers is thus accompanied by an increase in their segmental anisotropy, indicating the presence of orientational order in the side chains which significantly exceeds the conformational ordering of the usual Gaussian chains.

The degree of ordering Q, determined by the following value in the same way as for liquid crystals, is a measure of the orientational order in a chain molecule:

$$Q = (3\overline{\cos^2\theta} - 1)/2 \tag{3.5}$$

where θ is the angle formed by the chain element with the axis of the predominant orientation of the molecules, i.e., vector h, which joins the ends of the macromolecules; $\overline{\cos^2\theta}$ is not only averaged with respect to all elements of the chain but also with respect to all of its conformations with a given value of the ratio h/L (L is the contour length of the chain).

Since the contour chain length L and the mean square of the distance between the ends of the macromolecules are correlated by the ratio

$$h^2/2aL = 1 - (1 - x)(1 - e^{x}) \tag{3.6}$$

where $x = L/a$ (where a is the persistent length), and

$$\overline{\cos^2\theta} = \frac{1 - 3(h/L)}{L^*(h/L)} \tag{3.7}$$

where $L^*(h/L)$ is the inverse Langevin function, it is possible to reduce Eq. (3.5) to

$$Q(x) = \frac{6/5\{1 - [(1 - e^{-x})/x]\}}{x - 0.8[1 - (1 - e^{x})/x]} \tag{3.8}$$

Equation (3.8) indicates that the degree of the intramolecular orientational-axial order is a unique function of x.

The value of Q for chain molecules can be calculated from Eq. (3.8) if the molecular weight M ($L = \lambda M/M_0$, where M_0 and λ are the molecular weight and length of the unit in the direction of the chain, and a is the persistent chain length) is known. The value of Q is equal to 1 (ideal order) when $L \ll a$, i.e., for a rod-shaped molecule, and decreases with an increase in the chain length L or with a decrease in its equilibrium rigidity (i.e., a).

Of all the molecular characteristics, the optical anisotropy is the most sensitive measure of the axial ordering of the structural elements of a macromolecule. As demonstrated in the studies by Tsvetkov et al. [26], the specific anisotropy of a macromolecule $(\gamma_1 - \gamma_2)/M$ is equal to the degree of intramolecular order Q multiplied by the specific anisotropy of the chain element (monomeric unit or Kuhn segment) according to

$$\gamma_1 - \gamma_2 = (M/M_0)\Delta a Q \tag{3.9}$$

The calculation of the degree of intramolecular orientational order in the side chains of higher homologs of comb-shaped polymers with the values of s and the given values of x indicated above produced the following values of Q:

Polymer	s	x	Q
PMA-16	60	0.66	0.73
PA-18	80	0.50	0.77

With respect to the order, these values of Q are very close to the corresponding values of the order parameter of low-molecular-weight nematic liquid crystals (see Chapter 4).

The most perfect intramolecular order is developed in solutions of polymers containing mesogenic side chains capable of forming liquid crystals. The structural characteristics of one of these polymers, the polymethacryloyloxyphenyl ester of cetylhydroxybenzoic acid (PPMOB-16), are reported in Table 3.1 [24, 28].*

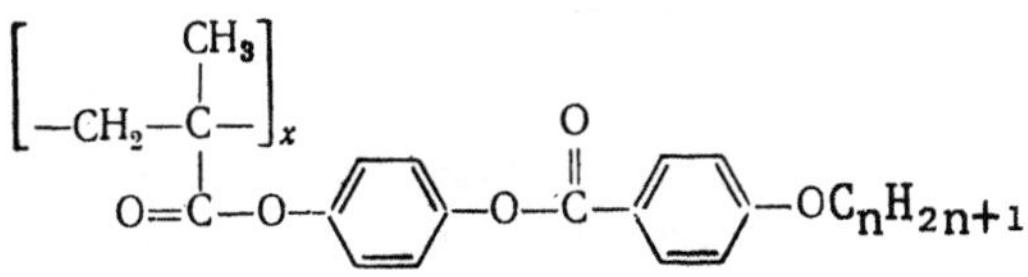

(PPMOB-n)

*A detailed examination of the structural features of this type of liquid-crystalline polymer in the solid phase is given in Chapter 4, and the properties of solutions of these polymers are described in the monograph by Tsvetkov [27].

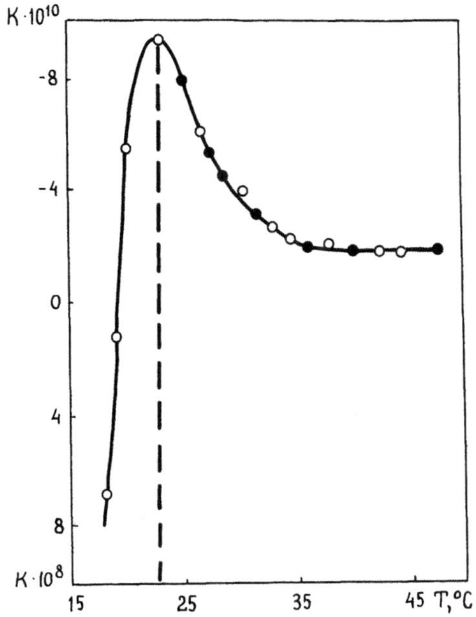

Fig. 3.4. Dependence of the Kerr constant K on the temperature for solutions of PPMOB-6 with a concentration of the polymer of 0.77 (o) and 1.4100 g per 100 cm³ (•) in benzene (the dashed line corresponds to the change in K while keeping the temperature of the solution constant at 24°C for 8 h) [14].

The presence of strongly interacting side groups in its molecules increases the rigidity of the main chain of the polymer (ν = 25). The negative segmental anisotropy of the molecule is more than one order of magnitude greater than the anisotropy of PMA-16 and is only comparable to the anisotropy of the "crystal-like" molecules of polypeptides in the helical conformation [29]. The anisotropy of the monomer of PPMOB-16 significantly exceeds the anisotropy of the monomeric units of polypeptide macromolecules. This means that the very high anisotropy of the macromolecules of PPMOB-16 is caused not only by the rigidity of their main chain (which is relatively high but nevertheless significantly lower than in molecules with a secondary structure, for example polypeptides and polynucleotides), but also by the very perfect orientational order in their side groups.

It is interesting to note that this order is not only axial but also polar: solutions of PPMOB-16 (as well as PPMOB-6 and PPMOB-9) [14] are characterized by high negative birefringence in an electric field. The value of the components of the dipole moment along the axis of the PPMOB-16 molecule is equal to 30D, which is one order of magnitude greater than the value of the dipole moments of the molecules of most flexible chain polymers. In contrast to flexible chain molecules, virtually the entire molecule, which rotates in an electric field as a unified whole, is the kinetic unit responsible for the observed birefringence in the case examined. This means that the PPMOB-16 macromolecule is characterized by very high kinetic rigidity in rotation in an alternating electric field which is comparable (like its anisotropy) to the stiffness of crystal-like polypeptides. All of these findings indicate that orientational order of the mesomorphic type, where the side chains of the molecule form a mobile liquid-crystalline structure, takes place in the molecules of PPMOB. The high degree of intramolecular order observed in solutions of PPMOB-n polymers attains the maximum value in conditions close to phase separation.

The temperature dependence of the Kerr constant $K = \Delta n/cE^2$ (where Δn is the electric birefringence, c is the concentration of the solution, and E is the field intensity) for solutions of PPMOB-6 in two different concentrations is shown in Fig. 3.4 [14]. The birefringence is negative in the 25-50°C temperature range, and the absolute value increases sharply when the experimental temperature approaches the phase separation temperature (i.e., with deterioration of the thermodynamic quality of the solvent). The Kerr constant is almost independent of the concentration, which indicates the molecular nature of the observed effect. A further decrease in the temperature (below 24°C) results in a decrease in the negative values of K to zero and the appearance of positive birefringence which is 100 times higher than the maximum negative values of the Kerr constant (Fig. 3.4, curve 1). Prolonged thermostating of the solution at 24°C results in the same high positive values of K (Fig. 3.4, curve 2).

This increase in the positive birefringence with time means that the observed effect is of a supermolecular nature and is caused by the formation of molecular aggregates whose structure in the condensed state, as indicated on p. 96, is characterized by the layered packing of the side chains of the macromolecules. The presence of a rigid linkage containing two phenyl groups in the macromolecules of PPMOB prevented, as we saw (cf. p. 19), the crystallization of the side methylene groups in the condensed state. At the same time, this is the determining factor for the realization of a high dipole-orientational order in dilute solutions of polymers of the PPMOB series. The similar values of the segmental anisotropy of PPMOB-9 and PPMOB-16, measured in tetrachloromethane [for PPMOB-9 $(\alpha_1 - \alpha_2) = -260 \cdot 10^{-30}$ m^3; for PPMOB-16, $(\alpha_1 - \alpha_2) = -270 \cdot 10^{-30}$ m^3], support the above statement [14].

As follows from the data reported, an increase in the side-chain length in a number of comb-shaped polymers is accompanied both by an increase in the optical anisotropy and by some decrease in the rigidity of the polymer chains, manifested by an increase in ν with an increasing number of carbon atoms in the side chains. The latter is the consequence of orientational ordering in the side chains, and this slightly decreases the equilibrium flexibility of the main chains of comb-shaped macromolecules.

The effect of the interaction of the side groups, which increases with increasing length, on the conformational stability of the main chains of the macromolecules is clearly illustrated in the case of comb-shaped polypeptides based on poly-L-lysine (PL-n) [30, 31]:

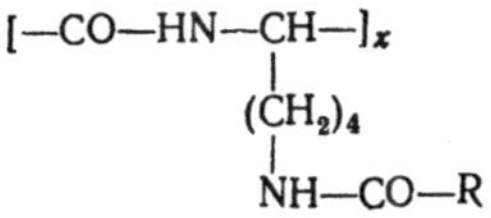

where R = C_5H_{11}, C_9H_{19}, $C_{17}H_{35}$.

According to the results of spectropolarimetric studies, all of these compounds in solutions in chloroform assume the α-helical conformation [32, 33]. However, as the calculation of parameter B_0 which characterizes the conformational state of the polypeptide chain according to the Moffit equation [32] ($B_0 = 630$ for the α-helix conformation and $B_0 = 0$ for the statistical coil conformation) indicates, an increase in the length of the substituent in PL-n results in a slight decrease in its value. This is due to an increase in the tension of the α-helical conformation of the polypeptide chain due to the strengthening of the interaction of the alkyl side groups with an increase in the number of

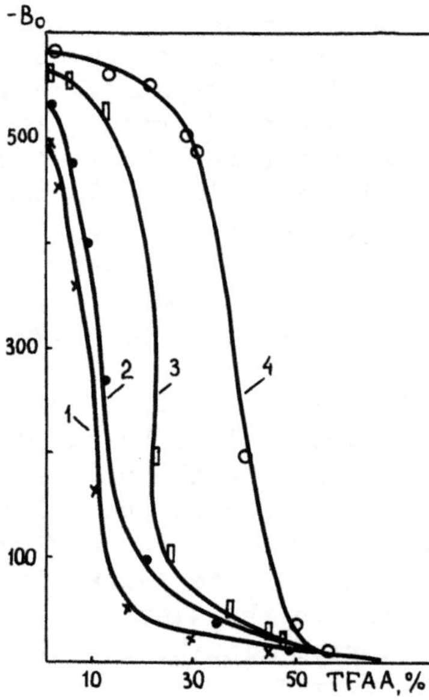

Fig. 3.5. Curves of helix—coil conformational transitions in comb-shaped polypeptides in the $CHCl_3$—TFAA (trifluoroacetic acid) system [32, 33]: 1) PL-17, 2) PL-12, 3) PL-9, 4) PL-5.

methylene units, as established for poly(n-alkyl acrylate)s in dilute solutions from the findings of dielectric measurements and birefringence, for example [5].

This interaction of the alkyl side groups in polypeptides can also result in some destabilization of the α-helix. As indicated in Fig. 3.5, an increase in the length of the R group in PL-n macromolecules also affects the character of the helix—coil conformational cooperative transition which takes place in solutions of PL-n in chloroform on addition of an uncoiling agent: trifluoroacetic acid (TFAA). With an increase in the length of the alkyl substituent, this transition takes place at a lower concentration of the uncoiling agent in the solution. As a consequence, the strengthening of the interaction in the side chains causes a decrease in the values of parameter B_0 in chloroform and facilitates the cooperative destruction of the α-helix of the polypeptide chain when the composition of the solvent changes.

TABLE 3.2. Dipole Moments and Correlation Parameters of Comb-Shaped Polymers and their Monomers in Toluene at 20°C

Polymer	Dipole moment		Correlation parameter	Reference
	Monomer μ_0, D	Polymer μ_{ef}, D		
	Poly(n-alkyl acrylate)s			
PA-1	1.63	1.45	0.79	[36]
PA-2	1.75	—	—	[36]
PA-7	1.74	1.57	0.81	[37]
PA-10	1.75	1.58	0.80	[37]
PA-16	1.79	1.48	0.68	[38]
	Poly(n-alkyl methacrylate)s			
PMA-1	1.70	1.33	0.60	[39]
PMA-2	1.82	1.35	0.55	[39]
PMA-4	1.84	1.40	0.58	[39]
PMA-8	1.81	1.43	0.62	[37]
PMA-16	1.81	1.40	0.60	[38]
	Poly(vinyl alkyl ester)s			
PVE-3	—	—	0.87*	[36]
PVE-17	1.59	1.53	0 93	[38]
	Poly(octadecyl acrylamide) and Poly(octadecyl methacrylamide)			
PAA-18	3.75	4.63	1.53	[40]
PMAA-18	3.60	5.10	2.0	[41]

*In benzene.

TABLE 3.3. Values of β and $\bar{z}$ for PAA-18 and PMAA-18 at 20°C [42]

Solvent	PAA-18		PMAA-18	
	β	$\bar{z}$	β	$\bar{z}$
Carbon tetrachloride	0.05	20±5	0.38	2.6±0.5
n-hexane	0.04	25±5	0.36	2.8±0.5
n-hexadecane	0.05	25±5	0.36	2.8±0.5
Chloroform	0.29	3.5±0.7	0.58	1.7±0.3
Polymer film	0.04	25±5	0.36	2.8±0.5

3.2. CONFORMATIONAL STATE AND INTRAMOLECULAR MOBILITY

The appearance of orientational order in the side chains of comb-shaped polymers, as the data on optical anisotropy show, is also reflected in the character of the relaxation behavior of these polymers in dilute solutions. Problems concerning the estimation of intramolecular interactions which determine the conformational state as well as some features of intramolecular mobility related to the orientational ordering of the side chains will be examined in the present section. The study of dipole moments and the temperature coefficients of the dipole moments is one of the methods of investigating the conformational state of the polymer chain; this provides information on the correlation of the orientations of neighboring polar groups along the chain which determine the conformation of the macromolecules. In addition, the study of the dipole polarization of polymers in solutions supplies information on the kinetic properties of the polar groups.

It is known [36] that for a freely joined chain, the total dipole moment M of the entire macromolecules is expressed by the relation

$$M^2 = N\overline{\mu_0{}^2} \tag{3.10}$$

where N is the number of polar groups in the chain, equal to the degree of polymerization in the general case; μ_0 is the dipole moment of the monomeric unit.

In real polymer chains, the dipole moment is dependent on the intramolecular interaction due to hindrance of internal rotation and is determined from the relation

$$\overline{M}^2 = N\mu_{ef}{}^2 = N\mu_0{}^2 g \tag{3.11}$$

The effective dipole moment of the monomeric unit μ_{ef} is a function of the parameters which describe the hindrance of internal rotation in the macromolecule. The correlation parameter $g = \mu_{ef}{}^2/\mu_0{}^2$ is a relative measure of the hindrance of internal rotation; a difference in the value of g from unity indicates the correlation of the orientations of neighboring dipoles caused by the presence of fixed valence angles and hindrance of internal rotation. In studying the dipole moments of polymers, the determination of the parameter g supplies direct information on the nature of the short-range interactions in the polymer chains.

The dipole moments of comb-shaped polymers and corresponding monomers and the correlation parameters for polymers of different homologous series are reported in Table 3.2. A comparison of the values of the correlation parameters for members of the three series PA, PMA, and PVE shows that the value of g is determined by the way in which the side chain is attached to the main chain regardless of the number of carbon atoms in the alkyl side chain.

Attachment of the side chain through the ester oxygen thus slightly increases g to 0.93 for PVE-17, in comparison to PA-16 and PMA-16 (g = 0.68 and 0.60, respectively), in which a polar carbonyl group is directly attached to the main chain. As a consequence, the interaction of the polar groups in the side chains in PVE is less pronounced than in the corresponding homologs of the PA and PMA series. Of the polymers examined, PMA is characterized by the lowest value of g in solutions. This is in good agreement with x-ray data on the structure of higher homologs of PMA in which the maximum defectiveness of the crystallites in the condensed state is undoubtedly due to the conformational state of the macromolecules, characterized by a strong intramolecular interaction of the polar groups which impedes crystallization of the alkyl side chains of the two-layer packing type. For all three types of polymers listed above, the value of g is thus less than 1, which indicates the presence of only a partial correlation in the orientation of the polar groups, resulting in some compensation of their dipole moments.

Of the polymers reported in Table 3.2, PAA-18 and PMAA-18 are sharply distinguished by the values of dipole moment and parameter g, as their dipole moments are 3-4 times higher than those of the comb-shaped polymers examined above, and their value of $g > 1$ [40, 41]. These g values indicate a significant correlation in the orientation of the polar amide groups of the neighboring units bound by intramolecular hydrogen bonds. According to IR spectroscopic data [42], the formation of so-called multimers, i.e., continuous sequences of amide groups bound in the macromolecule by hydrogen bonds, takes place in solutions of these polymers and other comb-shaped PA and PMA of more complex structure.

The results of calculating the average number of polymeric units $\bar{z}$ forming a continuous chain of hydrogen bonds and the fraction of free NH groups β for PAA-18 and PMAA-18 are reported in Table 3.3.

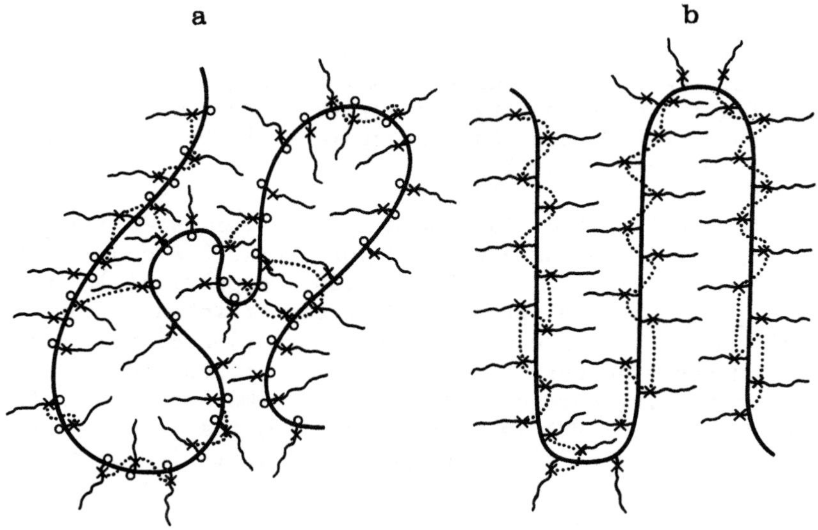

Fig. 3.6. Schematic illustration of the conformations of macromolecules of PAA-18 (a) and PMAA-18 (b) (the broken lines indicate the hydrogen bonds) [42].

Models corresponding to the conformation of these macromolecules were proposed based on the results of the IR spectroscopic analysis and the estimated values of parameter g in [42]. The presence of long sequences of hydrogen bonds in the case of PAA-18 ($\bar{z}$ = 20-25) results in the formation of a very perfect short-range order and stabilizes the primarily folded conformation of the macromolecules; the size of the fold is determined by the size of the multimer (Fig. 3.6a). The macromolecules of PMAA-18, whose multimers do not exceed three monomeric units in size, predominantly have the conformation of random coils deformed and compressed by a network of intramolecular hydrogen bonds (Fig. 3.6b). The difference in the conformational state of PAA-18 and PMAA-18 also determines the lower value of correlation parameter g for macromolecules of PAA-18. The presence of a predominantly folded chain conformation results in the mutual compensation of the dipole moments due to their antiparallel direction (Fig. 3.6). For this reason, the amide groups in the defective regions of the structure make the basic contribution to μ_{ef}, which results in lower values of g. A similar situation was observed in [44] for a synthetic polypeptide, poly-S-carbobenzoxymethylcysteine, whose β form was characterized by lower values of the dielectric increment and dipole moment in comparison to the values for this polypeptide in the coil form. The same situation also occurs in going from the coil conformation in the case of PMAA-18 to the folded intramolecular structure in PAA-18 (see Fig. 3.6). These two types of structures correspond to the limiting conformations obtained in solutions of these polymers in nonpolar solvents.

In going from solutions of nonpolar solvents to solutions of these polymers in chloroform (see Table 3.3), the number of bound amide groups decreases significantly due to the formation of hydrogen bonds between the chloroform, a weak proton donor, and the oxygen atom in the amide group, which causes a change in the conformational state of the macromolecules. It is important to emphasize here that the number of hydrogen bonds in the unit, as Table 3.3 indicates, remains the same as in solutions in nonpolar solvents; this indicates the predominant formation of intramolecular hydrogen bonds in block polymers as well. The data reported make the role and effect of the quality of the solvent on the process of crystallization of these polymers clear. It was emphasized previously (cf. p. 61) that the values of the d spacings for comb-shaped polymers, and particularly for PAA-18 [45, 46], can differ by 20-30% as a function of the type of solvent used in crystallization of the polymer. The quality of the solvent in this case determines the length of the multimers and consequently the conformational state of the macromolecules and the degree of perfection of the crystal structure.

By comparing the values of g for the polymers in Table 3.2, it can be concluded that the values of their correlation parameters are close to those of the first homologs (with the exception of polymers with amide groups: PAA-18 and PMAA-18), despite the existence of orientational order of the alkyl chains in the higher homologs of PA, PMA, and PVE. This is the basis for proposing that the internal rotations in the polar groups of these polymers are approximately the same as in the lower homologs, i.e., the presence of orientational order of the side chains has almost no effect on the value of the dipole moment of these comb-shaped polymers which contain no groups capable of a specific interaction.

However, if we turn to the data on the study of the temperature coefficients of the dipole moments $d\mu/dT$ and the

TABLE 3.4. Temperature Coefficient of the Correlation Parameter dg/dT of Polymers of the PMA-n Series [47]

Polymer	Solvent	[η], dl/g (25°C)	$(dg/dT)\cdot 10^3$, deg^{-1} (20-90°C)
PMA-1	Toluene	0.31	+0.5
	Xylene	0.16	+0.5
PMA-4	Toluene	2.80	−0.8
	Xylene	2.35	−0.8
PMA-10	Toluene	–	−1.7
PMA-16	Xylene	0.90	−2.8
	Toluene	0.75	−1.7

correlation parameters dg/dT, whose values are in some cases a more sensitive indicator of the conformational state of macromolecules than the absolute values, some conclusions can be formed concerning the differences in the conformational behavior of comb-shaped polymers of different chemical structure. According to the concepts elaborated in [35], the values and sign of $d\mu_{ef}/dT$ and dg/dT can indicate the preferred concentration of coiled or stretched conformations of the polymer chains. The values of dg/dT (in the 20-90°C temperature range) for polymeric homologs of the methacrylic series (toluene as the solvent) are reported in Table 3.4 [47]. These data show that the temperature coefficient of the correlation parameter changes sign when the side chains become longer. This is due to the fact that, beginning with PMA-4, the less-coiled chain conformations become energetically advantageous. Lengthening the side chain decreases the conformational stability of the macromolecules, as indicated by a sharper temperature dependence of g and by the effect of the thermodynamic quality of the solvent on this value (see p. 170 for more detail) [47].

A different situation is observed for polymers of the acrylic series. In this case, the dipole moment and correlation parameter change very slightly with an increase in the temperature. In the 20-90°C range, the temperature coefficient of the dipole moment does not exceed $1\cdot 10^{-3}$ deg^{-1} for all homologs of PA-n (where n = 1, 4, 7, and 16) [48]. The low and almost constant value of this coefficient (in hydrocarbon solutions) indicates the greater conformational stability and identical conditions of internal rotation in all homologs of the PA-n series. The positive value of the temperature coefficient of the dipole moment indicates the

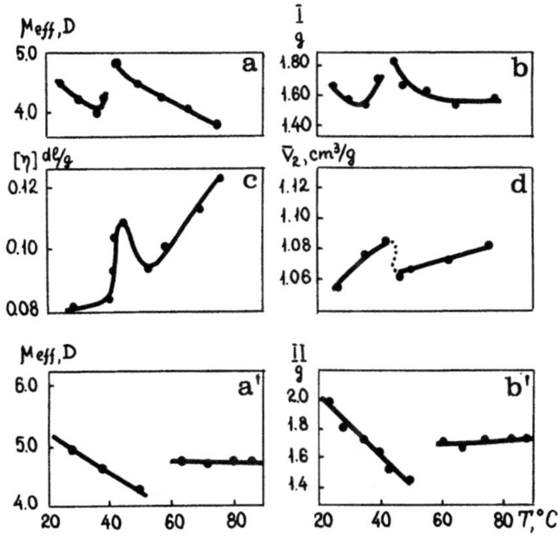

Fig. 3.7. Temperature dependences of the dipole moments (a, a') and correlation parameters (b, b') for PAA-18 (I) and PMAA-18 (II) and the dependence of the intrinsic viscosity (c) and partial specific volume (d) on the temperature for PAA-18 (I) (solvent: toluene) [41, 50].

existence of conformations with partially compensated dipole moments in all homologs of this series.* It is characteristic that the differences between the conformations of the PA-n and PMA-n series increase in going from the lower ($n = 4$) to the higher ($n \geq 16$) homologs.

In contrast to polymers of the acrylic and methacrylic series which contain no groups capable of specific interaction, the temperature dependences of the dipole moment and correlation parameter of PAA-18 and PMAA-18 are characterized by the presence of maxima and minima (Fig. 3.7) [47, 50]. In addition, similar extrema characterize the change in the viscosity and partial specific volume of solutions of

*We also note that a positive value of the temperature coefficient for macromolecules of PA-10 of unperturbed size was obtained in [49] using a series of n-aliphatic alcohols as θ solvents.

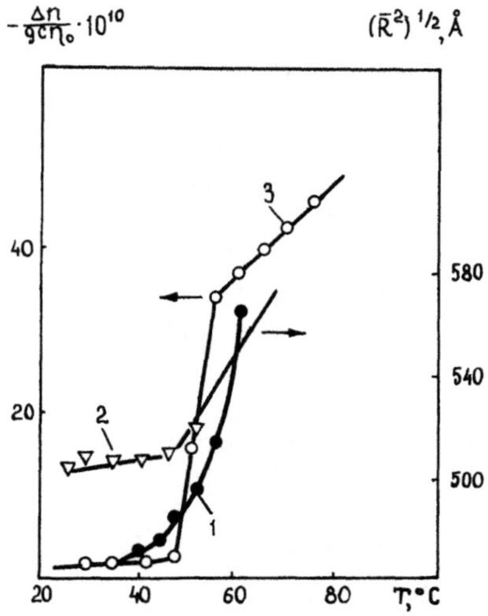

Fig. 3.8. Dependence of $(\bar{R}^2)^{1/2}$ (1, 2) and $\Delta n/gc\eta_0$ on temperature for solutions of PA-16 in n-hexyl (1) and n-octyl (2, 3) alcohols [53].

this polymer with changing temperature (Fig. 3.7). The data reported in Fig. 3.7 indicate that a sharp change in the conformational state takes place in these polymers in the 30-40°C range for PAA-18 and in the 50-60°C range for PMAA-18. Since no anomalous change in the ratio of the number of free and bound hydrogen bonds is observed in this temperature range based on the findings of IR spectroscopic studies [42, 43], the appearance of a conformational transition in these polymers can be correlated with the mutual rearrangement of the side chains, which affects the shape of the macromolecular coil. Conformational transformations have also been observed for other comb-shaped polymers in such systems as PMA-6—n-propyl alcohol [51, 52], PA-16—toluene [52], octyl alcohol [53], and others [41], for example.

The temperature dependences of the characteristic birefringence $\Delta n/gc\eta_0$ (where Δn is the birefringence, g is the shear rate gradient, η_0 is the viscosity of the solvent, and c is the concentration of the polymer) and size of PA-16 macromolecules in solutions in n-hexyl and n-octyl alcohols are shown in Fig. 3.8. The sharp changes in these parameters through a narrow temperature range (50-60°C) indicate a conformational transition caused by the cooperative interaction of the side chains of the PA-16 macromolecules according to [53]. The tendency of comb-shaped polymers toward conformational transformations in solution is apparently due to specific structural features of these macromolecules possessing chain segments of varying polarity; this makes it possible to consider these polymers as amphiphilic systems. The amphiphilic character of these compounds permits the selective interaction of the polymer segments due to their selective solvation. This is most clearly seen in the polymer—solvent systems where a specific interaction (of the hydrogen bond type) is possible both between the segments of the macromolecules (PAA-18, PMAA-18—nonpolar solvent) [50, 41] and between the solvent and the polymer chains (PA-16—octyl alcohol, PMA-6—propyl alcohol) [51-53].

The sharp change in the conformation of the macromolecules which takes place in a narrow temperature range is essentially the reflection of changes in the ratio of two types of interaction in polymer segments of different polarity with a change in temperature. The features of amphiphilicity in comb-shaped macromolecules will be examined in more detail in the next section. It should be emphasized here that the conformational transitions of comb-shaped polymers in solutions are apparently a relatively common event which predetermines the behavior of comb-shaped polymers both during gel formation and during their crystallization in the solid phase.

In turning to the evaluation of the relaxation behavior of comb-shaped polymers, it should be noted that the basic results were obtained from studies of the polymeric homologs of the methacrylic series using the methods of dielectric relaxation [38, 40, 54, 55] and polarized luminescence [56, 57]. The values for the activation energy and dipole polarization relaxation time of methacrylic polymers are reported in Table 3.4. These values for the lower homologs of PMA-n differ little up to n = 8. At the same time, a progression to higher homologs with n = 16 and 22 results in an increase in U and τ (see Fig. 3.9).

The observed dependence demonstrates the change in the kinetic properties of the polar side group, which is apparently due to the interaction of the hydrocarbon side chains and is most clearly illustrated beginning with values of n ~ 8-10. The lengthening of the side chains in the PMA-n series thus results in quantitative changes in the parameters char-

TABLE 3.5. Dipole Polarization Parameters of Methacrylic Polymers in Solutions in Toluene (2 mole %) [38-40]

Polymer	Activation energy, U, kJ/mole	Relaxation time, $\tau \cdot 10^9$, sec		Polymer	Activation energy, U kJ/mole	Relaxation time, $\tau \cdot 10^9$, sec	
		20 C	60 C			20 C	60 C
PMA-1	27.3	6.0	—	PMA-18	31.0	20	—
PMA-4	25.2	5.6	1.8	PMA-22	29.4	30	6.5
PMA-8	26.5	6.3	—	PMAA-18	42.0	250	—
PMA-10	31.5	15	3.0	PAA-18	69.3	450	—
PMA-16	32.8	18	—				

TABLE 3.6. Relaxation Times ($\tau_W \cdot 10^9$, sec) for PMA in Different Solvents at 25°C [56]*

Polymer	Toluene	Tetrahydrofuran	Heptane	Polymer	Toluene	Tetrahydrofuran	Heptane
PMA-1	4.3	3.3	—	PMA-16	21.0	25.0	27
PMA-4	6.7	7.7	—	PMA-22	22.0	25.0	29
PMA-6	11.0	6.7	—	PMAA-18	26.0	—	56
PMA-10	10.6	12.4	19				

*All values of τ_W are reduced to one viscosity value to exclude the values of τ_W related to a change in the viscosity of the solvent (for more detail see [56]).

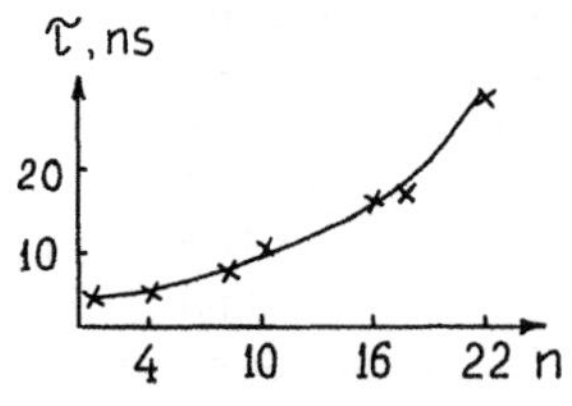

Fig. 3.9. Dependence of the dipole polarization relaxation time on the side-chain length of PMA-n [47].

acterizing the intramolecular mobility and interaction. The specific features of long side chains are reflected in a change in the conformational set and a decrease in the number of coiled conformations when the side chain lengthens.

In contrast to linear flexible-chain polymers, the dependence of the correlation parameter and its temperature coefficient on the thermodynamic quality of the solvent is a characteristic feature of comb-shaped polymers (Table 3.4). This indicates that a certain tendency of the side chains toward intramolecular aggregation in solution occurs in these polymers due to the dispersion interaction of the aliphatic groups.

The similar observation of the decrease in the intramolecular mobility with an increase in the length of the side chain also follows from the estimation of the relaxation time τ_w of a luminescent label, methacrylic acid anthrylmethyl ester

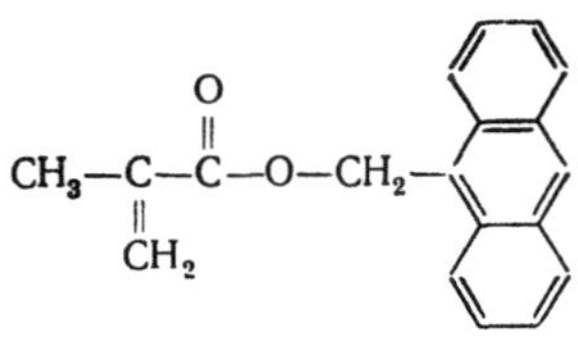

covalently introduced in PMA-n macromolecules by copolymerization with the corresponding n-alkyl methacrylates (Table 3.6) [56, 57]. The observed increase in τ and τ_w is determined by intramolecular crosslinking caused by the interaction and ordering of the side chains which inhibit the movement of the polar groups (see Table 3.3) or the luminescent label (Table 3.6).

The lessening of the increase in the values of τ and τ_w for polymers of the PMA-n series with increasing side chain length indicates the predominant role of the interaction of the methylene groups closest to the main chain of the polymer. As Tables 3.5 and 3.6 show, the maximum values of τ and τ_w (and U) are characteristic of polymers containing amide groups in the side chains and capable of forming hydrogen bonds (PAA-18 and PMAA-18).

A comparison of the values of U and τ for PMA-16 and PMAA-18 show that, with the same side-chain length, the presence of hydrogen bonds in PMAA-18 (and PAA-18) sharply increases the activation energy and the dipole polarization relaxation time by more than one order of magnitude. This circumstance and the data on the estimation of parameter g (see Table 3.2) indicate the significant inhibition of the intramolecular motion of the polar groups bound by hydrogen bonds. As a comparison of PAA-18 and PMAA-18 shows (see Table 3.5), the higher values of U and τ correlate well with the more uniform conformation of PAA-18, the model of which is shown in Fig. 3.6.

The role of hydrogen bonds in decreasing intramolecular mobility is also illustrated when comparing the relaxation times of the methyl esters of poly(N-methacryloyl-ω-hydroxy-undecanoic) and poly(N-methacryloyl-ω-aminolauric acid)s [47]:

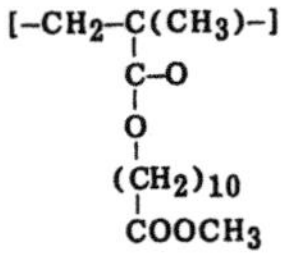

(PMM-10)

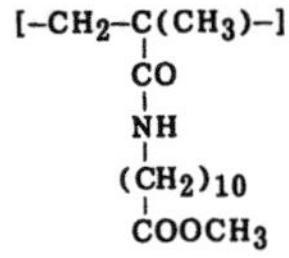

(PM-N-A-11)

Relaxation time	PMM-10	PM-N-A-11
τ, nsec	7.0	280

These values of τ are close to the corresponding values characteristic of polymers of the PMA-n and PMAA-18 series.

The appearance of orientational ordering of the side chains accompanied by an increase in the optical anisotropy thus causes a significant decrease in the intramolecular mobility of comb-shaped macromolecules with an increase in the side-chain length. In turn, an increase in the length of the side chains has an inhibiting effect on the mobility of the solvent molecules, as shown by the conformational polarization of the luminescence. The conformational depolarization of luminescence was studied [56] in solutions of PMA-n in toluene, and it was found that, in contrast to solutions of lower homologs (PMA-1, PMA-4), solvent (toluene) molecules with inhibited mobility exist in solutions of the higher homologs (PMA-10, PMA-16, PMA-22). These molecules are characterized by a relaxation time of 10^{-9} sec, which is almost three orders of magnitude higher than the rotational relaxation time of free molecules of the solvent. The existence of two types of solvent molecule with a different relaxation time in solutions of comb-shaped polymers is most clearly manifested in gel formation processes, examined in detail in Section 3.4.

3.3. UNPERTURBED DIMENSIONS OF THE MACROMOLECULES

Despite the possible appearance of orientational order in the side chains of comb-shaped polymers in dilute solutions and the corresponding chain stiffening, these polymers exhibit properties characteristic of flexible-chain polymers. The size of the macromolecules of comb-shaped polymers is very sensitive to the thermodynamic quality of the solvent (Table 3.7).

The specific amphiphilic character of the structure of the macromolecules, which permits selective solvation of the lyophobic methylene side chains, or vice versa, causes their interaction when the solvent preferentially solvates the polar linkages, and apparently plays a significant role here.

These circumstances are stimulating interest in investigations of the conformational structure of comb-shaped macromolecules. However, there are significant difficulties in interpreting the results due to the absence of rigorous quantitative relations between the structural parameters of the macromolecules and their behavior in solution. Let us examine the problems in determining the equilibrium rigidity of such macromolecules and the development of definite criteria for their evaluation.

Calculated results of the equilibrium flexibility of the macromolecules of some comb-shaped polymers in different solvents are reported in Table 3.8. The differences in their unperturbed dimensions are very significant. The unperturbed dimensions of PA-10 macromolecules $\langle\bar{r}_0\rangle^2$, determined from viscometric data obtained in decalin and ethyl acetate solutions ($\nu = 18$), significantly exceed the $\langle\bar{r}_0\rangle^2$ of this polymer in n-butyl and isoamyl alcohols, where $\nu = 9.5\text{-}10$ [23].

The difference in the values of $\langle\bar{r}_0\rangle^2$ for linear polymers in different solvents has long been debated [60-68]. This difference is usually attributed to the specific effect of the solvent. Even when the unperturbed dimensions of macromolecules in θ solvents are measured directly, the possible specific effect of the solvent should be taken into consideration in many cases. This is usually attributed to the fact that the rigidity of the polymer chain is a function of the frequency of the various repeating monomers. Redistribution of these monomers can be expected as a result

TABLE 3.7. Effect of the Solvent on the Intrinsic Viscosity of Some Comb-Shaped Polymers [22-24, 59]

Polymer	Molecular weight $M \cdot 10^{-6}$	Solvent	T, °C	[η], cm^3/g
		Decalin	21.0	1.43
	0.92	n-Heptane	21.0	1.38
PA-10		n-Butanol	24.5	0.35
		Ethyl acetate	10.0	0.62
		Decalin	21.0	0.97
		n-Butanol	24.5	0.28
	0.60	Ethyl acetate	10.0	0.46
		Decalin	21.0	3.50
PA-16	13.0	n-Heptane	21.0	3.20
		Butyl acetate	15.2	1.40
		Decalin	21.0	0.65
		n-Heptane	21.0	0.57
	0.50	Butyl acetate	15.2	0.36

TABLE 3.8. Unperturbed Dimensions (Kuhn Segment) of Some Comb-Shaped Polymers Determined in Different Solvents [23-25]

Polymer	Solvent	Temp., °C	Kuhn segment, (±2)	Method of determination
PA-10	Isoamyl alcohol	8.5 (θ)	9.5	Viscometry
	n-Butanol	24 (θ)	9.5	Viscometry
	n-Butanol	24 (θ)	10	Sedimentation, diffusion
	Decalin	21	14	The same
	Ethyl acetate	10 (θ)	14	The same
	n-Heptane	24	20	The same
	Ethyl acetate	10 (θ)	20	The same
PA-16	n-Pentanol	71 (θ)	19	Light scattering
	n-Heptane	21	22	Sedimentation, diffusion
	n-Heptane	37 (θ)	30	Light scattering

of the interaction of the individual segments of the macromolecules with the solvent molecules. As a result, the same polymer in different solvents at the same temperature can have a different flexibility and consequently different unperturbed dimensions. The fluctuations in the values of $\langle r_0 \rangle^2$ for linear polymers in different solvents in θ-conditions can be approximately 20% [64].

In the case of comb-shaped polymers, this difference, due to the amphiphilicity and relative lability of the side chains, is stronger, as should be expected, and the difference in size even attains 50%, as Table 3.8 indicates. This circumstance makes it possible for the solvent to actively affect comb-shaped macromolecules and to control the conformation in solution with its subsequent fixation in the condensed state. On the other hand, due to the dependence of $<\bar{r}_0>^2$ of comb-shaped polymers on the thermodynamic quality of the solvent, the question inevitably arises concerning the validity of using such concepts as θ-point and θ-solvent for describing the behavior of this type of polymer in solution and evaluating the equilibrium flexibility of the macromolecules. Actually, the attempts to use the usual criteria of the feasibility of θ-conditions have been unsuccessful. The satisfaction of a number of criteria, including the basic requirements of the second virial coefficient $A_2 = 0$ and $\alpha = 0.5$ in the Mark–Kuhn–Houwink relation $[\eta] - KM^{\alpha}$, corresponds to θ-conditions. At the same time, the data in Table 3.9 demonstrate the inapplicability of these criteria to the evaluation of the "ideality" of solutions of comb-shaped polymers.

The observed values of α, which differ from 0.5 in θ-conditions (when $A_2 = 0$), indicate the noncoincidence of ideal θ-conditions for intermolecular ($A_2 = 0$) and intramolecular ($\alpha = 0.5$) interactions in solutions of comb-shaped polymers. This behavior of solutions of comb-shaped polymers resembles the behavior of graft and block copolymers in single-component solvents [70-75] or some homopolymers in mixed solvents [76, 77]. The reasons for these observations may lie in the possibility of selective solvation of the individual segments of the macromolecules in the case of random and especially graft and block copolymers or the selective adsorption of one of the components of a mixed solvent.

The consequences of selective solvation have been clearly demonstrated for a series of copolymers studied by Rafikov et al. [78, 79]. The use of mixtures of cross-selective solvents for "heterogeneously soluble" copolymers made it possible to obtain structures in which the segments of chains of one type or another were "precipitated" depending upon which of the selective solvents predominated in the mixture.

In the case of comb-shaped polymers which are amphiphilic systems, the solvent can fulfill a dual function: it is simultaneously a solvent and a precipitant for different parts of the macromolecules. This should lead to unusual microstratification of the polymer–solvent system within the

TABLE 3.9. θ-Temperatures for Some Comb-Shaped Polymers [59]

Polymer	Solvent	θ-Temperature, °C	
		Method I ± 0.5°C	Method II ± 0.7°C
PA-16	n-Pentanol	70.0	75.6
	n-Hexanol	34.0	43.2
PMA-16	n-Hexanol	52.2	60.2
	n-Heptanol	20.5	30.1
PMA-6	n-Propanol	20.0	21.3

boundaries of each macromolecule with separation of the polar and nonpolar segments of the polymer chain in microvolumes of the homogeneous solution. This local microseparation results in the uneven distribution of the solvent (again maximally strong for an amphiphilic solvent) whose thermodynamic quality will be different with respect to different parts of the macromolecule. Solutions of comb-shaped polymers containing hydrogen bonds are systems with the most pronounced tendency toward microseparation previously discussed (see p. 161) with respect to the conformational transitions which reflect changes in the ratios of the energies of the intramolecular interaction between separate parts of macromolecules with changing temperature. For this reason, in the compensation of intermolecular forces ($A_2 = 0$), the intramolecular interactions can be uncompensated ($\alpha \neq 0.5$) due to attraction (Table 3.9, PA-10 in ethyl acetate and n-butyl alcohol) or repulsion (Table 3.9, PA-8 and PA-18) of the segments of comb-shaped polymers even in conditions which supposedly correspond to θ-conditions.

By varying the temperature, i.e., by changing the thermodynamic quality of the solvent, it is possible to obtain situations in which α will correspond to θ-conditions ($\alpha = 0.5$), but in this case the intermolecular interactions will not be completely compensated, and $A_2 \neq 0$. As a consequence, the θ-temperature for the same polymer, determined by different methods whose sensitivity to these interactions can differ, may not coincide. The values of the θ-temperatures for PA-16 and PMA-16 in different solvents, determined by light scattering (method I) (by interpolation of the temperature dependence of A_2 to the value of $A_2 = 0$) and by plotting the phase diagrams with estimation of the critical mixing temperatures and viscometric methods (method II), are reported in Table 3.9 [59]. The differences in the values of the θ-temperatures are very significant for polymers with long side chains, but decrease for the polymer with a short side chain.

The concept of the unperturbed dimensions of comb-shaped macromolecules thus loses its usual meaning, which implies that the repulsion (or excluded volume) between segments in the molecules is compensated by mutual attraction. The consequences of this behavior of comb-shaped macromolecules are extremely significant. If the dimensions of macromolecules determined under θ-conditions (either under the condition of $A_2 = 0$ or $\alpha = 0.5$) are not unperturbed, then the comparative evaluation of main chain flexibility in a series of comb-shaped polymers, even in the same homologous series of solvents, should be performed with regard to the change in the ratio of the hydrophilic and hydrophobic parts of the macromolecules, which varies with a change in the length of the side chain.

Consideration of this feature could require the reexamination of some data obtained in evaluating the equilibrium flexibility of comb-shaped polymers. The hypothesized presence of microvolumes with a different local concentration of the segments of the polymer chains due to their selective solvation by the solvent (which explains the data reported above) in dilute solutions of comb-shaped polymers undoubtedly requires further experimental verification. We can hope that the statistical studies of comb-shaped macromolecules to calculate their dimensions [80-84] together with experimental studies in this area will assist in solving this complex problem.

3.4. GEL FORMATION IN SOLUTIONS OF COMB-SHAPED POLYMERS

Many representatives of comb-shaped polymers of the acrylic and methacrylic series form crosslinked gels when mixed with aliphatic alcohols and hydrocarbons [85-89]. Gel formation in alcohols begins for the same polymer at a certain critical length of the solvent molecule (Table 3.10).

In the case of the alcohols $C_5H_{11}OH$ and $C_6H_{13}OH$, separation of the system into two phases takes place, and the value of the critical separation temperature T_{cr}, accordingly, is a function of the molecular weight of the sample. In

TABLE 3.10. Phase Separation in Solutions of Poly(hexadecyl acrylate) (PA-16) [87]

Molecular weight of PA-16	T_{cr}, °C		
	in amyl alcohol	in n-hexyl alcohol	in n-octyl alcohol
$5.1 \cdot 10^5$	68.0	38.2	24
$3.4 \cdot 10^5$	64.7	36.1	24
$1.4 \cdot 10^5$	61.0	33.8	24

going to higher alcohols of normal structure, $C_8H_{17}OH$, and even to $C_{12}H_{25}OH$, and when $C_{10}H_{22}$-$C_{16}H_{34}$ n-paraffins in a solution of PA-16 are used as the solvents, there is no separation as T_{cr} is reached, and the entire system forms a gel on cooling. The capacity to form a gel and the gel formation temperature are not dependent on the molecular weight (see Table 3.10, octyl alcohol).

Polymers with sufficiently long side groups, i.e., PA-16, PA-18, PA-22, PMA-16, PMA-18, etc., easily form gels. At the same time, the lower members of this series, PA-8 and PA-10, like PMA-8 and PMA-10, notwithstanding the lower members of the family, form the usual solutions in alcohols and hydrocarbons which separate into two phases on cooling.

The very low critical concentration for gel formation, which is 0.3% of the weight of the polymer, is a nontrivial feature of such gels. This finding has almost no analogies in other cases of gel formation, which is to some degree astonishing when a system containing 99.7% liquid solvent and only 0.3% dissolved substance forms a relatively stable gel. The critical concentration of gel formation is not dependent on the molecular weight of the polymer (this is demonstrated on the example of PA-16 in octanol, where fractions with M_w from $0.7 \cdot 10^5$ to $5.6 \cdot 10^6$ were studied) [88].

When the concentration of the polymer is below the critical concentration, the system does not form a continuous gel, but rather turbidity and the formation of 0.001-0.1 mm particles are observed at the gel formation point; these particles are none other than microgels which have the same temperature of dissolution as the T_m of the microgel.

The gels of comb-shaped polymers are thermally reversible; melting and gel formation take place at a point like

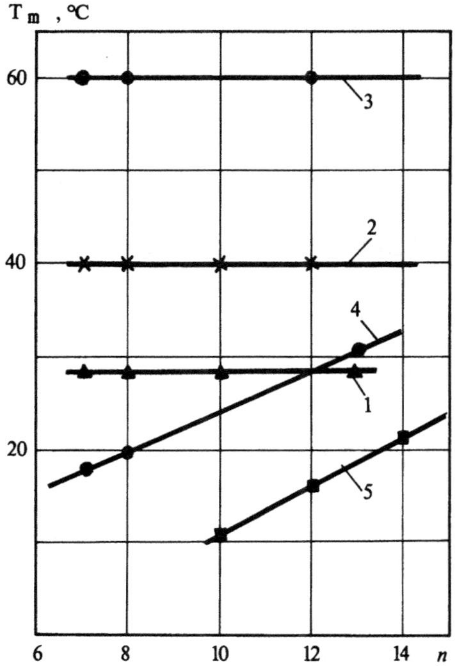

Fig. 3.10. Dependence of the melting point of gels of PA-16 (1, 5), PA-18 (2, 4), and PA-22 (3) in n-aliphatic alcohols (1-3) and hydrocarbons (4, 5) on the number of carbon atoms in the solvent molecule [91].

a phase transition and can be repeated an infinite number of times when the system is heated and cooled. The gels are stable in time, and even holding for a long time (several years) at a temperature below T_m does not cause syneresis and separation of the system into two phases.

The basic factors which cause gel formation in solutions of polymers and affect the properties of the gels have now been formulated. It is well known that the formation of such systems is due to processes of phase separation and that incomplete phase separation is a distinctive feature of these systems [90]. However, the causes of the incomplete phase separation and the mechanism of the formation of the gel network are still being debated.

The dependence of the melting points of gels on the length of the solvent molecule is shown in Fig. 3.10 [91].

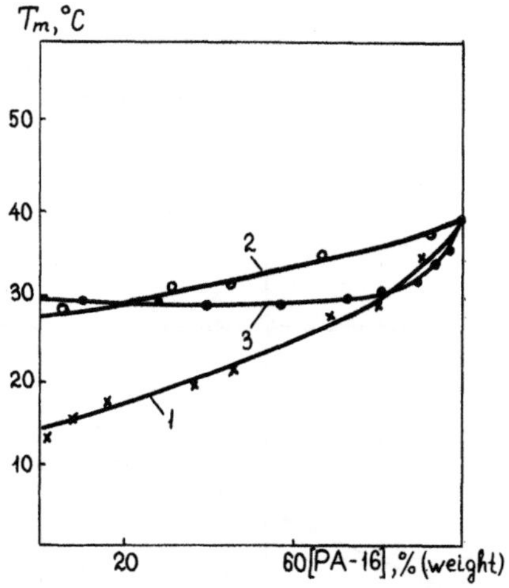

Fig. 3.11. Concentration dependence of the melting point of gels of PA-16 in n-decane (1), cetane (2), and alcohols (3) [87].

Note that for gels in alcohols, T_m is not dependent on the length of the alcohol molecule. At the same time, the dependence of T_m on the number of carbon atoms in the solvent molecule is linear for hydrocarbons, and going from PA-16 to PA-22 results in an increase in T_m in the same solvents, although the character of the observed mechanisms does not change.

The course of the concentration dependence of the melting points indicates the differences in the thermal behavior of gels in alcohols and hydrocarbons. Although a gradual depression of T_m is observed in hydrocarbons with dilution of the system (Fig. 3.11, curves 1 and 2), these dependences have a totally different character in the case of alcohols. In a broad concentration range, the T_m of the gels is not dependent on the concentration, and it increases to the T_m of the pure polymer only as some critical concentration (approximately 70 wt. %) is attained (Figs. 3.11 and 3.12). These results are due to the different chemical nature of the solvents studied and, as a consequence, the different character of the polymer—solvent interactions.

The changes in the gel formation temperatures are similar to the change in T_m. T_{gel} is almost always 4-5°C lower than T_m.

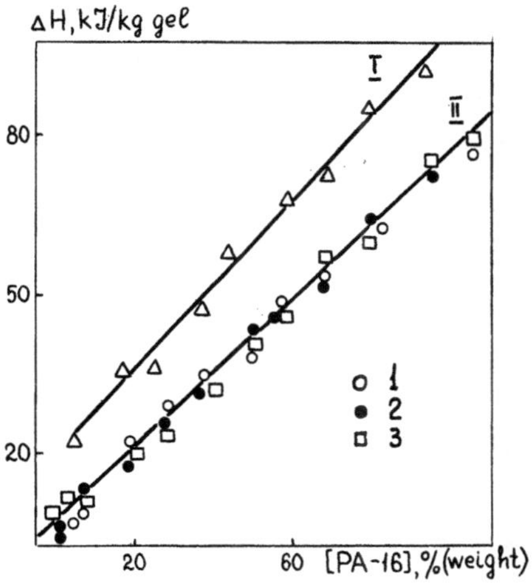

Fig. 3.12. Dependence of the heat of fusion of PA-16 gels in alcohols (o: n-octanol; □: n-decanol; •: n-dodecanol), and in cetane (Δ) on the concentration of the polymer in the gel [91].

The calorimetric study of the gel—solution transition process and the calculation of the heat of fusion of the gels in a broad range of concentrations show that the dependence of the heat of fusion on the concentration of the polymer in the gel is linear for the n-aliphatic alcohols octyl, decyl, and dodecyl. All points lie on one line regardless of the length of the aliphatic portion of the alcohol molecule. Extrapolation of this dependence to zero dilution (100% concentration of the polymer) produces a value of 80 J/g, close to the heat of fusion of the polymer crystallized from a melt (84 J/g). These results suggest that melting of gels in alcohols is due to melting of crystallites of the polymer [92]. This is also confirmed by the data from structural analysis.

The structure of low-concentration gels is characterized by the presence of only short-range order in the arrangement of the methylene chains, and there is only one diffuse maximum in the region of 4.6 Å in the x-ray patterns of PA gels corresponding to the short-range order in liquid alcohols and hydrocarbons. An increase in the concentration of the polymer in the gel to 2-3% results in the appearance of a sharp maximum at 4.19 Å on the periphery of the diffuse halo

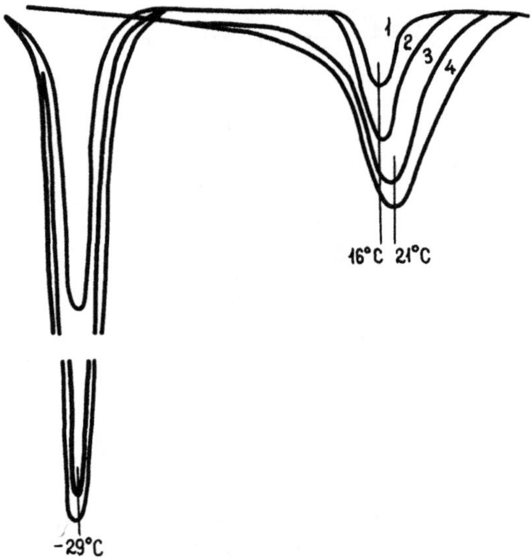

Fig. 3.13. Thermograms of PA-16 gels in decane [91]: 1) 10, 2) 30, 3) 50, and 4) 70 wt. % PA-16.

[93], characterizing the hexagonal packing of the side chains of PA-16. Since amorphous PA (PA-6, PA-8) do not exhibit a tendency toward gel formation, crystallization of the polymer in the gel is naturally considered one of the factors of the formation of gels. The layered ordering of gels in alcohols is identical to the ordering of the side groups of the macromolecules in a block copolymer (see Chapter 1).

The behavior of gels of comb-shaped PA in n-aliphatic alcohols and hydrocarbons differs significantly, as do the values of the d spacings calculated for gels of the same polymer in alcohols and hydrocarbons. Although the values of the d spacing coincide and are 42 ± 2 Å for block PA-16, as for PA-16 gels in n-octanol and n-dodecanol in the 1-50 wt. % range of concentrations, an increase in the d spacing to 48 ± 2 Å is observed for PA-16 gels in cetane (1-50 wt. %) (Table 3.11). Hydrocarbons have a marked effect on the layered order of the side chains of the polymer molecules. The absence of a small-angle reflection in the x-ray diffraction patterns of the gels in decane indicates the important imperfection of the packing of the macromolecules in these systems.

TABLE 3.11. Interplanar Distances for PA-16 Gels in Hydrocarbons and Alcohols

Solvent	Conc., wt. %	Interplanar distances, Å					
		$d_1 \pm 2$ Å	$d_2 \pm 0.3$ Å	$d_3 \pm 0.3$ Å	$d_4 \pm 0.1$ Å	$d_5 \pm 0.04$ Å	$d_6 \pm 0.05$ Å
Octanol	1	—	15.8	—	4.5	—	—
	3	—	15.6	—	4.5	4.20	—
	5	—	15.2	—	4.6	4.20	—
	20	—	15.2	—	4.5	4.19	—
	30	40	14.9	—	4.5	4.19	—
	50	42	14.9	—	4.6	4.18	—
	70	42	14.7	—	—	4.19	—
	90	42	14.7	—	—	4.19	—
Decanol	1	—	16.6	—	4.6	—	—
	2	—	16.6	—	4.6	4.19	—
	3	—	16.6	—	4.7	4.19	—
	5	—	16.3	—	4.5	4.20	—
	20	—	15.6	—	4.6	4.18	—
	30	40	15.2	—	4.6	4.20	—
	50	42	15.0	—	—	4.19	—
	70	42	14.9	—	—	4.19	—
	90	41	14.7	—	—	4.19	—
Dodecanol	1	—	17.6	—	4.6	—	—
	3	—	17.6	—	4.6	4.19	—
	5	—	17.2	—	4.6	4.19	—
	20	42	16.5	—	4.5	4.20	—
	30	42	16.0	—	4.5	4.19	—
	70	42	15.1	—	—	4.17	—
	90	42	14.7	—	—	4.19	—
Decane	1	—	—	—	4.6	—	—
	5	—	—	—	4.6	4.19	—
	10	—	—	—	4.6	4.19	—
	20	—	—	—	4.7	4.20	—
	30	—	—	—	4.7	4.18	—
	50	42	14.9	—	4.7	4.19	—
	60	42	14.8	—	—	4.17	—
	70	42	14.7	—	—	4.19	—
	90	42	14.7	—	—	4.19	—
Cetane	1	48	16.6	—	4.6	4.19	—
	5	48	16.4	—	4.6	4.16	—
	10	48	16.3	—	4.6	4.18	—
	30	48	16.6	10.6	4.7	4.19	—
	40	48	16.4	10.2	4.7	4.19	2.48
	50	46	16.4	9.9	4.6	4.20	2.44
	60	47	16.5	9.8	—	4.19	2.44
	70	46	16.4	9.7	—	4.19	2.46
	80	46	16.2	7.9	—	4.19	2.44

Going to higher homologs of the hydrocarbons not only involves an increase in the T_m of the gels (Fig. 3.10, curves 4 and 5) but also an important improvement in their structure. Maxima of higher orders, absent for the gels in decane, appear in the x-ray patterns of gels in cetane. The appearance of these reflections is characteristic of a block copolymer undergoing prolonged annealing. In addition, the value of the d spacing, as indicated above, increases in comparison to the block copolymer. The heat of melting of the gels in cetane exceeds the heat of melting of the gels in alcohols (Fig. 3.12), and the extrapolated value of the heat of fusion for the pure polymer is 113 J/g based on these data.

A comparison of the dissolving capacity in a series of n-aliphatic alcohols and hydrocarbons with respect to comb-shaped polymers (PA-16, PA-18, PA-22) shows that hydrocarbons are better solvents than alcohols. This is indicated by the values of the second virial coefficients A_2 in dilute solutions of PA-16 in octanol and cetane reported below [87]:

Temp., °C	$A_2 \cdot 10^4$, $cm^3 \cdot mole \cdot g^{-2}$
Cetane	
60	1.92
30	2.10
26	2.10

Temp., °C	$A_2 \cdot 10^4$, $cm^3 \cdot mole \cdot g^{-2}$
n-Octanol	
60	0.78
30	0.66
26	0.62

The "more active" solvation of the aliphatic groups of the polymer in hydrocarbons in comparison to alcohols is a consequence of this. Cetane, which is structurally similar to the side chains of the polymer, eliminates the steric hindrances in crystallization of the polymer to the greatest degree of all hydrocarbons. It causes the incorporation of a larger number of methylene groups in the crystal, with a consequent increase in the d spacing.

Gels in hydrocarbons clearly show the presence of two phases in the system (Fig. 3.13). One of the endothermic peaks corresponds to melting of the gel or, more precisely, the polymer crystallites in it. The low-temperature transition corresponds to melting of the pure solvent. An increase in the concentration of the polymer in the gel results in an increase in the T_m of the gel and redistribution of the area of these two peaks. The pure solvent disappears with a con-

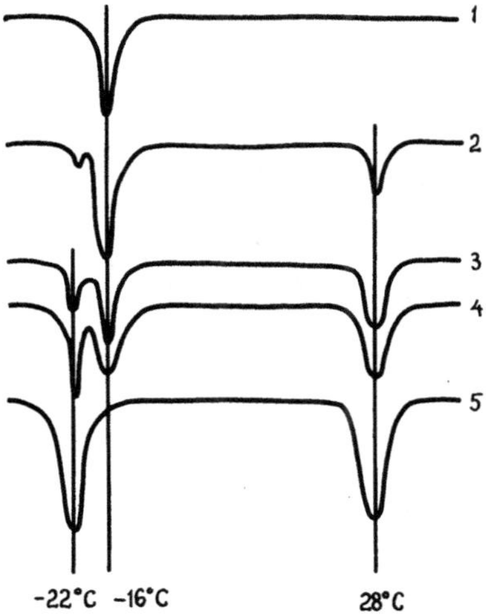

Fig. 3.14. Thermograms of PA-16 gels in octanol [87]: 1) all octanol; 2) 10, 3) 30, 4) 35, and 5) 60 wt. % PA-6.

centration of polymer greater than 70 wt. %, judging by the thermograms.

The picture is different in the case of gels in alcohols. There are two phases with less than a 10% concentration of PA-16: polymer and solvent; i.e., the situation is the same as in hydrocarbons. However, with concentrations of the polymer of 15, 30, 50, and 60%, a third peak begins to appear in the thermograms, the intensity of which increases (Fig. 3.14). This peak, which corresponds to the appearance and melting of a new phase in the system and which is located 6°C below the melting point of the pure solvent, totally changes the peak of the ordinary solvent in a 60% gel [87]. A decrease in the melting point of the pure solvent is also observed for gels in decanol and dodecanol.

The x-ray structural study of PA-16 gels in octanol permits identification of a new low-temperature phase in the thermograms shown in Fig. 3.14 as a new phase of the solvent. It follows from Fig. 3.15 that the x-ray pattern of solid n-octanol is characterized by several diffraction maxima (Fig. 3.15, curve 1). An increase in the concentration of the polymer results in a change in the character of the diffraction patterns: two new maxima appear, corresponding to in-

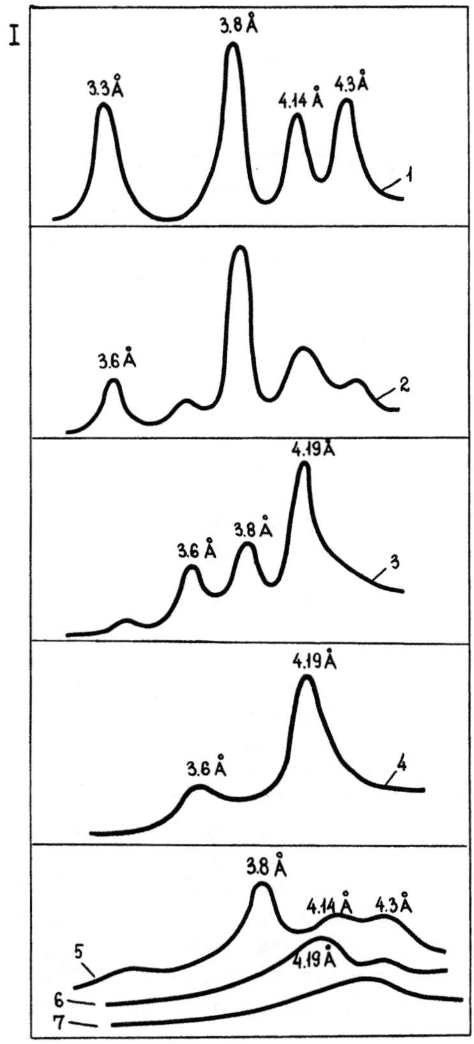

Fig. 3.15. Diffraction patterns of PA-16 gels in octanol [88]: 1) pure octanol at −30°C; 2) 10 wt. % at −30°C; 3) 30 wt. % at −30°C; 4) 60 wt. % PA-16 at −30°C; 5) 30 wt. % PA-16 at −19°C; 6) the same, at −10°C; 7) the same, at −30°C.

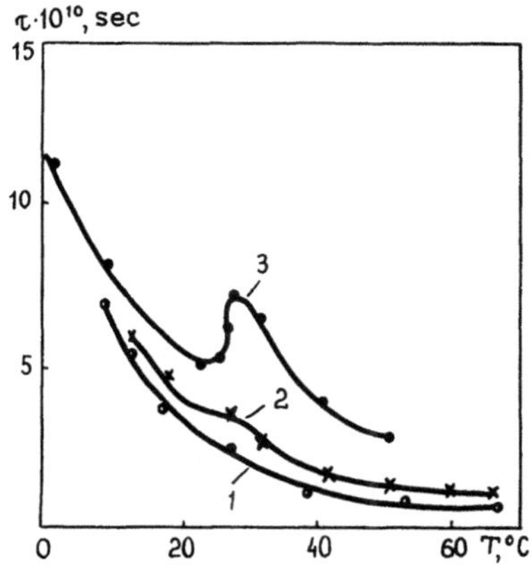

Fig. 3.16. Temperature dependence of the correlation time of rotational diffusion of the paramagnetic probe in octanol (1) and in 10% (2) and 70% (3) PA-16 gels [94].

terplanar distances of 3.6 and 4.19 Å. A further increase in the concentration is accompanied by an increase in the fraction of the new structure in the system (Fig. 3.15, curves 2-4), which corresponds to the appearance of a new endothermic peak in the thermograms whose value increases with an increase in the concentration of the polymer (see Fig. 3.14). The size of the peak corresponding to the melting of pure octanol decreases. The disappearance of the diffraction maxima with interplanar distances of 3.6 and 4.19 Å corresponds to the appearance of the "new phase" of the solvent at a temperature of −22°C. Two poorly resolved maxima (4.14 and 4.3 Å) appear, and the pattern as a whole is identical to the diffraction pattern of pure crystalline octyl alcohol (see Fig. 3.15) which melts at −16°C. Its melting is in turn accompanied by the disappearance of several reflections (see Fig. 3.15, curve 5) and the separation of a low-intensity maximum (d = 4.19 Å) against the background of a broad, diffuse halo. Melting of the gel at 28°C is accompanied by the sample becoming totally amorphous (see Fig. 3.15, curve 7).

The addition of a comb-shaped polymer to an aliphatic alcohol thus not only results in the formation of a gel network but also in the appearance of a new phase of the solvent, characterized by a lower melting point and having different structural parameters of the crystal lattice in com-

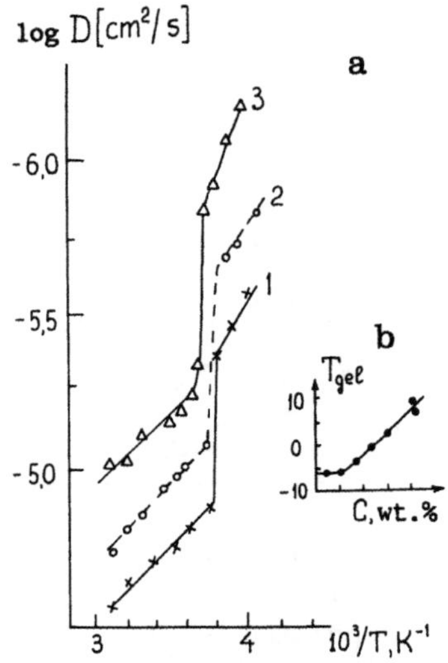

Fig. 3.17. a) Reciprocal temperature dependence of the autodiffusion coefficient of toluene in the PA-16—toluene mixture with a concentration of polymer of 5 (1), 15 (2), and 25 wt. % (3) of PA-16 [97]. b) Dependence of the gelation temperature on the concentration of PA-16 in deuterated toluene [97].

parison to the usual solvent. The change in the structure of the solvent is due to the polymeric nature of the gel-forming substance, since the phase analysis of the octanol—cetyl propionate model system (an analog of PA-16) does not reveal any evidence of a new modification of the solvent [88]. What is the solvent in such a system? The study of the structure of gels using a paramagnetic probe [94] showed that the EPR spectrum of a stable radical probe in a 70% gel is the superposition of two different spectra, one corresponding to inhibited rotation of the probe in the pure polymer and the other corresponding to the pure solvent, octanol. In gels with an equimolar concentration of PA-16 with octyl alcohol (70% of the weight of the polymer), a significant fraction of the probe molecules is located in octanol-filled cavities which are so large that the presence of the polymer does not significantly affect the polarity of the environment of the probe. The dependence of the correlation time of the rotational diffusion of the probe (τ) on the concentration of polymers in the gel shows (Fig. 3.16) that the rate of rotation of the probes in a 10% PA-16 gel is higher than in a 70% gel.

In melting, a two-phase gel turns into a one-phase gel, as indicated by x-ray, DTA, and paramagnetic probe data, as well as data on the dielectric properties of PA-16 gels in hydrocarbons (Fig. 3.17) [95, 53].

The estimation of the translational autodiffusion coefficient D of the solvent by the spin-echo method and the temperature characteristics of the nuclear magnetic relaxation for the PA-16—deuterated toluene system unambiguously indicate that the formation of a gel is accompanied by a relatively sharp decrease in the translational mobility of the solvent [97]. As Fig. 3.17a shows, at T_{gel}, the decrease in D is greater the higher the concentration of the polymer. For comparison, the dependence of the gel formation temperature on the concentration of the polymer is shown in Fig. 3.17b.

All of these results thus clearly indicate the coexistence of two phases in the gel: the polymer backbone (or crystallites) and cavities of important volume filled with the solvent. The solvent is capable of crystallization even in a highly concentrated gel.

The simultaneous presence of two different phases of the solvent (octanol) in PA-16 gel indicates that some of the solvent in these cavities (apparently near the polymer molecules) behaves differently than an isotropic liquid crystallizable in the pure octanol phase. It is logical to call this solvent "bound"; its crystal structure, as the data reported above show, differs from the structure of the pure solvent.

It was previously noted in the discussion of the molecular mobility of the molecules of comb-shaped polymers (see p. 168) that even in dilute solutions of toluene, a fraction of the toluene molecules with a different (smaller) relaxation time than the pure solvent appears, and these "special" molecules are in layers adjacent to the aliphatic combs of the molecule, i.e., they also behave like "bound" solvent molecules.

The combination of segments of different polarity in the polymer and alcohol molecules in gels in alcohol solutions apparently results in an additional bonding effect (in addition to the classic van der Waals interaction): the appearance of hydrogen bonds between the OH groups of the alcohol and the C—O groups of the polymer.

Since alcohols have no effect on the crystal structure of the polymer (judging by the data from the structural analysis and estimation of the heat of fusion of the gels), it should be assumed that "binding" of the solvent does not take place in the crystalline but instead in the amorphous phase of the gel, i.e., near the basic backbone of the macromolecules where, once again, there are carbonyl groups. The appearance of a low-temperature modification of octanol in the amorphous PA-8—octanol system, which has the same T_m as PA-16 gels, confirms the above hypothesis.

One of the unexpected features of the behavior of comb-shaped polymers in aliphatic alcohols is thus their "organizing" role with respect to the solvent molecules. This opens up new possibilities in the creation of gel-like systems in which the molecules of a low-molecular-weight substance are an organized ensemble with the simultaneous preservation of elevated mobility. In essence, this system is similar to the usual type of liquid crystals, and if the solvent molecules are simultaneously molecules of a reagent, then the chemical consequences of the possible reactions in this system can be nontrivial with respect to the rates, selectivity, yields, etc.

When and how does a gel appear? It is clear that crystallization of the polymer is a necessary but not sufficient condition for the formation of a thermally reversible gel. A certain balance in the polymer—polymer and polymer—solvent interactions is also necessary. In a gel-forming solvent, for example in octanol, PA-16 cannot attain the θ-condition, since aggregation of the polymer long before the θ-point is attained, with high positive values of the second virial coefficient A_2, is observed with a decrease in the temperature [53]. We note that for polymers which are only soluble with difficulty, precipitation of the polymer above the θ-temperature has been observed in the literature [1]. For cetane, gel formation is also observed with high positive values of A_2 [87]. The capacity of the polymer to form thermotropic gels is due to the tendency toward aggregation of the parts

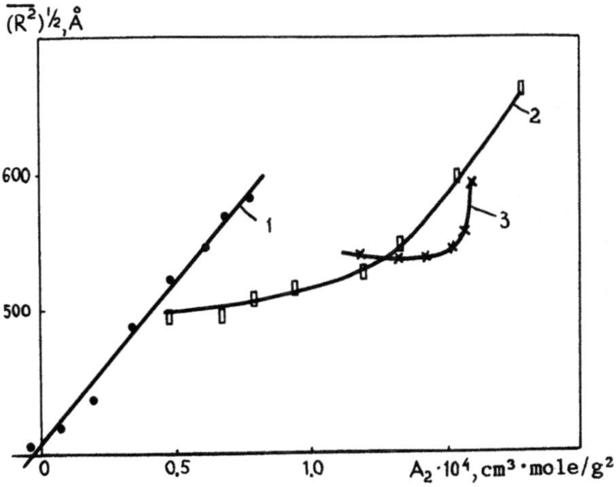

Fig. 3.18. Dependence of $(\overline{R^2})^{1/2}$ on A_2 for PA-16 solutions in n-pentanol (1), n-hexanol (2), and n-octanol (3) [53].

of the macromolecules separated from the solution and the presence of a certain thermodynamic affinity of the polymer and the solvent. The actual mechanisms of the transition from a molecularly disperse solution to a gel could be different. It is known that either the formation and development of supermolecular structures or transition of the macromolecules into a globular conformation can be observed in solutions of polymers when the quality of the solvent deteriorates [51, 57]. These are the nuclei of a new phase near the separation point. No conformational transitions in dilute solutions are observed in a non-gel-forming solvent (n-pentanol) (Fig. 3.18).

At the same time, a pronounced inflection in the curve of the dependence of $(\overline{R^2})^{1/2}$ on A_2 is observed in octanol (Fig. 3.18, curve 3), and the dimensions of the macromolecule no longer change in the 24-45°C temperature range. The value of the segmental anisotropy simultaneously changes sharply in the 50-60°C range [53], which once again shows that a conformational transition of the PA-16 macromolecules takes place in this temperature range (see p. 164). The constancy of the values of $(\overline{R^2})^{1/2}$ and $(\alpha_1 - \alpha_2)$ below this conformational transition (the solution is molecularly disperse) up to the gel formation temperature, which is 20°C lower, shows that the intramolecular aggregates formed are

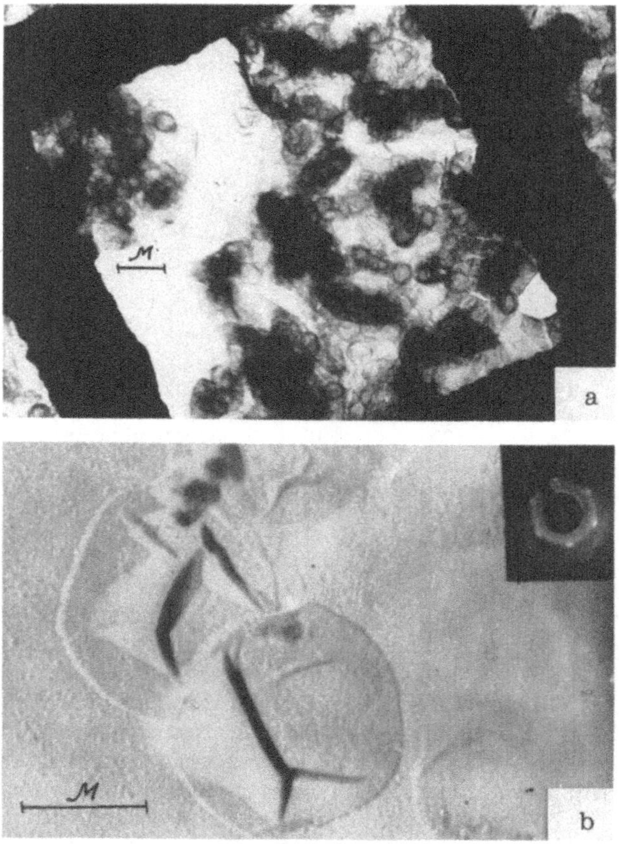

Fig. 3.19. Electron microscopic photographs of 1% PA-16 gel in octanol [99]: a) in a gas cell; b) from solution in octanol after evaporation of the solvent; a picture of the electronic microdiffraction from a single crystal is shown in the corner.

totally stable and their appearance could be the initial stage of gel formation. In conditions of phase separation, the presence of these aggregates results in intermolecular aggregation and subsequent crystallization in the gel.

The crosslinking of the system on the molecular level causes the appearance of very perfect supermolecular formations. The structure of a 1% gel in octanol, which can be observed in the gas chamber of the electron microscope, is a collection of discrete formations, monodisperse in size, with a radius of approximately 1 μm (Fig. 3.19a) [98]. Pictures of the small-angle scattering of polarized light in crossed and parallel polaroids obtained for the same system are shown in Fig. 3.20 [99]. This scattering indicates the optical anisotropy of the structural elements of the gel. The dimensions of the formations directly observed in the electron microscope coincide with the mean diffraction dimensions calculated from the diffraction patterns based on the position of the "large cross" maxima with the Stein equation [100].

Figure 3.19b shows that each of these formations is a plate or disk with a fold ("spine" or crease) in the middle. The picture of the microdiffraction (Fig. 3.19) taken from one disk indicates that it is a single crystal in which the side chains of the macromolecules are arranged in a hexagonal cell perpendicular to the plane of the support. This packing should correspond to the thickness of a single plate equal to twice the length of the side chain (≈56-58 Å), which is in good agreement with the experimentally determined values of the thickness of single crystals (50-60 Å). The character of the "wrinkled" plates formed and the microdiffraction results are the basis for proposing that, as in the crystallization of polyethylene from a suspension [101], pyramidal crystals of PA-16 are also involved in this case; they collapse with the formation of "spines" in slow evaporation of the solvent from the gel. One of the factors in their formation could be the slow rate of evaporation of the solvent, but all of the factors which affect the formation of hollow pyramidal single crystals (as in the case of polyethylene) have not been definitively established.

These monocrystalline formations are bound to the single gel network either by their direct contact or due to the joining of the individual monocrystalline structures or their aggregates by shared chains. The estimation of the average distances between the centers of gravity of the supermolecular formations reported in [88] suggests that binding of the aggregates in the gel network probably takes place as a result of direct contact between them, especially since the value of the average distance (1.2 μm) is in good agreement with the radius of these particles. "Direct contact" should be understood as contact of the type where any part of the surface of one of the aggregates touches the edge of a disk of a neighboring aggregate. The bond between them can arise both due to van der Waals interaction and due to short connecting chains with a length on the order of 0.1 μm.

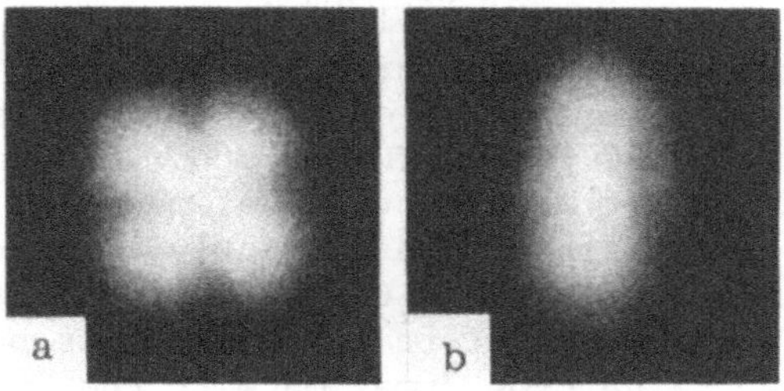

Fig. 3.20. Optical diffraction patterns of PA-16 gels in octanol [99]: a) H_v, b) V_v.

All of the above makes it possible to postulate the following sequence of events which takes place in gel formation when a solution of a comb-shaped polymer is cooled: first, stable intramolecular nuclei of a new phase are formed as a result of the interaction of the aliphatic side chains. A second stage appears at the phase separation point: intermolecular association, which takes place at a high rate and results in the formation of perfected supermolecular structures with three-dimensional crystalline order: individual crystals. Then a volume network is formed from these crystals, and the solvent molecules are positioned in the cells of this network. The size of the cells is relatively large in order to hold the solvent in the form of a separate phase. The specific interaction of the polymer with the solvent results in the appearance of a new modification of the solvent.

The tendency of the macromolecules of comb-shaped polymers to associate both on the intramolecular and intermolecular levels and their amphiphilic character indicate that these compounds can play the role of active crosslinking (gel-forming) agents. The specific structure of branched macromolecules which associate side-chain mobility with ordering in their arrangement forms the basis for considering this class of compounds as systems with a strong crosslinking effect both with respect to the macromolecules themselves and with respect to the molecules of low-molecular-weight compounds.

REFERENCES

1. V. N. Tsvetkov, V. E. Eskin, and S. Ya. Frenkel', Structure of Macromolecules in Solutions [in Russian], Nauka, Moscow (1964).
2. V. N. Tsvetkov, Vysokomol. Soedin., 4, 894-900 (1962); 7, 1468-1475 (1965); Dokl. Akad. Nauk SSSR, 165, 360-363 (1965).
3. V. N. Tsvetkov, S. Ya. Magarik, T. Kadyrov, and G. A. Andreeva, Vysokomol. Soedin., A10, 943-954 (1968).
4. W. Philippoff and E. S. Tornqvist, J. Polym. Sci., C, No. 23, 881-899 (1968).
5. V. N. Tsvetkov, L. P. Andreeva, E. V. Korneeva, and P. N. Lavrenko, Dokl. Akad. Nauk SSSR, 205, 895-897 (1972).
6. V. N. Tsvetkov, G. Hardy, I. N. Shtennikova, E. V. Korneeva, G. F. Pirogova, and K. Nitrai (Nyitrai), Vysokomol. Soedin., A11, 349-358 (1969).
7. V. N. Tsvetkov, Usp. Khim., 38, 1674-1709 (1965).
8. V. N. Tsvetkov, Vysokomol. Soedin., A11, 132-151 (1969).
9. V. N. Tsvetkov and N. N. Boitsova, Vysokomol. Soedin., 2, 1176-1187 (1960).
10. V. N. Tsvetkov, M. G. Vitovskaya, and S. Ya. Lyubina, Vysokomol. Soedin., 4, 577-582 (1962).
11. A. E. Grishenko, M. V. Vitovskaya, V. N. Tsvetkov, and L. N. Andreevna, Vysokomol. Soedin., 8, 800-803 (1966).
12. S. N. Chinai and R. A. Gurri, J. Polym. Sci., 41, 475-487 (1959).
13. H. T. Lee and D. W. Levi, J. Polym. Sci., 47, 449-458 (1960).
14. V. N. Tsvetkov, E. I. Ryumtsev, I. I. Konstantinov, Yu. B. Amerik, and V. A. Krentsel', Vysokomol. Soedin., A14, 67-71 (1972).
15. V. N. Tsvetkov, I. N. Shtennikova, E. I. Ryumtsev, G. F. Kolbina, I. I. Konstantinov, Yu. B. Amerik, and B. A. Krentsel', Vysokomol. Soedin., A11, 2528-2536 (1969).
16. M. Matsumoto and I. Ohyanagi, J. Polym. Sci., 50, 1-2 (1961).
17. V. N. Tsvetkov, E. V. Korneeva, P. N. Lavrenko, G. Hardy, and K. Nitrai (Nyitrai), Vysokomol. Soedin., A12, 426-434 (1970).
18. V. S. Skazka, G. A. Fomin, G. V. Tarasova, I. G. Kirillova, V. M. Yamshchikov, A. E. Grischchenko, I. A. Shefer, and I. S. Alekseeva, Vysokomol. Soedin., A15, 2561-2569 (1973).

19. E. V. Korneeva, V. N. Tsvetkov, and P. N. Lavrenko, Vysokomol. Soedin., A12, 1369-1373 (1970).
20. I. N. Shtennikova, G. Hardy, V. N. Tsvetkov, E. V. Korneeva, and K. Nitrai (Nyitrai), in: International Symposium on Polymers, Proceedings, Vol. 4, Budapest (1969), pp. 109-112.
21. V. N. Tsvetkov, N. N. Boitsova, and M. G. Vitovskaya, Vysokomol. Soedin., 6, 297-303 (1964).
22. L. N. Andreeva, A. A. Gorbunov, S. A. Didenko, E. V. Korneeva, P. N. Lavrenko, N. A. Platé, and V. P. Shibaev, Vysokomol. Soedin., B15, 209-212 (1973).
23. V. N. Tsvetkov, L. N. Andreeva, E. V. Korneeva, P. N. Lavrenko, N. A. Platé, V. P. Shibaev, and B. S. Petrukhin, Vysokomol. Soedin., A14, 1737-1745 (1972).
24. V. N. Tsvetkov, L. N. Andreeva, E. V. Korneeva, P. N. Lavrenko, N. A. Platé, V. P. Shibaev, and B. S. Petrukhin, Vysokomol. Soedin., A13, 2226-2235 (1971).
25. T. I. Garmonova, Vestn. Leningr. Gos. Univ., 22, 721-723 (1962).
26. V. N. Tsvetkov and L. N. Andreeva, Adv. Polym. Sci., 39, 98-207 (1981).
27. V. N. Tsvetkov, Rigid-Rod Polymer Molecules [in Russian], Nauka, Leningrad (1985).
28. V. N. Tsvetkov, E. I. Ryumtsev, I. N. Shtennikova, H. V. Korneeva, B. A. Krentsel', and Yu. B. Amerik, Eur. Polym. J., 9, 481-492 (1973).
29. V. N. Tsvetkov, I. N. Shtennikova, E. I. Ryumtsev, and G. I. Okhrimenko, Vysokomol. Soedin., 7, 1111-1116 (1965).
30. V. P. Shibaev, M. Palumbo, and E. Peggion, Biopolymers, 14, 73-78 (1975).
31. V. P. Shibaev, V. V. Chupov, and N. A. Platé, Zh. Org. Khim., 10, 412-413 (1974).
32. V. P. Shibaev, V. V. Chupov, V. M. Laktionov, and N. A. Platé, Vysokomol. Soedin., B16, 332-333 (1974).
33. V. V. Chupov, V. P. Shibaev, and N. A. Platé, Vysokomol. Soedin., A21, 218-228 (1979).
34. W. Moffit, J. Chem. Phys., 25, 467-474 (1956).
35. T. M. Birshtein and O. B. Ptitsyn, Conformations of Macromolecules [in Russian], Nauka, Moscow (1964).
36. G. P. Mikhailov and L. L. Burshtein, Vysokomol. Soedin., 4, 720-725 (1962).
37. T. I. Borisova, L. L. Burshtein, V. N. Malinovskaya, T. P. Stepanova, V. P. Shibaev, and N. A. Platé, Vysokomol. Soedin., B13, 621-625 (1971).

38. T. I. Borisova, L. L. Burshtein, V. N. Malinovskaya, T. P. Stepanova, V. P. Shibaev, and N. A. Platé, Vysokomol. Soedin., A14, 2106-2111 (1972).
39. G. P. Mikhailov and L. L. Burshtein, Vysokomol. Soedin., 4, 270-275 (1962).
40. T. I. Borisova, L. L. Burshtein, T. P. Stepanova, V. P. Shibaev, V. M. Moiseenko, and N. A. Platé, Vysokomol. Soedin., B18, 229-232 (1976).
41. T. I. Borisova, L. L. Burshtein, T. P. Stepanova, and V. P. Shibaev, Vysokomol. Soedin., B21, 186-190 (1979).
42. N. A. Kuznetsov, V. M. Moiseenko, Z. A. Roganova, A. L. Smolyanskii, and V. P. Shibaev, Vysokomol. Soedin., A19, 399-408 (1977).
43. A. L. Smolyanskii and V. P. Shibaev, Vysokomol. Soedin., A19, 2221-2228 (1979).
44. V. I. Frolov, Vysokomol. Soedin., B17, 45-48 (1975).
45. N. A. Platé and V. P. Shibaev, J. Polym. Sci., Macromol. Rev., 8, 117-253 (1974).
46. T. I. Borisova, L. L. Burshtein, N. A. Nikonorova, and V. P. Shibaev, Vysokomol. Soedin., A19, 1218-1225 (1977).
47. L. L. Burshtein and V. P. Shibaev, Vysokomol. Soedin., A24, 3-19 (1982).
48. T. I. Borisova, L. L. Burshtein, T. P. Stepanova, V. P. Shibaev, and N. A. Platé, Vysokomol. Soedin., 17B, 692-695 (1975).
49. V. E. Eskin, A. I. Kipper, D. Kh. Khalikov, N. A. Platé, and V. P. Shibaev, Vysokomol. Soedin., 13B, 257-259 (1971).
50. T. I. Borisova, L. L. Burshtein, T. P. Stepanova, V. M. Moiseenko, and V. P. Shibaev, Vysokomol. Soedin., B18, 550-553 (1976).
51. V. A. Kasaikin, A. B. Zezin, N. A. Platé, and V. P. Saraeva, Vysokomol. Soedin., A15, 1581-1586 (1973).
52. T. I. Borisova, L. L. Burshtein, T. P. Stepanova, V. P. Shibaev, and N. A. Platé, Vysokomol. Soedin., A19, 286-290 (1977).
53. V. P. Shibaev, R. V. Tal'roze, V. A. Kasaikin, E. A. Terent'eva, O. A. Agranova, and V. G. Baranov, Vysokomol. Soedin., B17, 124-129 (1975).
54. T. I. Borisova, L. L. Burshtein, T. P. Stepanova, Ya. S. Freidzon, and V. P. Shibaev, Vysokomol. Soedin., B21, 829-832 (1979).
55. T. I. Borisova, L. L. Burshtein, T. P. Stepanova, Ya. S. Freidzon, V. P. Shibaev, and N. A. Platé, Vysokomol. Soedin., B19, 552-555 (1977).

56. E. V. Anufrieva, V. D. Pautov, Ya. S. Freidzon, V. P. Shibaev, and N. A. Platé, Vysokomol. Soedin., A17, 586-592 (1975).
57. E. V. Anufrieva, V. D. Pautov, Ya. S. Freidzon, and V. P. Shibaev, Vysokomol. Soedin., A19, 755-758 (1977).
58. T. I. Borisova, L. L. Burshtein, T. P. Stepanova, Ya. S. Freidzon, and V. P. Shibaev, Vysokomol. Soedin., A20, 1380-1384 (1978).
59. V. P. Shibaev, Doctoral Dissertation, Moscow State University, Moscow (1974).
60. U. Bianchi, F. Patrone, and E. Pedemonte, J. Phys. Chem., 70, 3057-3068 (1966).
61. U. Bianchi, J. Polym. Sci., A2, 3083-3091 (1964).
62. L. Crescenzi and P. Flory, J. Am. Chem. Soc., 86, 141-150 (1964).
63. T. Orofino and J. Mickag, J. Chem. Phys., 38, 2512-2515 (1963).
64. H. Morawetz, Macromolecules in Solution, Wiley, New York (1971).
65. G. Moraglio, G. Giannotti, E. Loppi, and U. Bonicelle, Eur. Polym. J., 7, 303-309 (1971).
66. U. Biscup and H. Cantow, Makromol. Chem., 168, 315-322 (1973).
67. V. E. Eskin, Light Scattering by Polymer Solutions [in Russian], Nauka, Moscow (1973).
68. P. Flory, Statistical Mechanics of Chain Molecules, Wiley, New York (1969).
69. V. E. Eskin and A. I. Kipper, Vysokomol. Soedin., A16, 1862-1866 (1974).
70. M. Huggins, J. Polym. Sci., C., No. 4, 445-452 (1965).
71. N. A. Platé and V. P. Shibaev, Zh. Vses. Khim. Ova. D. I. Mendeleeva, 7, 147-156 (1962).
72. A. Dondos, P. Rempp, and H. C. Benoit, Polymer, 13, 97-112 (1972).
73. J. R. Urwin and M. Girolamo, Makromol. Chem., 150, 179-189 (1971).
74. M. Girolamo and J. R. Urwin, Eur. Polym. J., 8, 299-312 (1972).
75. A. Dondos, J. Polym. Sci., B9, 871-882 (1971).
76. A. Dondos and H. Benoit, J. Polym. Sci., B7, 335-342 (1969).
77. A. Dondos and H. Benoit, Macromolecules, 6, 242-250 (1973).
78. S. R. Rafikov, Izv. Akad. Nauk Kazakh. SSR, Ser. Khim., No. 5, 67-75 (1967).

79. S. V. Bereza, E. A. Bekturov, and S. R. Rafikov, Vysokomol. Soedin., A11, 1681-1688 (1969).
80. K. Solc, Macromolecules, 5, 705-713 (1972).
81. I. N. Tverdokhlebova, Usp. Khim., 46, 1279-1301 (1977).
82. A. R. Khokhlov, Vysokomol. Soedin., A20, 1860-1866 (1978).
83. H. Galina, Macromolecules, 16, 1479-1483 (1983).
84. J. Freire, R. Prats, J. Pla, and de la Torre, Macromolecules, 17, 1815-1821 (1984).
85. R. V. Tal'roze, V. P. Shibaev, and N. A. Platé, Proceedings of the 1st Scientific Conference on Liquid Crystals [in Russian], Ivanovo (1970), pp. 48-49.
86. V. P. Shibaev, R. V. Tal'roze, and N. A. Platé, Vysokomol. Soedin., B13, 4-5 (1971).
87. R. V. Tal'roze, V. P. Shibaev, and N. A. Platé, J. Polym. Sci., Polym. Symp., No. 44, 35-47 (1974).
88. R. V. Tal'roze, Candidate's Dissertation, Moscow State University, Moscow (1973).
89. T. I. Borisova, L. L. Burshtein, T. P. Stepanova, and V. P. Shibaev, Vysokomol. Soedin., A26, 2582-2587 (1984).
90. S. P. Papkov, Physicochemical Principles of Processing of Polymer Solutions [in Russian], Khimiya, Moscow (1971).
91. N. A. Platé, V. P. Shibaev, and R. V. Tal'roze, in: Advances in Polymer Chemistry and Physics [in Russian], Khimiya, Moscow (1973), pp. 127-161.
92. V. P. Shibaev, R. V. Tal'roze, and N. L. Strusovskaya, Proceedings of the 18th All-Union Conference on Macromolecular Compounds [in Russian], Nauka, Moscow (1973), p. 119.
93. R. V. Tal'roze, Yu. K. Ovchinnikov, L. A. Shteinberg, V. P. Shibaev, and N. A. Platé, Vysokomol. Soedin., 15B, 289-292 (1973).
94. A. N. Kuznetsov, R. V. Tal'roze, B. G. Tenchov, and V. P. Shibaev, Vysokomol. Soedin., A17, 1332-1336 (1975).
95. A. I. Artyukhov, T. I. Borisova, L. L. Burshtein, D. A. Dmitrochenko, and V. A. Shevelev, Vysokomol. Soedin., A17, 2552-2560 (1975).
96. T. I. Borisova, M. I. Lifshits, E. R. Chicagova, V. A. Shevelev, and V. P. Shibaev, Vysokomol. Soedin., A22, 875-880 (1980).
97. V. E. Eskin and I. N. Serduk, Vysokomol. Soedin., 9A, 2431-2436 (1969).

98. V. P. Shibaev, R. V. Tal'roze, L. V. Vladimirov, and N. A. Platé, Proceedings of the 2nd All-Union Conference on Liquid Crystals [in Russian], Ivanovo (1972), p. 92.
99. V. P. Shibaev, R. V. Tal'roze, L. V. Vladimirov, N. A. Platé, S. I. Banduryan, and M. M. Iovleva, Vysokomol. Soedin., A17, 298-302 (1975).
100. E. S. Stein and M. B. Rhodes, J. Appl. Phys., 31, 1873-1881 (1960).
101. D. H. Reneker and P. H. Geil, J. Appl. Phys., 31, 1916-1921 (1960).

Chapter 4

Thermotropic Liquid-Crystalline Polymers

Investigators in the field of physical chemistry of macromolecular compounds have recently turned to problems in the creation and study of polymeric liquid-crystalline (LC) systems [1-25]. The great interest in these systems is due to the significant advances in the study and practical application of low-molecular-weight liquid crystals in such new fields of technology as electronics, electrooptics, holography, etc., as well as in medicine, chemistry, and biology [26-32]. On the other hand, the interest in this field is also due to the possibility of creating polymer systems which successfully combine the unique properties of low-molecular-weight liquid crystals and low-molecular-weight compounds, which allows for the preparation of films, fibers, and coatings with unusual properties. The use of low-molecular-weight thermotropic liquid crystals in most cases involves the creation of special hermetic membranes (electro-optical cells, microcapsules, etc.) which form the required shape and protect the LC compound from external effects. In the case of thermotropic LC polymers, the creation of these shapes can be and is optional, since the properties of low- and high-molecular-weight liquid crystals are combined in a single polymer, and this opens up new prospects for their use by significantly extending the number of LC compounds.

On the other hand, the study of this type of polymer is also of significant independent interest related to the clarification of the nature and specific features of the LC state of macromolecular compounds. The study of this third state of condensed systems, which occupies an intermediate position between the amorphous and crystalline states, is very important for perfecting the theoretical concepts of the structure of polymers and low-molecular-weight liquid crystals.

The study of both lyotropic and thermotropic LC polymers is also directly related to the solution of a number of practical problems in the creation of polymer materials with given properties. For example, the use of the anisotropy of the LC state in the processing and molding of polymers widens the prospects for significantly improving the physical and mechanical qualities of polymer materials.

As in the case of low-molecular-weight liquid crystals, LC polymers can be divided into two large classes: lyotropic and thermotropic liquid crystals.

Lyotropic polymeric LC systems (solutions of synthetic polypeptides, proteins) have only been investigated for approximately the last 13 years, beginning with the studies by Elliot and Ambrose [33] and Robinson [34]. However, in the past 10 years, the studies of lyotropic synthetic rigid-chain polymers has obtained a powerful new stimulus both due to the synthesis of an extensive group of rigid-chain polymers such as polyamides, polyamide esters, etc., and due to the study of the structure of the liquid-crystalline state to obtain new types of high-strength, high-modulus fibers [35-37].

Extremely important results, both scientific and practical, have now been obtained in this area of physical chemistry of polymers, as indicated in many books [8, 13, 18, 38-40] and surveys [9, 41-43].

The field of research concerned with thermotropic polymeric liquid crystals is currently developing so intensely that a new scientific direction has emerged in the physical chemistry of polymers, related to the development of methods for the preparation of thermotropic LC polymers by studying their physicochemical properties and controlling their structure with electric and magnetic fields.

The further successful development of this field of research will inevitably involve the development of methods for the preparation of new types of thermotropic polymeric LC systems and control of crosslinking processes, which is directly related to the preparation of polymer materials with a required set of physicochemical properties.

The use of comb-shaped polymers for the preparation of LC polymers is very promising, as these polymers have served as very convenient and unusual "matrices" for their synthesis.

The autonomous nature of the behavior of the side chains of comb-shaped polymers, manifested in their capacity to form ordered layered structures and even to crystallize regardless of the configuration of the main chains, makes it possible to create LC polymers by incorporating mesogenic groups in the side chains; these mesogenic groups usually have elevated anisotropy of polarizability which determines the appearance of mesomorphic properties in comb-shaped polymers. The method of introducing the mesogenic units in the side chains of the macromolecules was subsequently extended to linear polymers to obtain LC systems consisting of macromolecules with alternating flexible and rigid (mesogenic) segments.

Several hundred LC polymers with comb-shaped and linear structures have now been synthesized and described.

The present chapter basically concerns LC polymers containing mesogenic groups only in their side chains and only partially touches on questions concerning the LC state of linear polymers. The generalization of all of the relatively extensive material in the literature on comb-shaped thermotropic LC polymers is not attempted; rather, the most important questions concerning some aspects of the synthesis of nematic, smectic, and cholesteric polymeric liquid crystals and the structure and properties of these substances are examined, along with their behavior in dilute solutions, and data are reported concerning the effect of electric and magnetic fields on LC polymers.

4.1. GENERAL INFORMATION ON THE FORMATION AND STRUCTURE OF LOW-MOLECULAR-WEIGHT LIQUID CRYSTALS

More than 90 years have passed since the discovery of liquid crystals. They were first discovered by the Austrian botanist Reinitzer [44], who observed two melting points for cholesteryl benzoate in the region of 145 and 179°C. At 145°C, the solid crystalline substance was transformed into a turbid <u>anisotropic liquid</u>, which become transparent and isotropic at 179°C. The state of cholesteryl benzoate in the 145-179°C temperature range was called <u>liquid crystalline</u>,

TABLE 4.1. Some Low-Molecular-Weight Liquid-Crystalline Compounds

Substance and structural formula	Phase transition temperature*
Nona-2-4-dienoic acid $C_3H_7CH_2-CH=CH-CH=CH-COOH$	Cr $\xrightarrow{23}$ N $\xrightarrow{53.5}$ I
Octoxybenzoic acid 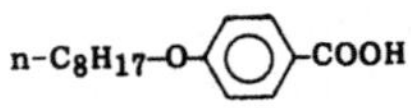	Cr $\xrightarrow{100.7}$ S $\xrightarrow{107.5}$ N $\xrightarrow{147.3}$ I
4'-n-Pentyl-4-cyanobiphenyl 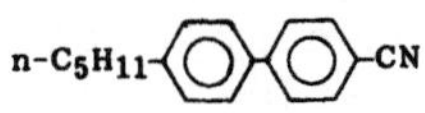	Cr $\xrightarrow{22.5}$ N $\xrightarrow{35}$ I
p-Azoxyanisole 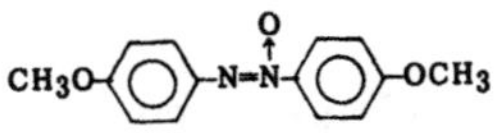	Cr $\xrightarrow{118.2}$ N $\xrightarrow{135.3}$ I
2-p-n-Decyloxybenzylidene aminofluorene 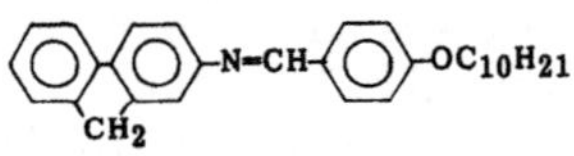	Cr $\xrightarrow{170.5}$ N $\xrightarrow{181}$ I
Hexahydroxybenzene hexa-n-octanoate 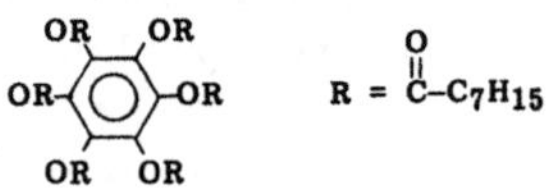	Cr $\xrightarrow{81.2}$ D $\xrightarrow{87.0}$ I
Cholesteryl myristate 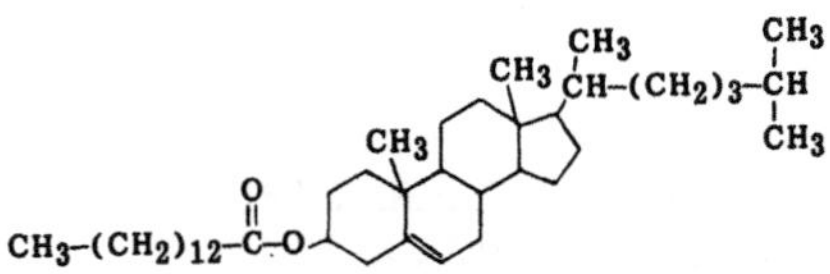	Cr $\xrightarrow{71.0}$ S $\xrightarrow{79.1}$ C $\xrightarrow{84.6}$ I

*Cr: crystalline state; N, S, C: nematic, smectic, and cholesteric mesophases; I: isotropic melt; D: discotic mesophase.

and substances capable of simultaneously combining the properties of liquids (fluidity, capacity to form drops) and the properties of crystalline substances (anisotropy of mechanical, optical, electrical, and other properties) above the melting point T_m (in a certain temperature range) were called liquid crystals. Despite the unusual combination of these two words, the term persisted and is still used together with such terms as "mesophase," "mesomorphic substance," and "mesomorphic state."*

Several tens of thousands of organic substances which can exist in the LC state are now known, and the number of these compounds is increasing continuously [45-46]. The anisodiametric rod, lamellar, or disk shape of the molecules is a basic feature of the structure of LC compounds.

The most characteristic representatives of organic compounds capable of forming a LC phase in a certain temperature range are reported in Table 4.1. They are basically aromatic organic compounds with one or more benzene groups containing a number of substituents. The elongated shape of the molecules, responsible for the anisotropy of polarizability, determines the primarily parallel arrangement of the molecules, which is perhaps the basic distinctive feature of low-molecular-weight liquid crystals.

Three basic types of LC compounds are distinguished as a function of the arrangement of the molecules according to the classification proposed by Friedel [47]: 1) smectic, 2) nematic, and 3) cholesteric (Fig. 4.1).

Smectic liquid crystals are closest to naturally crystalline substances. The molecules are arranged in ordered (Fig. 4.1a) or unordered (Fig. 4.1a') layers, and their centers of gravity are mobile in two dimensions (in the smectic plane).

The long axes of the molecules in each layer can be positioned both perpendicular to the plane of the layer and at a slight angle, which makes the formation of different polymorphous modifications possible within the limits of the smectic mesophase (smectic A, B, C, D, E, F, etc.) [48].

The nematic form is characterized by the orientation of the long axes of the molecules along a certain direction with an unordered arrangement of the centers of gravity of the molecules (Fig. 4.1b).

*Applied to polymers, these terms are not always synonymous, as we will discuss below; see also [4].

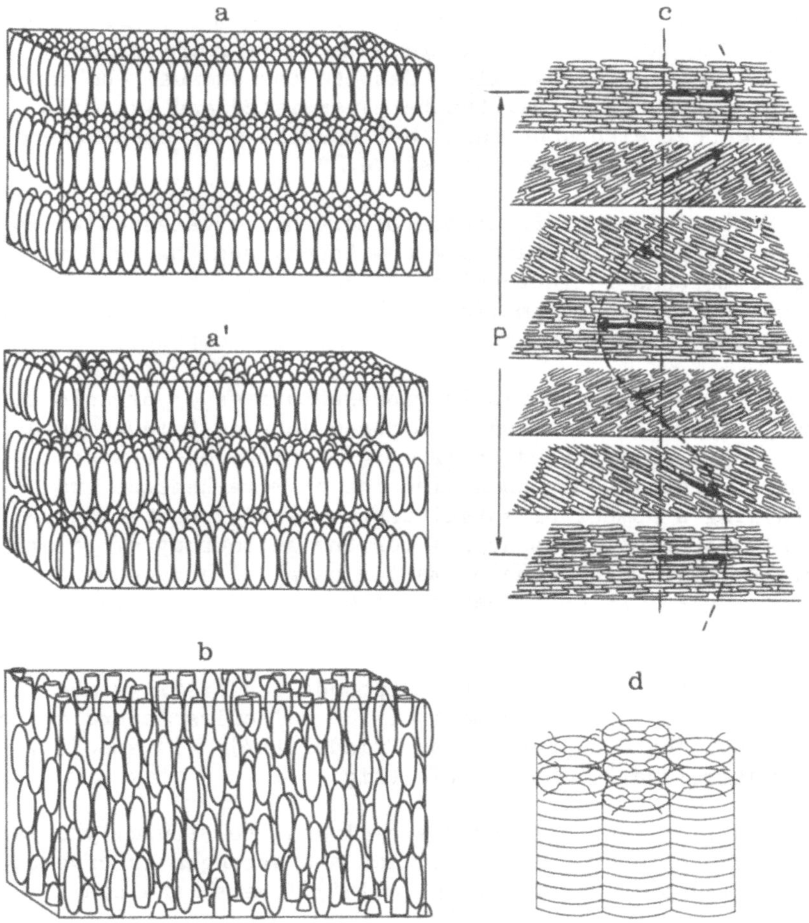

Fig. 4.1. Schematic representation of the different types of mesophases. Smectic with ordered (a) and unordered (a') arrangement of the molecules in layers; b) nematic; c) cholesteric; and d) discotic.

The most complex type of ordering of the molecules is the cholesteric type, formed by derivatives of cholesterol or optically active chiral nematic liquid crystals, for example. The molecules are assembled in layers in which their arrangement is similar to the arrangement in the nematic phase, but each layer is rotated by a certain angle with respect to the preceding layer so that some helical twisting of the molecules occurs on the whole, describing a helix with pitch p (Fig. 4.1c).

An unusual type of ordering was recently discovered for disk-shaped molecules, called discotic (Fig. 4.1d, Table 4.1). In this case, the molecules lie in the plane of the layers, forming close hexagonal packing [49-50].

These types of mesophases exist in thermotropic liquid crystals, which are only formed when the substance is heated or cooled.

Another type of mesophase occurs in lyotropic liquid crystals, formed when a number of compounds are dissolved in certain solvents; they have a more complex structure than thermotropic liquid crystals. In this case, molecular complexes, and not molecules, are the structural units; they are distributed in the medium of the solvent and can have a nematic, smectic, or cholesteric structure, assuming a lamellar, cylindrical, or spherical shape, for example.

However, despite the fact that the structure of liquid crystals, responsible for the combination of orientational ordering with the high lability of the molecules, has only been known for several decades and that a large amount of experimental material had already accumulated by the 1940-1950s, most investigators did not consider the study of the LC state to be of any interest other than theoretical.

This attitude changed sharply at the beginning of the 1960's, when extensive practical applications began to be found for the unique properties of low-molecular-weight LC compounds (primarily thermotropic compounds), determined by the elevated anisotropy and lability of the molecular structure. The capacity of LC compounds to rapidly "respond" to a change in the temperature, mechanical stress, electromagnetic radiation, and even the chemical environment, has been widely used for creating temperature indicators, displays in special recording instruments, and data displays in microelectronics and electrooptics. Liquid crystals began to be used in medicine as temperature indicators for the diagnosis of vascular diseases, different types of tumors, and skin neoplasms. A detailed examination of the practical application of low-molecular-weight liquid crystals can be found in [51-56], for example. It can now be said without exaggeration that research in a number of the most important fields in technical physics would be impossible without the use of liquid crystals.

The ways of using liquid crystals in practice in turn significantly stimulated a surge in scientific research. The data reported below on the number of published articles and patents in this field (based on data from the British information service Locus) indicates the unusual increase in these studies at the end of the 60s and beginning of the 70s.

Year	1968	1969	1970	1971	1972	1973
No. of publications	180	300	330	500	600	700

The studies of liquid crystals began to include macromolecular compounds from the beginning of the 70s. At this time, different approaches to the creation of thermotropic polymeric LC systems were beginning to be developed, and the term "liquid-crystalline" or "mesomorphic" entered the vocabulary of polymer chemists.*

However, the polymeric systems studied were frequently called liquid-crystalline and their special state was called liquid-crystalline with inadequate justification.

We will give the term "liquid-crystalline state" to the thermodynamically stable phase state of polymers or polymeric systems characterized by spontaneously appearing anisotropy of properties (particularly the optical anisotropy) in the absence of a three-dimensional crystal lattice. At the same time, it is necessary to note that, in addition to the term which describes the LC state as a phase state, the term "liquid-crystalline structure," which only indicates the formation of orientational order in polymers, will also be frequently used. Despite the narrower meaning of the latter term, the concept of LC structure (or order) is widely used in structural studies of polymers, and we will attempt to show the difference and similarity of these terms below.

*It should be emphasized that this basically concerns the development of research in the area of thermotropic liquid crystals, since lyotropic LC systems, for example, based on natural and synthetic polypeptides, had already been investigated for a relatively long time (see, e.g., [33, 34]). Extensive material on lyotropic polymeric LC systems is given in [8] and [57-59].

If the term "LC state" in polymers is approached from this point of view, it is now apparently possible to mention the creation of a LC or mesomorphic state where orientational LC order can arise in the following systems: 1) melts of some crystalline polymers, 2) lyotropic LC systems, 3) mesomorphic structures of block copolymers in gels, and 4) polymers with mesogenic groups which simulate the structure of the molecules of low-molecular-weight liquid crystals contained in the main or side chains of the macromolecules.

The LC state of melts of some nonlinear polymers will be briefly examined below, and the results of the study of the structure and properties of polymers containing mesogenic groups in the side chains, i.e., comb-shaped LC polymers, will be reported in the greatest detail.

4.2. THE LIQUID-CRYSTALLINE STATE IN LINEAR POLYMERS

The appearance of anisotropic properties in low-molecular-weight LC compounds is the result of a certain mutual orientation of anisodiametric molecules primarily positioned parallel to each other in separate regions called domains or clusters, regardless of the type of mesophase, and consisting of 10^4-10^6 molecules, according to the different data.* Since polymers consist of long-chain, highly asymmetrical molecules, their presence should apparently predetermine the possibility of the formation of the LC state for any polymeric substances. Kargin and Slonimskii first drew attention to this possibility in 1941 [60]. However, in addition to the presence of polymers, certain conditions must also be fulfilled for the formation of the LC state. It was found that, for the overwhelming majority of flexible-chain polymers, the size of the regions of spontaneously ordered segments of macromolecules is too small for the manifestation of anisotropy of properties by the polymeric material (in

*The continuum theory [65, 66] is an alternate theory of clusters; according to this theory, the mesophase is considered an essentially continuous state with a successive change in the directions of the long axes of the molecules. The continuum theory is the basis for examining the thermodynamics and dynamics of low-molecular-weight liquid crystals; those wishing to become familiar with its basic hypotheses are referred to the corresponding literature [28, 31].

the absence of the effect of an external orienting field). Nevertheless, a number of investigators [61-64] draw an analogy between this structure and nematic liquid crystals based, for example, on the electron diffraction studies of melts of a series of crystalline polymers (polyethylene, polytrifluorochloroethylene, polyethylene sebacate), finding aggregates (regions) of the macromolecules with parallel stacking of the segments of the polymer chains. Long-range order in the orientation of the molecules is preserved within the boundaries of these regions (size up to 25 Å), while there is no long-range order in the position of their centers of gravity in the projection in the plane perpendicular to the axes of the molecules. In the opinion of these investigators, the greater length of chain molecules and the difference in the molecular weights excludes the possibility of formation of a smectic modification in this case. The LC order which exists in a melt can also be preserved in the solid phase of a polymer by abrupt cooling of the melt (quenching), as found in [68] with quenched isotactic polypropylene. The existence of a metastable form with a mesomorphic structure was later confirmed by pulsed NMR and IR spectroscopy [69]. In this partially ordered structure of polypropylene, the macromolecules, with helical conformation 3_1, are arranged parallel to each other with azimuthal disorder in the plane perpendicular to the axis of the macromolecules. This type of structure, called smectic in [68], is apparently an example of packing characteristic of the gas-crystalline state [70, 71], which is a variant of the LC state.

However, in drawing an analogy between melts of polymers and low-molecular-weight liquid crystals, attention was only focused on the similarity of the structural orientational ordering of these compounds in the above studies, and the question of their phase state was not considered. However, Smith [72] obtained thermodynamic confirmation of these structural data; he studied the temperature dependence of the specific heat C_p of polyethylene and isotactic polypropylene and found several reversible transitions in their melts at temperatures 50-100°C above their melting points. Two typical curves of the temperature dependence of the specific heat for polyethylene and isotactic polypropylene are shown in Fig. 4.2. Despite the fact that the number and precise temperatures of these transitions and the absolute values of C_p are dependent on the prehistory and heat treatment of the sample to a significant degree, their existence apparently reflects the special properties of their melts.

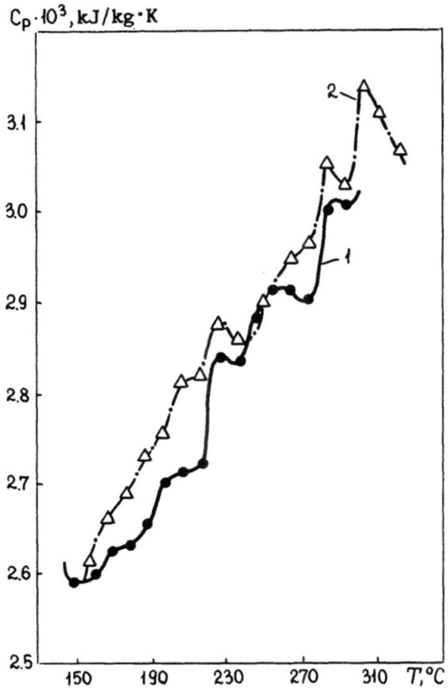

Fig. 4.2. Temperature dependence of the heat capacity of polyethylene (1) and isotactic polypropylene (2) [72].

In comparing the values of ΔC_p with a number of values of the specific heat for LC compounds of the higher aliphatic acid salt type, Smit directly mentions the formation of several types of smectic modifications for both polyethylene and polypropylene. Actually, the value of ΔC_p between a supercooled isotropic melt of polypropylene and one of its smectic modifications is equal to 0.50 kJ/kg, which is close to ΔC_p = 0.46 kJ/kg for a typical LC compound, p-azoxyphenetole, in its transition from the smectic modification to an isotropic liquid. A comparison of the transition temperatures found for polyethylene and polypropylene with the corresponding known transition temperatures for aliphatic acid salts made it possible to estimate the number of monomeric units in the ordered regions of polyethylene and polypropylene. In Smit's opinion, these regions in the smectic state consist of lamellae 20-50 Å thick and contain approximately six and nine monomeric units of polypropylene and

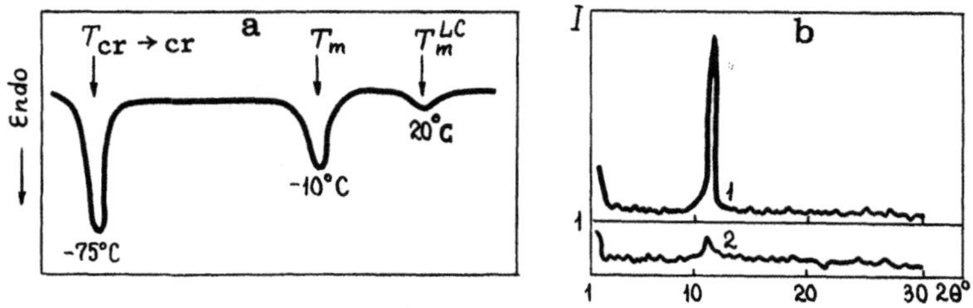

Fig. 4.3. DSC curve of polydiethylsiloxane (a) and wide-angle x-ray diffraction patterns (b) of the partially ordered phase (1) and the isotropic melt (2) at 15 and 30°C, respectively [78].

polyethylene, respectively. It is interesting to note that segments of macromolecules in a helical conformation containing at least five monomeric units was also found in a melt of isotactic polypropylene by IR spectroscopy [73].

The formation of the so-called high-pressure modification of polyethylene during its crystallization or annealing under a pressure above 2 tons/cm^2 [74] and the polyethylene which exists in the 260-290°C temperature range (where 260°C is the phase transition of polyethylene from the rhombic to the hexagonal form under pressure and 290°C is melting of the hexagonal form) can apparently also be considered formation of LC order [75]. In [76], which presents a detailed examination of the results of a structural study of polyethylene with straight chains, Bassett also indicates the possibility of drawing an analogy between the hexagonal form of polyethylene and the structure of liquid crystals.

The study of the structure of melts of ultrahigh-molecular-weight polyethylene with $M_w \geq 2 \cdot 10^6$ showed the preservation of anisotropy up to a temperature of 345°C [77]. It is indicated that the anisotropy of the melt cannot be due only to the preservation of superheated crystalline segments or orientation of the amorphous phase; the anisotropy of the melt is correlated with the formation of a smectic LC phase.

The existence of a LC phase in melts of polydiethylsiloxane was first hypothesized in [78]. Based on the results of x-ray, optical, thermographic, and NMR studies, it was concluded that there is a "partially ordered phase" in melts of polydiethylsiloxane at temperatures of −10 to +20°C, which is similar to the LC phase of low-molecular-weight compounds

TABLE 4.2. Transition Temperatures and Heats of Transition of Syndiotactic Cyclolinear Polysiloxanes [81]

Polymer			T_g, °K	Region of transition to the mesomorphic state		Region of transition to the isotropic state	
n	R	R		T, °K	ΔH_{tr}, J/g	T, °K	ΔH_{tr}, J/g
1	CH_3	CH_3	222	330-345	18.8	350-355	6-8
1	CH_3	C_6H_5	278	356-366	12.0	380-420	1.0
2	C_6H_5	CH_3	318	415-430	2.4	460-478	0.4
2	CH_3	CH_3	218	210-220*	2.5	500-600	10

*Apparently a transition of the crystal—crystal type.

(Fig. 4.3a). The polymer exhibits optical anisotropy in this temperature range and its diffraction pattern is characterized by the presence of one sharp maximum corresponding to the distance between the polydiethylsiloxane chains (8.7 Å) (Fig. 4.3b). As Fig. 4.3a shows, the transition to an isotropic melt is accompanied by the appearance of a small endothermic peak in the DSC thermogram and a sharp decrease in the intensity of scattering of the x rays (Fig. 4.3b). The existence of the mesophase was later found for other syndiotactic cyclolinear polysiloxanes with a more complex structure in [79-81], whose characteristics are reported in Table 4.2:

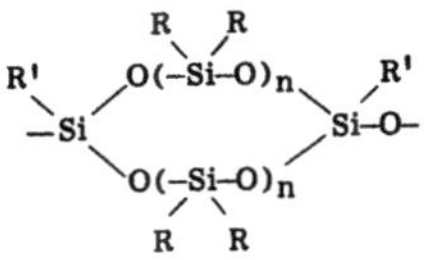

However, despite the large number of studies of the thermodynamic and dynamic properties of polysiloxane derivatives conducted in recent years [79-81], the question of the causes of the formation of the LC phase for these flexible-chain polymers is still unclear.

The polymers capable of existing in the LC state include both organic and heteroorganic polymers; the latter include an extensive group of derivatives of polyorganophosphazenes with the general formula $[-\overset{OR}{\underset{OR}{P}}=N-]$; some of the members of this group are reported in Table 4.3 [82, 83].

TABLE 4.3. Properties of Some Polyorganophosphazenes $[-P(OR)_2=N-]$ [82, 83]

R	T_m, °C	T_m^{LC}, °C	ΔH_f, kJ/kg	ΔH_f^{LC}, kJ/kg	ΔT^{LC}	$T_{chem\ dec}$, °C
CF_3CH_2O	92	240	36.0	3.3	148	360
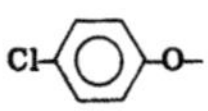	169	366	27.6	0.8	196	410
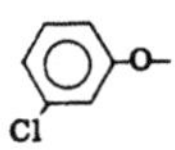	81	370	–	–	289	380
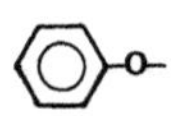	160	390	–	–	230	400
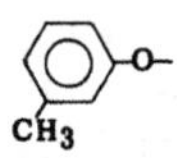	90	348	–	–	258	350

Most of the polymers in this class of compounds are characterized by the presence of two endothermic transitions of the first kind with T_m and T_m^{LC} values which differ by more than 100-200°C in most cases (Table 4.3). T_m corresponds to the transition of the crystalline polymer from the orthorhombic modification to the mesomorphic state characterized by hexagonal packing of the macromolecules (Fig. 4.4) with the parameters a_h shown below [82]:

R	T, °C	a_h, Å
$CF_3CH_2O^-$	100	11.7
	120	11.8
p-ClC_6H_4O	180	14.0
	199	14.0
	238	14.2
m-ClC_6H_4O	180	14.0
p-$C_6H_5CH_2C_6H_4O$	120	19.4

The LC state of polyorganophosphazenes, which exists through a wide temperature range (ΔT^{LC}), is characterized by the very stable structural organization of the macromolecules, since an increase in the temperature causes almost no change in the values of the hexagonal lattice parameters a_h, as the data reported above show.

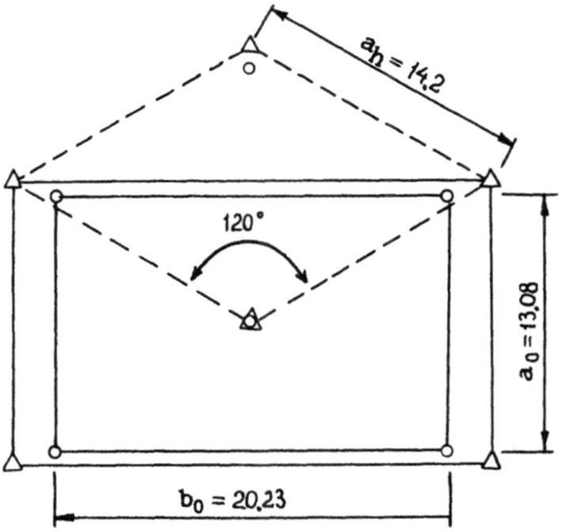

Fig. 4.4. Schematic representation of the structural transition of poly-p-(bis-chlorophenoxy)phosphazenes from the orthorhombic (o) to the pseudohexagonal (Δ) structure [82].

Although inclined to support the smectic organization of the macromolecules, Desper et al. [82, 83] nevertheless find it difficult to apply the existing classification for low-molecular-weight liquid crystals to polymers. The study of the molecular dynamics [84] and rheology of polyphosphazenes [85] revealed at least two relaxation processes related to the intense motion of the side groups (β process) and the main chain (α process) of the polymer. How and why a relatively stable mesophase of polyorganophosphazenes is preserved over a wide range of temperatures is still not clear.

The results of the studies examined above thus clearly indicate the possible formation of both a LC structure and a LC state in melts of flexible-chain polymers, despite the fact that they frequently differ with respect to the determination of the type of LC modification.

The studies indicated above are basically concerned with comparatively flexible-chain polymers such as polyethylene, polypropylene, and polydiethylsiloxane. The hypotheses concerning the existence of a LC state in these polymers came as a result of studying their structure. Most investigators tend to conclude that <u>LC structural order</u>, but not a

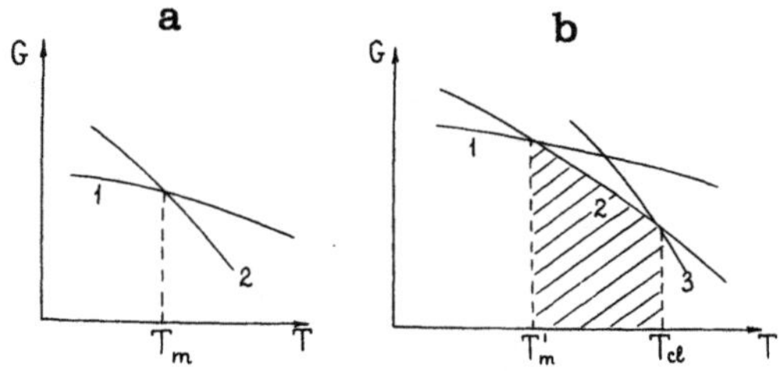

Fig. 4.5. Temperature dependence of the free energy. a) Initial state of the system: 1) crystalline phase, 2) isotropic melt; b) system under mechanical stress: 1) crystalline phase, 2) liquid-crystalline state, 3) isotropic melt [38].

thermodynamically stable LC phase, can be realized for flexible-chain polymers, and prefer to call these polymers partially ordered, viscocrystalline, or mesomorphic. For this reason, the examples given above are more of an exception than a rule, since the isotropic phase in which the macromolecules possess the static coil conformation is thermodynamically stable for melts of flexible-chain polymers. The relatively high rigidity of the macromolecules is one of the most important conditions which favors the formation of the LC state in polymers. The required chain rigidity can be both thermodynamically stable, that is, inherent in a macromolecule of a given chemical structure, and "induced," caused, for example, by the effect of an external mechanical field on the flexible-chain macromolecules. In applying an external force effect (stretching, pressure) to melts of flexible-chain polymers, it is possible to "force" the molecular coils to uncoil so that some of the macromolecules will have more stretched "linear" segments, which is somewhat equivalent to an increase in their rigidity.

According to the thermodynamic analysis conducted by Flory [86-88] and the thermokinetic approach developed by Frenkel', Baranov, and El'yashevich in examining the phase transitions in the orientations of polymers [89, 91], the described process is accompanied by the appearance of one-dimensional orientational order corresponding to the nematic LC phase.

The appearance of a region corresponding to the liquid-crystalline phase can be seen in Fig. 4.5, where the change in the free energy of the polymer (Gibbs energy) as a function of the temperature is shown. In the absence of a field of force, the polymer goes from the crystalline state to an isotropic melt at the T_m. The effect of the field on the melt of the polymer results in straightening of the molecules (which is equivalent to a decrease in their flexibility) and stimulates the formation of a new LC phase. T_{cl} corresponds to the transition of the polymer from the isotropic to the liquid-crystalline state, and T_m' corresponds to the transition of the polymer from the liquid-crystalline to the crystalline state.

The diagram in Fig. 4.5 essentially shows the theoretically possible formation of an "induced" LC phase, although the experimental evidence of its existence is significantly complex. This is due to the complexity of conducting precise structural studies of mechanically stressed, oriented melts of polymers in the extremely narrow temperature range of mesophase existence. However, the study of oriented melts of polymers and the processes which precede melting is not only of significant scientific interest with respect to the physics and mechanics of the condensed state of a substance, but is also important practically. The uncoiling of macromolecular coils induces the formation of a LC phase actually constructed of macromolecules with straightened chain conformations. The ordering (or orientational order) of such a melt is significantly greater than the "liquid" order of an isotropic melt. The use of previously ordered melts creates the most favorable conditions for the subsequent crystallization of a polymer with straightened (and not folded) chains and the preparation of fibers and films with unique physical and mechanical properties which cannot be obtained by any other methods.

This method, called "orientational crystallization" [92], is the basis for preparation of high-strength, high-modulus fibers and films from flexible-chain polymers (polyethylene, polypropylene).

Polyethylene fibers and films with a breaking strength of up to 1.4 GPa for the fibers and 0.4 GPa for the films and a 30-GPa modulus of elasticity for the fibers and a 6-GPa modulus for the films have thus been prepared by this method; these values are approximately one order of magnitude greater than the same properties of standard samples of polyethylene prepared from isotropic melts [93].

These results are a clear example of the effect of preliminary LC ordering of the macromolecules in melts on the physical and mechanical properties of the polymers.

The attention of scientists in different countries (primarily the USSR, USA, U.K., Japan, and FRG) has recently turned to the study of processes of orientational crystallization for developing the technology and production of ultrastrong polymeric materials.

The next problem for investigators in this area lies in finding the conditions under which known polymers exist in the LC state in order to use the structural features of the mesophase for increasing their mechanical properties. Determining the temperature range of the existence of the mesophase in polymer melts and creating the conditions for the maximum appearance of LC order (due to external effects) are all means of inducing unusual structural modifications of polymers in order to create new polymeric materials with the required set of properties.

The formation of a mesomorphic structure for these flexible-chain polymers generally does not require the presence of special mesogenic groups, an obligatory condition for the formation of the LC state in low-molecular-weight compounds. However, the question of the nature and structure of these mesomorphic polymers is still being debated.*

With respect to linear polymers which contain relatively rigid segments (benzene rings, short conjugated systems) which essentially play the role of mesogenic groups, their tendency to form a LC structure and LC state should be manifest to the maximum degree. Only high melting points and chemical decomposition temperatures close to the melting points apparently limit or totally exclude the temperature range of the existence of the mesophase in melts of common rigid-chain polymers.

This circumstance probably narrows significantly the group of polymers potentially capable of existing in the LC state. Due to the high tendency of such polymers to self-orga-

*Problems related to the thermotropic mesophases and classification of linear polymers are examined in the most detail by Wunderlich and Grebowicz [94, 95].

nize, they exhibit LC properties only when dissolved in a series of organic solvents, forming lyotropic liquid crystals. The examination of these compounds is beyond the scope of the present discussion, and detailed information on these systems can be found in [8, 38] and [41-43, 57-59].

The data presented above concern the possibility of the formation of the LC state in such common linear polymers as polyethylene, polypropylene, and polysiloxanes. The discovery of the LC state in these polymers, as indicated above, was a result of structural studies. However, as soon as this state was found, studies began to appear in which LC polymers were deliberately synthesized by introducing mesogenic groups similar to the molecules of low-molecular-weight liquid crystals in the macromolecules. In the initial studies, the mesogenic groups were added in the form of side substituents in the macromolecules. The principles of preparation and the properties of these polymers will be examined in detail in the following sections. We will examine data here dealing with the preparation methods of linear LC polymers with mesogenic groups in the main chains.

Polymers of this type are obtained by polycondensation (A) or copolycondensation (B) reactions of different bifunctional derivatives which usually consist of flexible (I) and rigid (mesogenic) (II, II') fragments

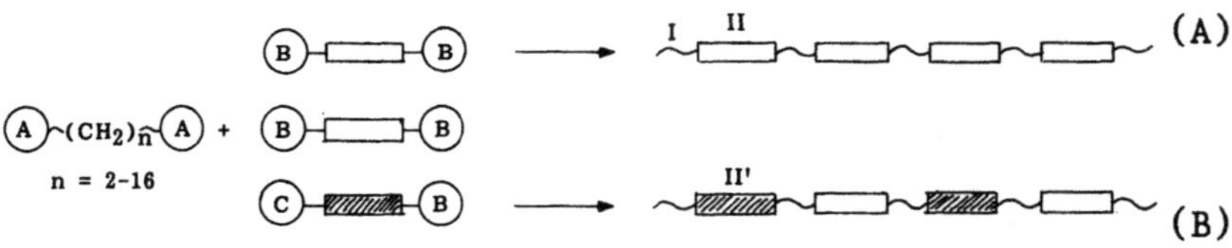

A, B, and C are functional groups;  ,  are mesogenic groups: 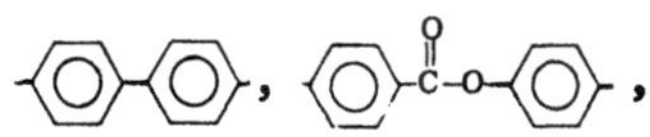 , etc.

Polymers of this type can also be obtained by modification of polycondensation polymers arising from chemical reactions in the main chains and the occurrence of structural and chemical irregularity in the macromolecules. The most intensely studied process is the reaction of polyethylene terephthalate with aromatic acids (for example, with acetoxybenzoic acid), resulting in the formation of a copolyester containing units with a heterogeneous chemical composition [96]:

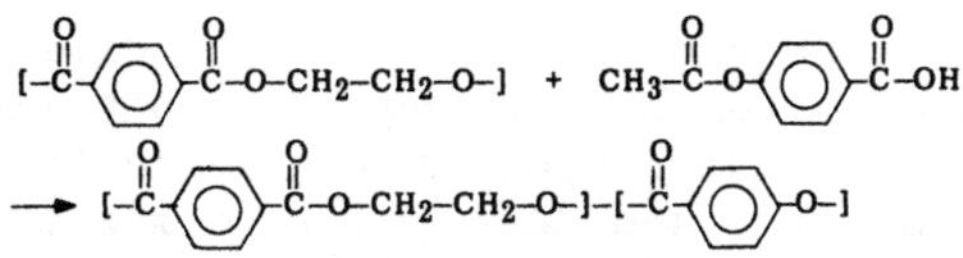

Some examples of the structure of macromolecules of LC polycondensation copolymers with different types of perturbations in the regular arrangement of the monomeric units are schematically shown below using the example of polyacrylates [97]

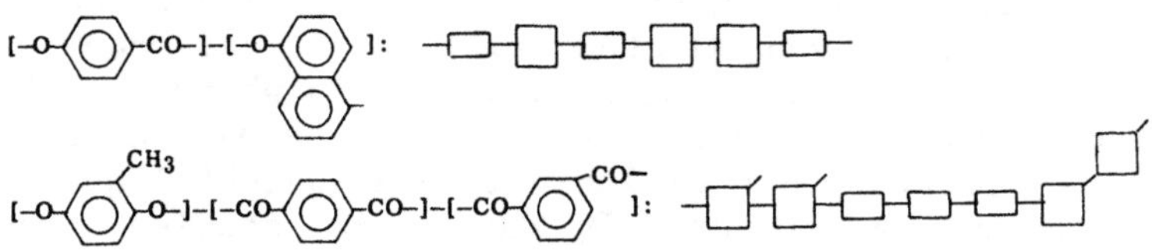

The construction of macromolecules of linear thermotropic LC polymers thus requires the presence of relatively rigid aromatic nuclei either separated by flexible segments or by monomeric units which "perturb" the regularity of the structure of the chain. This structure apparently determines some critical flexibility of the chain which is necessary for the appearance of mobility in the rigid fragments and their segregation with the formation of a mesophase.

Examples of some thermotropic LC polymers synthesized with mesogenic groups in the main chain are reported in Table 4.4. The interest in this type of compound primarily arises from the possibility of using the anisotropy of the LC state to significantly increase the physical and mechanical characteristics of polymers formed from melts and create high-strength materials based on them.

The discovery of the LC state in lyotropic systems — solutions of rigid-chain aromatic polyamides (fibers of the Kevlar type from DuPont) — led to the creation of a new type of high-strength polymer by the formation of fibers from the ordered mesomorphic state in the presence of a solvent rather than from an isotropic solution. The studies and synthesis of thermotropic LC polymers with mesogenic groups in the main chain, conducted more intensely in recent years, have essentially pursued the same goals [8, 9, 96]. Actually, the use of preliminarily ordered LC melts, as indicated above for polyethylene, should apparently create the most favorable conditions for the subsequent crystallization of the polymer from the oriented state for the preparation of fibers and films with elevated physical and mechanical properties. For example, for the copolyester prepared from polyethylene terephthalate, such important characteristics as the breaking strength and rigidity are 4-5 times higher than the corresponding characteristics for the usual polyesters [96]. However, despite the increasing number of newly synthesized linear LC polymers, which already number several hundred, only a limited number are used commercially [9, 106, 107]. Of the most interesting structural and rheological studies of LC polymers with mesogenic groups in the main chain, we cite the studies by Blumstein et al. [108-111], Krigbaum and Cifferi [112-115], Roviello and Sirigu [116-118], Noel et al. [119-120], Skorokhodov and Bilibin [121-123], Lenz et al. [37, 124-125], Windle et al. [126-129], etc. Studies of the same type have been widely conducted by companies in the U.S. (Eastman Kodak, Celanese, Carborundum, DuPont), Great Britain (ICI Ltd.), France (Rhône-Poulenc Textile), and Japan (Torey Industries), but the basic information on these studies is in the form of patents (see, e.g., the bibliography in the article by Dobbs and McIntyre [36]). Unfortunately, the newly synthesized LC polymers are not always macromolecular compounds; a large number of oligomeric products have been obtained, and the MWD of the compounds obtained and their compositional homogeneity are usually unknown. All of the above significantly complicates the establishment of a correlation between the structure and properties of thermotropic linear LC polymers and the chemical structure and flexibility of their macromolecules. There is no doubt that the recently developed theoretical concepts on LC ordering in melts of LC polymers [88, 93, 130-132] should be used to explain the experimental data and predict the limiting thermodynamic conditions of the formation of the mesophase based on the molecular structure of the newly synthesized polymers.

4.3. SYNTHESIS OF LIQUID-CRYSTALLINE POLYMERS WITH MESOGENIC SIDE GROUPS AND SOME FEATURES OF THE FORMATION OF MESOPHASES

As the previous section showed, LC order can occur in linear polymers, and the existence of the LC state is indicated by a certain temperature region in which macromolecular compounds combine the anisotropy of the properties of a solid and a viscous liquid. This is especially clear for

TABLE 4.4. Some Types of Thermotropic Liquid-Crystalline Polymers with Mesogenic Groups in the Main Chain

No.	Structure of the monomeric unit	Melting point	Clearing point	Proposed type of mesophase*	Reference
1	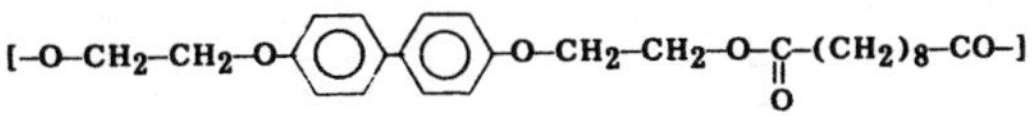	155	163	N	[98]
2	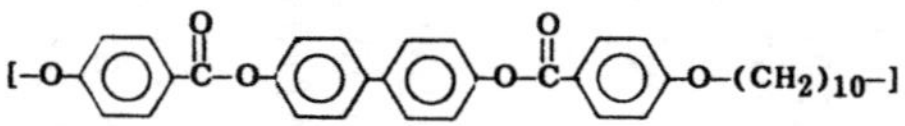	258	374	N	[99]
3	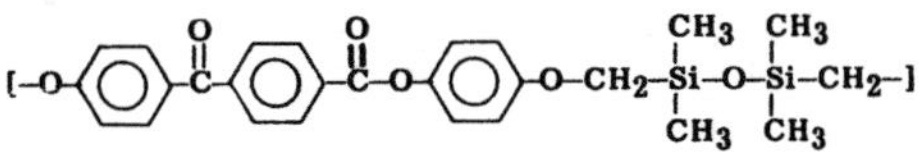	199	208	N	[100]
4	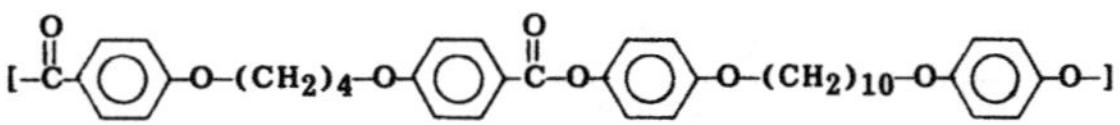	222	265	N	[101]
5	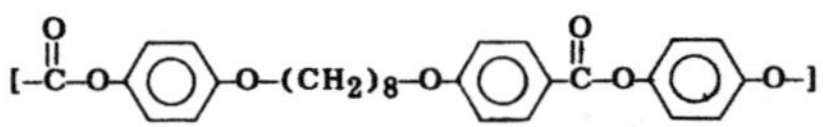	240† 300‡	350	S N	[102]

6	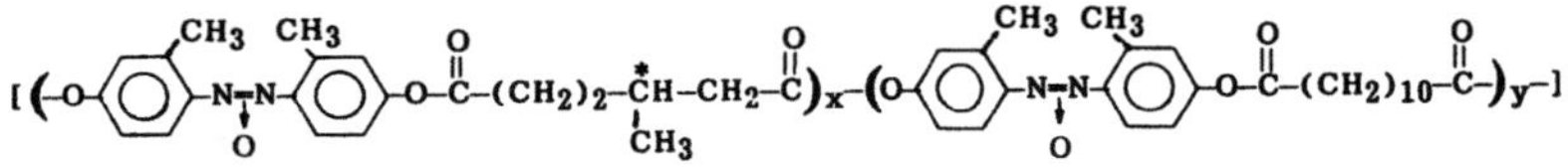 x/y = 1	76	178	Chol.	[103]
7	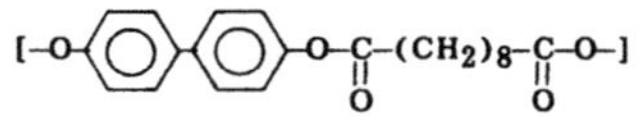	209	280	S	[98]
8	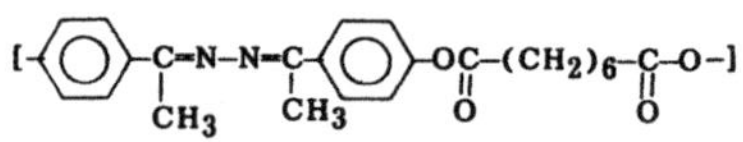	208	322	N	[104]
9	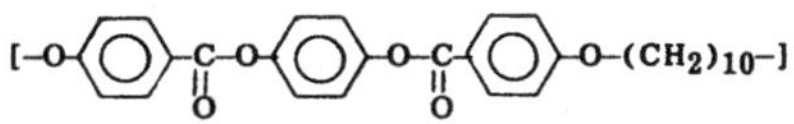	210-220	255-275	N	[105]

*Type of mesophase: N: nematic; S: smectic; Chol: cholesteric.
†Melting point of crystalline phase with formation of smectic mesophase.
‡Melting point of smectic phase.

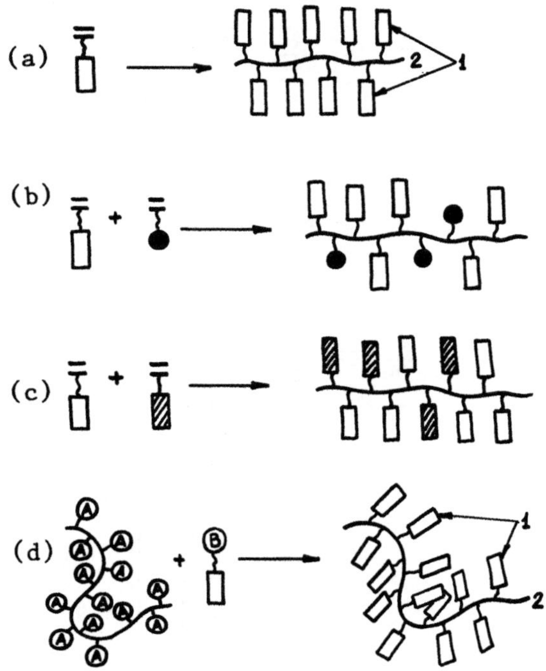

Fig. 4.6. Synthesis of liquid-crystalline (LC) polymers with mesogenic side chains: a) homopolymerization; b) copolymerization of mesogenic and nonmesogenic monomers; c) copolymerization of mesogenic monomers; d) polymer-analogous reaction; 1) mesogenic groups; 2) main chain; A, B) functional groups.

polymers with mesogenic groups in the main chain. By using the LC orientational order previously created in melts of these polymers, it is possible to deliberately affect the physical and mechanical characteristics of polymeric materials, as clearly demonstrated in [36-39, 91, 93]. Attempts to combine the properties of polymers (with their capacity to form films and fibers) with the properties of low-molecular-weight LC compounds were instituted a relatively long time ago and were primarily focused on the synthesis and study of polymers containing mesogenic groups in the side chains and not in the main chain (examined in the previous section).

The preparation of polymers with mesogenic side groups involves either synthesis of monomers with liquid-crystalline (mesogenic) groups and their subsequent homopolymerization or copolymerization with mesogenic or nonmesogenic compounds (Fig. 4.6), or addition of molecules of low-molecular-weight liquid crystals to the polymer chain by polymer-like transformations. The latter is still used relatively rarely, and the number of LC polymers synthesized by this method is very small [133-134]. Examples of the known reactions between polymers and mesogenic compounds are shown below:

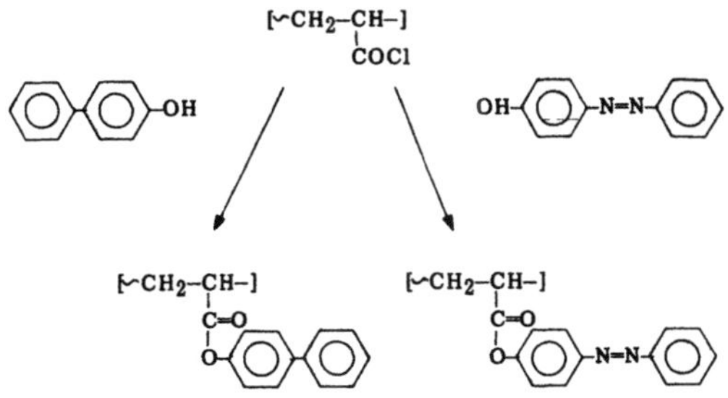

Poly(p-biphenyl acrylate) Poly(n-acryloyloxyazobenzene)

A variant of this method is the method proposed in [135] which consists of using an unusual mesogenic monomer of the olefin series as the mesogenic compound:

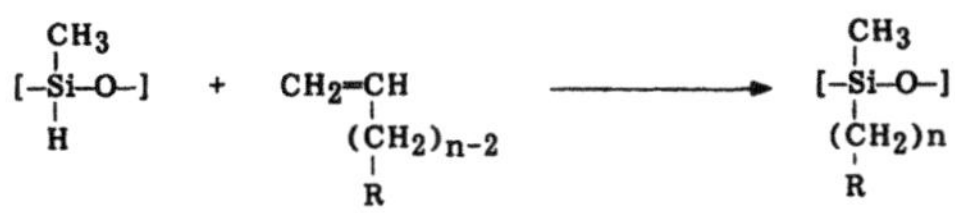

where R is the mesogenic group.

Several hundred polymeric compounds with mesogenic groups which simulate the structure of smectic, nematic, and cholesteric liquid crystals have now been synthesized; their structure and properties will be examined in detail in the following sections. The preparation of such exotic systems as discotic systems [136] and LC polymers with several mesogenic groups in each monomeric unit (so-called "en bloc" systems) [137] will also be described (Fig. 4.7). Examples of compounds of this type and their transition temperatures are shown below:

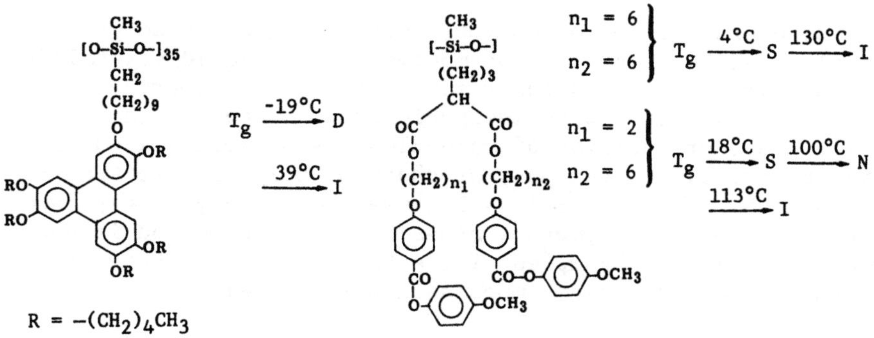

Crosslinked LC polymers can also be prepared, for example, based on polysiloxanes, using the reaction of addition of the crosslinking agent to the polymer chain [138]

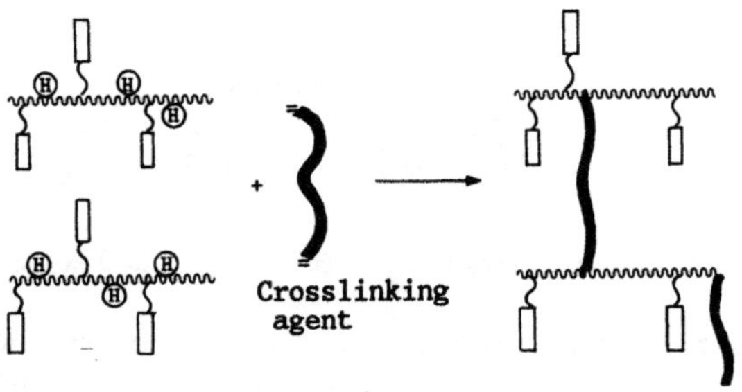

The preparation of unusual block copolymers consisting of oligomeric blocks of poly-L-lysine bearing mesogenic groups (hydrophobic block) and hydrophilic blocks of polyethylene glycol has also been described [139]:

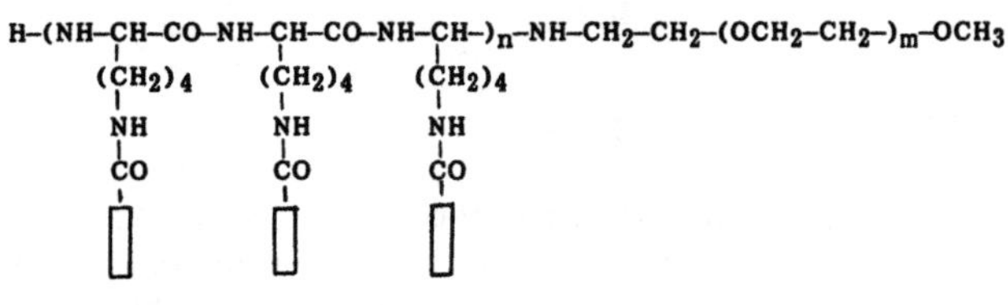

Of the synthesized liquid-crystalline polymers, polymers of the acrylic and methacrylic series and polysiloxanes containing different types of common mesogenic segments of low-molecular-weight liquid crystals (Schiff bases, cyanobiphenyl groups, alkoxybenzoic acid esters, cholesterol esters, etc.) are most common. A large group of copolymers of mesogenic monomers with alkyl acrylates and methacrylates has been syn-

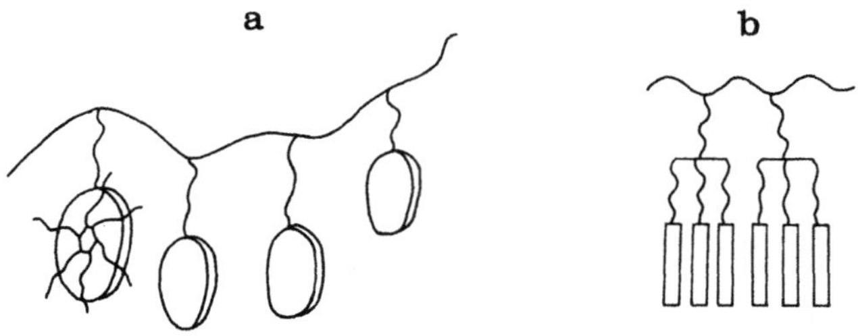

Fig. 4.7. Schematic diagrams of the macromolecular structure in polymers with disc-shaped mesogenic groups (a) and block arrangements of the mesogenic groups (b) in each monomeric unit [136, 137].

thesized, and heteroorganic compounds and linear and crosslinked LC polysiloxanes have been prepared.

The attempts to obtain liquid-crystalline polymers based on inorganic polymers – polyphosphazenes with mesogenic side groups (cholesterol) – merit attention, but the first publications have not yet produced the desired results. In the synthesis of new systems, attention should probably be turned to other classes of chemical compounds, for example heterocyclic compounds. It is very clear that here the possibilities of synthesis are far from exhausted.

In the first studies of LC polymers, most of the attention was focused on methods of synthesis of <u>LC monomers</u> capable of existing in one or more LC modifications and the study of the kinetics of their polymerization.* In conducting polymerization in the corresponding form or forms of a liquid crystal (as well as in bulk or solution), most investigators compared the properties of the polymers obtained by examining the effect of the preliminary ordering of the monomer on the configurational structure and conformation of the macromolecules without focusing special attention on the structural organization of the polymer. Several types of monomers either capable of existing in the LC state or bearing mesogenic groups are reported in Table 4.5. The poly-

*We will not examine the features of polymerization in the LC phase here, as it is of independent interest and is examined in detail by Amerik and Krentsel' [5] and Hardy et al. [20, 21, 24].

TABLE 4.5. Mesogenic Monomers and the Polymers Prepared from Them Not Exhibiting LC Properties

Monomer No.	Monomer and structural formula	Phase transition temperatures (°C) and type of mesophase of the monomers	Polymerization conditions (bulk solution and T_{pol}) and structure of polymer (interplanar distance, T_m, and T_g)	Reference
1.	Vinyl oleate (VO) $CH_2{=}CH{-}O{-}C({=}O){-}(CH_2)_7{-}CH{=}CH{-}(CH_2)_7{-}CH_3$	$C \underset{-45}{\overset{-32}{\rightleftarrows}} N \overset{-18}{\rightleftarrows} I$	In polymerization in solid and LC phase, a crystalline polymer is formed (T_m = 34-38°C); in polymerization in an I melt, an isotropic viscous liquid is formed	[140, 141, 142]
2.	Cetylvinyl ether (VE-16) $CH_2{=}CH{-}O{-}(CH_2)_{15}{-}CH_3$	$C \xrightarrow{-36} S_1 \xrightarrow{-26} S_2 \xrightarrow{-8} I$ $C_1 \xrightarrow{-16} C_2 \xrightarrow{-2.5} S \xrightarrow{16.5} I$	Polymerization in the LC phase yields a crystalline polymer Polymerization at −26, −10, and 12°C yields a crystalline polymer (T_m = 36°C)	[143]
3.	p-Alkoxybenzylidene-p-aminostyrenes 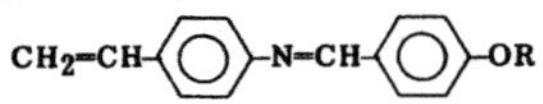			
3.1.	R = $-CH_3$	$C \xrightarrow{97.3} N \xrightarrow{110.6} I$	–	[144]
3.2.	R = n-C_4H_9	$C \xrightarrow{88.3} N \xrightarrow{120.6} I$	–	[144]
3.3.	R = n-$C_{18}H_{37}$	$C \xrightarrow{98} S \xrightarrow{104} I$	–	[144]
4.	p-Allyloxybenzylidene-p-butylaniline 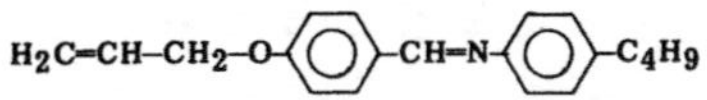	$C \xrightarrow{49.6} N \xrightarrow{68.5} I$	–	[145]

5.	p-Allyloxybenzylidene-p-butoxyaniline 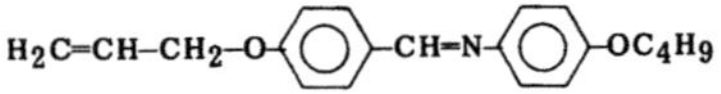	$C \xrightarrow{114} N \xrightarrow{119.6} I$	–	[145]
6.	p-Acryloyloxybenzylidene-p-alkyl-anilines 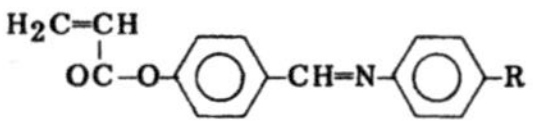			
6.1.	R = $-CH_3$	$C \xrightarrow{114} I$	–	[146]
6.2.	R = n-C_2H_5	$C \xrightarrow{96} I$	–	[146]
6.3.	R = n-C_4H_9	$C \xrightarrow{48.5} N \xrightarrow{56.5} I$	–	[146]
		$C \xrightarrow{28} S \xrightarrow{52} I$	Amorphous polymer	[147]
6.4.	R = n-C_5H_{11}	$C \xrightarrow{47} N \xrightarrow{72} I$	–	[147]
6.5.	R = n-C_6H_{13}	$C \xrightarrow{67} N \xrightarrow{68.5} I$	–	
6.6.	R = n-C_7H_{15}	$C \xrightarrow{50} N \xrightarrow{75.5} I$	–	[148]
6.7.	R = n-C_8H_{17}	$C \xrightarrow{46.5} N \xrightarrow{71} I$	–	
6.8.	R = n-C_9H_{19}	$C \xrightarrow{45} N \xrightarrow{77.5} I$	–	
6.9.	R = n-$C_{10}H_{21}$	$C \xrightarrow{55} N \xrightarrow{75} I$	–	[148]
6.10.	R = n-$C_{12}H_{25}$	$C \xrightarrow{63} N \xrightarrow{76.5} I$	–	

TABLE 4.5 (continued)

Monomer No.	Monomer and structural formula	Phase transition temperatures (°C) and type of mesophase of the monomers	Polymerization conditions (bulk solution and T_{pol}) and structure of polymers (interplanar distance, T_m, and T_g)	Reference
7.	p-Acryloyloxybenzylidene-p-alkoxyanilines 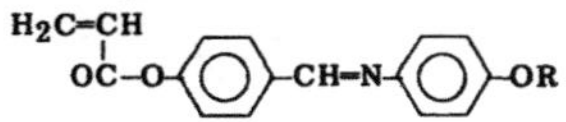	$C \xrightarrow{63} N \xrightarrow{76.5} I$	–	[148]
7.1.	R = $-CH_3$	$C \xrightarrow{97} I \xrightarrow{117} I$	–	[145]
7.2.	R = n-C_4H_9	$C \xrightarrow{74} N \xrightarrow{122} I$	–	[145]
		$C \xrightarrow{74} N \xrightarrow{105} I$; $I \xrightarrow{104} C\text{–}N \rightarrow C$ (dashed arrows)	Amorphous polymer	[149]
7.3.	R = n-C_5H_{11}	$C \xrightarrow{68} N \xrightarrow{111.5} I$	–	[148]
7.4.	R = n-C_6H_{13}	$C \xrightarrow{70.5} N \xrightarrow{116.5} I$	–	[148]
7.5.	R = n-C_7H_{15}	$C \xrightarrow{82} N \xrightarrow{111} I$	–	[148]
7.6.	R = n-C_8H_{17}	$C \xrightarrow{79} N \xrightarrow{114} I$	–	[148]
7.7.	R = n-C_9H_{19}	$C \xrightarrow{85.6} N \xrightarrow{110.5} I$	–	[148]
7.8.	R = n-$C_{10}H_{21}$	$C \xrightarrow{83} N \xrightarrow{112.5} I$	–	[148]

8.	Derivatives of p-acryloyloxybenzyl-idene 			
8.1.	R = $-CN$, p-cyanoaniline (ABCA)	$C \xrightarrow{136} N \xrightarrow{164.5} I$	T = 150°, amorphous polymer	[147]
8.2.	R = $-N{=}N-$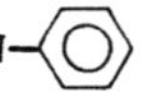, p-aminoazobenzene	$C \xrightarrow{118} N \xrightarrow{196} I$	–	[149]
8.3.	R = $-CH-CO-CH_3$, p-aminoacetamide	$C \xrightarrow{197} I$	–	[149]
8.4.	R = $-CO-CH_3$, p-aminoacetophenone	$C \longrightarrow I$; C $\xrightarrow{80}$ N $\xrightarrow{108}$ I (dashed)	–	[149]
8.5.	R = $-COOC_2H_5$, p-aminobutyl benzoate	$C \xrightarrow{73} I$	–	[149]
8.6.	R = $-COOC_4H_9$, p-aminobutyl benzoate	$C \xrightarrow{76} I$	–	[149]
9.	p-Methacryloyloxybenzylidene-p-alkylanilines 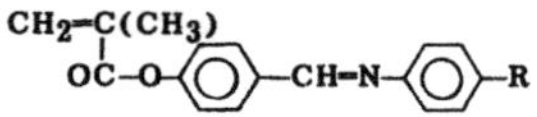			
9.1.	R = $-CH_3$	$C \xrightarrow{68} I$	–	[146]
9.2.	R = n-C_2H_5	$C \xrightarrow{62} I$	–	[146]
9.3.	R = n-C_4H_9	$C \xrightarrow{58} I$	–	[146]
9.4.	R = n-C_5H_{11}	$C \xrightarrow{56} I$	–	[148]
9.5.	R = n-C_6H_{13}	$C \xrightarrow{59} I$	–	

TABLE 4.5 (continued)

Monomer No.	Monomer and structural formula	Phase transition temperatures (°C) and type of mesophase of the monomers	Polymerization conditions (bulk solution and T_{pol}) and structure of polymers (interplanar distance, T_m, and T_g)	Reference
9.6.	R = n-C_7H_{15}	$C \xrightarrow{63.5} I$	–	
9.7.	R = n-C_8H_{17}	$C \xrightarrow{67.5} I$	–	
9.8.	R = n-C_9H_{19}	$C \xrightarrow{68} I$	–	[148]
9.9.	R = n-$C_{10}H_{21}$	$C \xrightarrow{74} I$	–	
9.10.	R = n-$C_{12}H_{25}$	$C \xrightarrow{80} I$	–	
10.	p-Methacryloyloxy-2-methoxybenzylidene-p-substituted anilines 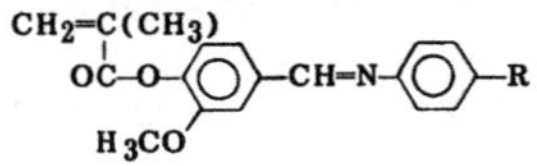			
10.1.	R = H	$C \xrightarrow{38} N \xrightarrow{39} I$		[150]
10.2.	R = $-CH_3$	$C \xrightarrow{41} N \xrightarrow{42} I$		[150]
10.3.	R = $-OCH_3$	$C \xrightarrow{48} N \xrightarrow{52} I$		[150]
10.4.	R = $-Cl$	$C \xrightarrow{45} N$		[150]

11.	p-Methacryloyloxybenzilidene-p-alkoxyanilines 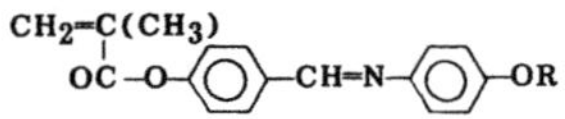			
11.1.	R = $-CH_3$	$C \xrightarrow{130.5} I$	–	[146]
11.2.	R = n-C_4H_9	$C \xrightarrow{126} I$	–	[146]
11.3.	R = n-C_5H_{11}	$C \xrightarrow{91.5} I$	–	
11.4.	R = n-C_6H_{13}	$C \xrightarrow{89} N \xrightarrow{96.5} I$	–	
11.5.	R = n-C_7H_{15}	$C \xrightarrow{98} I$	–	[148]
11.6.	R = n-C_8H_{17}	$C \xrightarrow{95} N \xrightarrow{99} I$	–	
11.7.	R = n-C_9H_{19}	$C \xrightarrow{99.5} I$	–	
11.8.	R = n-$C_{10}H_{21}$	$C \xrightarrow{97} N \xrightarrow{100} I$	–	[148]
12.	Di-(p-acryloyloxybenzylidene)-1,4-diamino-2-chlorobenzene 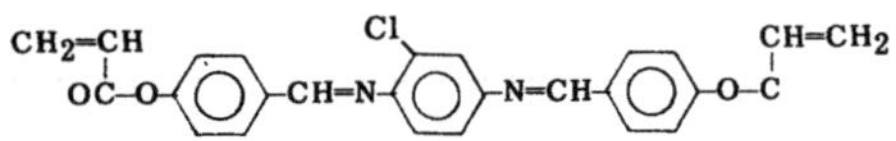	$C \xrightarrow{185} S \xrightarrow{189} I$	–	[147]
13.	Diacryloyloxy-4-4-tolane 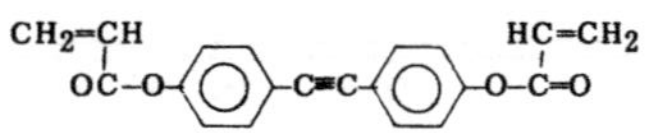	$C \xrightarrow{135} N \xrightarrow{141} I$	–	[151]

TABLE 4.5 (continued)

Monomer No.	Monomer and structural formula	Phase transition temperatures (°C) and type of mesophase of the monomers	Polymerization conditions (bulk solution and T_{pol}) and structure of polymers (interplanar distance, T_m, and T_g)	Reference
14.	N-methacryloyl-ω-aminolauric acid methyl ester $H_2C{=}C(CH_3)$ $O{=}C{-}NH{-}(CH_2)_{11}{-}COOCH_3$	$C \xrightarrow{45} I$	Amorphous polymer	[152]
15.	N-methacryloyl-ω-aminolauric acid benzyl ester (PBMAA-11) $H_2C{=}C(CH_3)$ $OC{-}NH{-}(CH_2)_{11}{-}COO{-}CH_2{-}$ 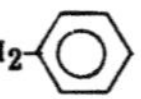	$C \xrightarrow{54} I$	Amorphous polymer	[152]
16.	N-methacryloyl-ω-aminolauric acid cetyl ester (PCMAA-11) $H_2C{=}C(CH_3)$ $OC{-}NH{-}(CH_2)_{11}{-}COO{-}(CH_2)_{15}{-}CH_3$	$C \xrightarrow{58} I$	The polymer crystallizes in a hexagonal cell, T_m = 40°C	[152]
			Above T_m, the polymer is optically isotropic	[152]

Note. T_{pol} is the temperature of polymerization; C) crystalline state; C_1, C_2 and S_1, S_2) different types of crystalline and smectic modifications; N, S) nematic and smectic types of mesophase; I) isotropic melt; the symbol "–" indicates no data on the structure of the polymer is available.

mers prepared from them do not exhibit LC properties, although the corresponding polymers were not synthesized from all of the monomers. We included these monomers in Table 4.5 to show that the group of previously unstudied potential mesogenic polymeric materials is relatively large. This could draw the more intense interest of investigators to the study of the physicochemical properties of these compounds.

Before turning to an examination of some of the properties of the polymers reported in Table 4.5, we will discuss some questions of terminology concerning the concept of the term liquid-crystalline as applied to polymers. In our opinion, all thermotropic macromolecular compounds can be divided into two groups. The first group consists of thermotropic macromolecular compounds for which the LC state, existing in a certain temperature range, is a thermodynamically stable phase state and is characterized by spontaneous anisotropy of the properties in the absence of a three-dimensional crystal lattice. Like low-molecular-weight liquid crystals, thermotropic LC polymers in this state exhibit both the properties of solids (anisotropy of properties) and the properties of liquids. The latter is possible in polymers only at temperatures above their glass transition temperature (or flow temperature). In other words, the transition to the LC phase is only possible for polymers at temperatures above the glass transition temperature. In this case, the melting point of the LC phase will henceforth be called the clearing point (T_{cl}) by analogy with low-molecular-weight liquid crystals, and should be higher than the glass transition temperature ($T_g < T_{cl}$).

However, due to the slowness of the relaxation processes which take place in polymers and their tendency to form nonequilibrium structures, "freezing" of the LC phase of polymers in the glassy state is possible. Such a polymer exhibits anisotropy of properties and can remain in the glassy state up to the temperature of its chemical decomposition (T_{dec}) if $T_{cl} > T_{dec}$. Such LC polymers, which have a LC structure (or LC order), are essentially glassy liquid crystals which do not have the distinctive feature of liquid crystals: the mobility of the mesogenic groups. In this case, since it remains glassy through a wide range of temperatures, the polymer cannot have a T_{cl} corresponding to its transition into a true LC state. Such polymers form a second group of LC polymers [10].

While acknowledging the arbitrariness of this division, we nevertheless would like to turn attention to the necessity for a clear separation of the concepts used in describing LC polymers. Although there is only an insignificant number of glassy LC compounds among low-molecular-weight liquid crystals, there is an extremely large number among polymers.

It is necessary to note that the first studies on the synthesis and structure of LC polymers basically concerned glassy macromolecular compounds with optical anisotropy but without the set of properties characteristic of liquids (flow, in particular), which raised doubt as to the correctness of their assignment to LC compounds in many cases. Basing conclusions on the results of optical studies alone and attributing the occasionally observed optical anisotropy caused by polymer orientation or the internal stresses to the LC state of the polymer, investigators were not always correct in calling these polymers liquid-crystalline.

The situation in this area of research is also complicated by the fact that not all due precautions, to exclude both the possible formation of oriented structures (assigned to the LC phase) and the presence of LC monomers in the polymerization products, are always taken in conducting bulk polymerization of LC monomers. For this reason, some polymers are called liquid-crystalline without sufficient justification, and this significantly complicates the study and determination of the specific features and role of the LC state in polymeric compounds. Establishing the thermodynamic limits of existence, characterized by certain heats and temperatures of transition of the polymer, is the most reliable criterion for assigning a polymer to the LC state, as in the case of low-molecular-weight liquid crystals, but this is only possible for polymers of the first type. Extreme care is required in all other cases in establishing the reality of the spontaneous appearance of anisotropy of the properties, which would indicate formation of the LC state and not LC ordering.

The data reported in Table 4.5 are intended to show that polymerization of LC monomers and monomers containing mesogenic groups does not automatically result in polymers with LC properties or LC structures, as hypothesized in the first studies of synthesis of LC polymers.

Vinyl oleate (VO) (Table 4.5, polymer 1), which forms a nematic mesophase in the temperature range of −32 to −18°C, was the first LC monomer studied [140, 141]. In bulk polymerization of VO, a transparent polymer is formed as a viscous liquid, while the polymer obtained by polymerization in the LC phase or solid state is crystalline, resembles a fat in external appearance, and melts in the 34-38°C region. Despite the fact that there are no experimental data in [140, 142] which indicate the LC structure of polyvinyl oleate, the same investigators call this polymer liquid-crystalline in subsequent surveys [153-154].

Hungarian investigators [142] subsequently studied the polymerization of vinyl oleate and showed that polymers prepared in the solid, liquid-crystalline, and isotropic states differ slightly in structure. While unfortunately not citing the melting points, they nevertheless note that the polymers prepared by polymerization in the solid phase and isotropic melt have a type S_A and S_C smectic structure (S), respectively. A crystalline polymer is formed in the polymerization of vinyl oleate in the LC phase, as in polymerization of cetylvinyl ether [143] (Table 4.5, polymer 2). The usual comb-shaped polymers which crystallize due to the packing of the side chains, as examined in detail in Chapter 1, were apparently involved in the examples cited above.

A wide group of monomers capable of forming a mesophase was synthesized by addition of groups of the Schiff base type to the side chains of vinyl, allyl, and acrylic compounds (Table 4.5, monomers 3-12) [140, 144-154]. The first communication of Paleos et al. [144] was primarily a description of the synthesis of these monomers and an examination of some mechanisms of their polymerization kinetics (Table 4.5, monomers 3.1-3.4). The structure of the polymers formed was either not studied, or the polymerization products were amorphous polymers with no LC properties and of no interest to research.

The effect of the initial mesomorphic monomer structure on the structure of the polymer formed was examined in detail, although not clearly enough, in a series of studies by Strzeleski and Liebert [147-149] using diacrylic compounds and monomeric Schiff bases (Table 4.5, monomers 6.3-6.10 and 7.2-9.10). Since a rigorously determined method of arrangement of the molecules corresponds to each of the three types of thermotropic liquid crystals, in the opinion of these investigators the initial mesophase in the polymeric compound can be established by conducting polymerization in each of

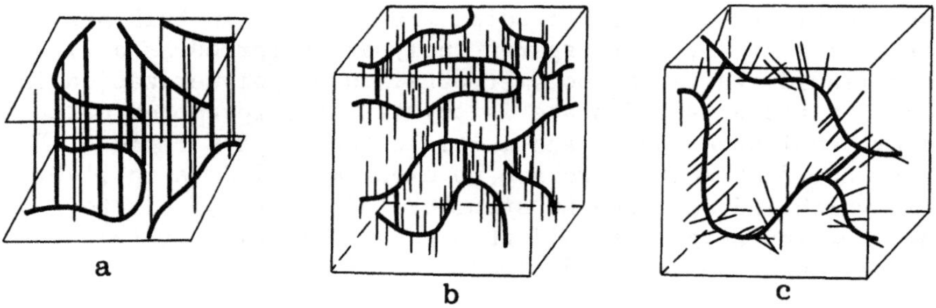

Fig. 4.8. Arrangement of the side chains of macromolecules in crosslinked polymers obtained by copolymerization of mono- and diacrylic monomers in the smectic (a), nematic (b), and cholesteric (c) mesophases [147].

the mesomorphic phases. From this, and assuming that in the case of linear polymers it is impossible to preserve an "organized" structure in a melt of the polymer above the melting point, Strzeleski and Liebert proposed using De Gennes' idea [155] on fixing the LC structure by polymerization with crosslinking in anisotropic media.

In conducting thermal polymerization of monoacrylic compounds or copolymerization with diacrylates (Table 4.5, monomers 4-6), it was concluded that a nematic, smectic, and even a cholesteric structure could be established in a polymer based on direct observation of the polymerization process on the polarization stage of a microscope or based on a study of optical photomicrographs obtained from sections of bulk samples.

However, instead of describing the observed morphological pictures and optical photomicrographs which are the basis for assigning a polymer to a certain type of mesophase, almost no structural data were reported on the polymers and copolymers studied. All of the polymers obtained were brittle, insoluble, and nonmelting, and resembled a brown glass.

The schematic drawings corresponding to the arrangement of the macromolecules of the polymers in the nematic, smectic, and cholesteric modifications (Fig. 4.8) shown in these studies include almost no concrete information on the packing of the side branches of the macromolecules. In addition, no data are cited on the degree of conversion of the monomers during their polymerization or on controlling the purity of the polymers; for this reason, the observed optical pictures could correspond to mixtures of polymers and LC monomers, so that the question of the LC state and actual molecular and supermolecular structure of these polymeric compounds remains open. Despite the possibility of obtaining anisotropic copolymers by copolymerization of the ABCA monomer (Table 4.5, monomer 8.1) with diacrylic derivatives in the nematic phase in a magnetic field (5000 G field intensity) [147], the properties of these systems with a fixed mesophase structure indicated above make them unsuitable for practical use and "inconvenient" for structural studies.

Nevertheless, the approach based on the use of crosslinking agents for fixing the LC structure was of definite interest and it was later used by Blumstein et al. [156] and Finkelmann [138], who selected the same and other polymers and conducted more detailed studies of the preparation of crosslinked polymers.

As Table 4.5 shows, most of the polymers studied have an amorphous structure (Table 4.5, polymers 6.3, 7.2, 8.1, etc.), and only if the side chain length exceeds n_{cr} (cf. Section 1.4) do the side chains of such polymers crystallize, regardless of the phase state of the medium in which polymerization is conducted (Table 4.5, polymers 1, 2, 15-17). The absence of data on the structure of a series of polymers based on the monomers listed in Table 4.5 unconditionally makes the problem of their study urgent, but there is still no basis for assigning them to the class of LC polymeric compounds.

The absence of LC polymers among the compounds examined thus indicates that the presence of mesogenic groups as side chains of the macromolecules is still not a sufficient condition for the formation of a LC phase by these polymers.

We specially listed the polymeric compounds which do not form LC phases in Table 4.5 to show how complex the question is of predicting the possible formation of the LC state in polymers with mesogenic side groups. The absence of a LC phase for some of these polymers is apparently due to the significant steric hindrance from the main chain on the packing of the mesogenic groups; this prevents the possible formation of a mesophase by the side groups. In addition, as previously noted, some of the polymeric compounds indicated

TABLE 4.6. Phase Transition Temperatures of Monomers and Properties of the Polymers Obtained, Most Exhibiting a LC Structure

Monomer No.	Monomer and structural formula	Phase transition temperatures	Polymerization conditions (bulk, solution, T_{pol}) and structure of polymer (interplanar distance d, T_g, T_m) (°C)	Reference
1.	p-Acryloyloxybenzylidene-p-ethoxyaniline 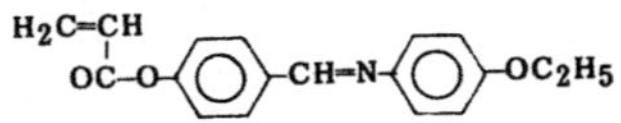	$C \xrightarrow{78} N \xrightarrow{136.5} I$	a) In polymerization in the N phase, I melt, and anionic polymerization at −78°C, a polymer of the S type is formed, $T_g \cong 180$°C; BF disappears at 308-310°C; d = 33 Å	[146]
			b) In polymerization in solution at 20°C, an amorphous polymer is formed	[146]
2.	p-Methacryloyloxybenzylidene-p-ethoxyaniline 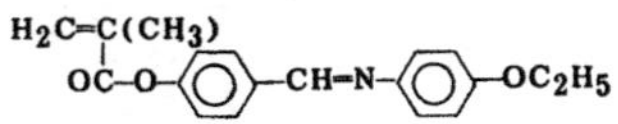	$C \xrightarrow{86.5} N \xrightarrow{100} I$	In polymerization in an I melt and the N phase, a polymer of the S type is formed; BF disappears at 308°C	[146]
3.	Di-N-(p-acryloyloxybenzylidene)-p-diaminobenzene (di-ABDAB) 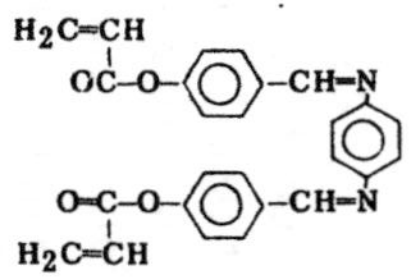	$C \xrightarrow{181} S \xrightarrow{250} N$	a) T_{pol} = 200-205°C, S-type polymer (d = 22 Å, side chains tilted 60° toward the plane of the end groups); in polymerization in the N phase a polymer of the N type is formed; S and N structures of the polymer are preserved up to 360°C	[147, 148]

		$C \xrightarrow{180} S \longrightarrow$ Polymerization	b) T_{pol} = 194°C, S-type polymer (d = 43.3 Å m; 22.2 Å s; 4.88 Å; halo ~4.4 Å)	
			c) T_{pol} = 200°C, S-type polymer (d = 43 Å w; 22 Å s; halo 4.4 Å)	[156]
			d) T_{pol} = 250°C, S-type polymer (d = 43.5 Å w; 22.2 Å s; halo 4.4 Å)	
4.	Di-N-(p-acryloyloxybenzylidene)-hydrazine 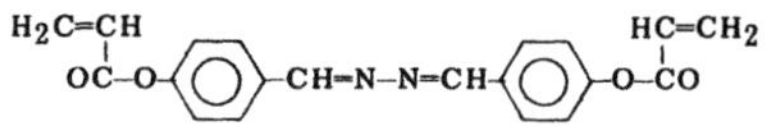	$C \xrightarrow{140} I$	a) T_{pol} = 140°C, N-type polymer	[157]
			b) T_{pol} = 170°C, amorphous polymer	
		$C \xrightarrow{138} N \longrightarrow$ Polymerization	c) N-type polymer	[156]
5.	p-Phenylene-bis(N-methylene-p-aminostyrene) 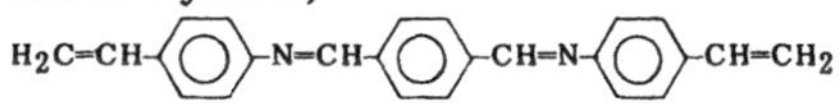	$C \xrightarrow{150} N \longrightarrow$ Polymerization	N-type polymer	[158]
6.	4[1]-Acryloyloxybenzoic acid 4-n-hexoxyphenyl ester 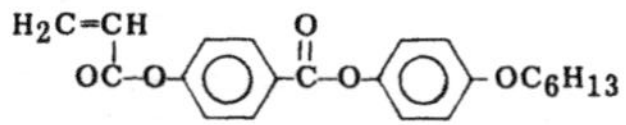	$C \xrightarrow{55-58} N \xrightarrow{97.6} I$	a) T_{pol} = 102°C, isotropic polymer	
			b) Polymerization in N phase without application of magnetic field and in a magnetic field (0.5 T) produces a highly oriented anisotropic polymer	[159]
7.	p-Acryloyloxybenzylidene-p-aminobenzoic acid (ABAA) 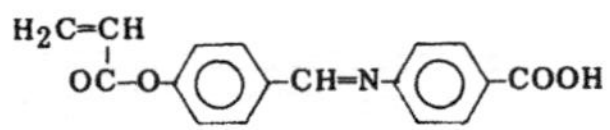	$C \xrightarrow{218} S \longrightarrow$ Polymerization	a) T_{pol} = 230°C; S-type polymer	
			b) T_{pol} = 350°C; N-type polymer. Both polymers insoluble	[160]

TABLE 4.6 (continued)

Monomer No.	Monomer and structural formulas	Phase transition temperatures	Polymerization conditions (bulk, solution, T_{pol}) and structure of polymer (interplanar distance d, T_g, T_m) (°C)	Reference
8.	p-Methacryloyloxybenzylidene-p-aminobenzoic acid (MABAA) 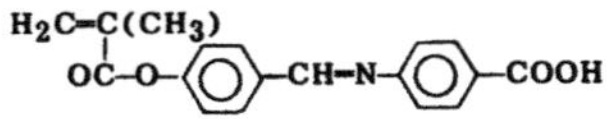	$C \xrightarrow{178} S_3(?) \xrightarrow{182} S_2$; $\xrightarrow{201} S_1 \xrightarrow{205} N \xrightarrow{?} I$	a) T_{pol} = 201°C, N-type polymer, T_g = 154°C	[161]
			b) T_{pol} = 205°C, N-type polymer	[162]
			c) T_{pol} = 213°C, S-type polymer (d = 32 Å, T = 124); in polymerization in DMF solution at 120°C, an N-type polymer is formed (T_g = 87°); in all cases, the BF is preserved up to T_{dec} of the polymer	[163]
		$C \xrightarrow{182} C_2 \xrightarrow{201} S(?)$; $\xrightarrow{206} N(?) \longrightarrow$ Polymerization	–	[164]
9.	p-Acryloyloxybenzoic acid (ABA) 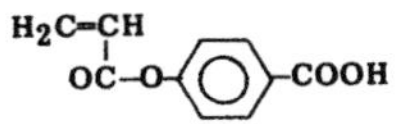	$C \xrightarrow{201} I$	a) T_{pol} = 210°C; S-type polymer is formed during fast bulk polymerization (T_g = 125°C; d = 17.7 Å)	[163]
			b) T_{pol} = 210°C; a crystalline polymer is formed during polymerization of the monomer or from form II by pouring a film from solution in DMF. Polymerization in DMF solution at 115°C, amorphous polymer, T_g = 90°.	

10.	p-Methacryloyloxybenzoic acid (MBA) 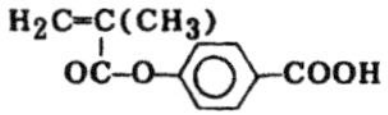	$C \xrightarrow{182} I$	a) T_{pol} = 185°C, S_B-type polymer (d = 19 Å, T_g = 110°, T_{dec} = 220°, b) Polymerization in solution in heptoxybenzoic acid at 150°C, amorphous polymer	[163, 165]

Note. Intensity of reflections: s) strong; m) medium; w) weak; C) crystalline state; C_1, C_2 and S_1, S_2) different types of crystalline and smectic modifications; N, S) nematic and smectic mesophases; I) isotropic melt; S_A, S_B, S_E) different types of smectic structure; BF) spontaneous birefringence.

in Table 4.5 have generally not been studied, and the question of their phase state remains unanswered.

At the same time, there is a relatively large number of polymers with mesogenic side groups directly bound to the main chain for which the formation of a LC phase is possible. Some compounds of this type are reported in Table 4.6. While not examining the structural features of these polymers in detail here, as this will be the subject of the following sections, we will only attempt to answer generally the question of why some of the polymers with direct linkage of mesogenic groups to the main chain are capable of forming a LC phase and others are not.

Since the formation of the LC phase is basically determined by the packing of the anisodiametric (mesogenic) side groups, the possibility of existence in the LC state in polymers is naturally primarily dependent on three basic factors: the chemical structure of the mesogenic group, the structure and conformation of the main chain, and the linkage which binds the mesogenic groups to the main chain. The identity of the chemical structure of the mesogenic side groups and the molecules of low-molecular-weight liquid crystals is only one of the necessary but insufficient conditions for the formation of a polymeric LC phase. A significant role in this process is attributed to the polymer main chain, linkage, conformation, and the intramolecular mobility, which should ensure the required arrangement of the mesogenic groups in the formation of the mesophase. The required conformation can be created directly during polymerization, and the polymer formed then has a LC structure (as this takes place in the polymerization of monomer in a melt or in a LC matrix, for example; cf. Table 4.6). Polymerization in solution usually results in the formation of polymers which do not have a LC structure (Table 4.6, polymers 1b, 10b) and only subsequent treatment accompanied by conformational changes can cause (but not always!) the appearance of LC order in the arrangement of the side groups.

This very clearly follows from the data in [166-168] on the study of the structure of comb-shaped polymers based on N^{α}-acyl derivatives of N^{ε}-methacryloyl-L-lysine (PML-n).

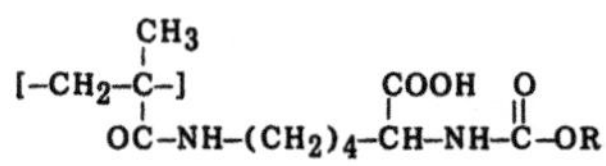

where R = $n\text{-}C_{15}H_{31}$, $n\text{-}C_{17}H_{35}$, and $n\text{-}C_{21}H_{43}$. Since chemical bonding of the asymmetric side chains only takes place at their end groups, the side chains with polar groups can be considered as a structurally organized set of long-chain molecules joined with the main chain, for which the properties characteristic of compounds capable of forming the LC state can be expected. Actually, the study of films of polymers of the PML-n series in polarized light revealed the presence of optically anisotropic structures over a wide range of temperatures, including both the regions of the glassy and the viscoelastic states. It is interesting to note that the method of treatment for this series of polymers essentially determines the phase state. Precipitation of polymers of the PML-15, -17, and -21 series from chloroform solutions results in amorphous systems, while preparation of polymer films from solutions in polar solvents (precipitating agents) results in the crystallization of these polymers due to the packing of the aliphatic side chains in a hexagonal, although more defective, cell, as in the previously studied comb-shaped polymers of the acrylic and methacrylic series ([169], Chapter 1). The melting points of these polymers are 32°C (PML-15), 34°C (PML-17), and 56°C (PML-21).

Nevertheless, the very pronounced amphiphilic nature of the PML-n macromolecules is responsible for the special position of these systems among comb-shaped polymers. The presence of two types of interactions — hydrogen bonds between functional groups and the dispersion forces between the methylene side chains — results in the appearance of microseparation which takes place in homopolymers of PML-n, especially under the effect of specifically acting solvents. As a result of microseparation, regions are formed which are capable of the molecular rearrangements and phase transitions typical of LC compounds; the formation of these regions takes place as a result of ordering of the anisodiametric side groups, which tend to form liquid crystals, and the stabilization of the structure is due to reinforcement of its matrix of "rigid" side chains. As a result, crystalline polymers PML-15, -17, and -21 enter the optically anisotropic state and exhibit LC properties above the melting points. Since the glass transition temperatures of these polymers are much higher than the melting point of the regions formed by the main chains, the processes of crystallization and melting only occur in microregions of the glassy matrix, permitting molecular rearrangements only within the boundaries of these regions. This character of the structural organization of the macromolecules covers a broad range of temperatures lying within the boundaries of the melting point—chemical decomposition temperature of the polymer (~200°C).

The analysis of the data in Table 4.6 results in the following conclusions.

LC polymers are usually obtained from mesogenic compounds (i.e., monomers bearing mesogenic groups), but the presence of a mesophase for the monomer is not an obligatory condition for the formation of a LC polymer. There are a number of monomers which do not form mesophases, but polymerization of these monomers results in mesomorphic polymers (Table 4.6, polymers 9 and 10, for example). This is most clearly seen in comparing low-molecular-weight and high-molecular-weight biphenyl derivatives. It is known that acyl derivatives of n-hydroxybiphenyl (compound I)

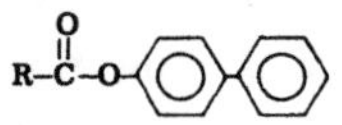

(I)

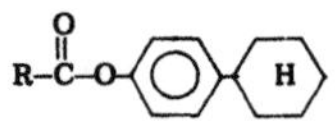

(II)

and its partially hydrogenated analog (II) are not mesogenic (since there are no substituents at position 4 in the benzene and cyclohexane rings) and do not form a mesophase. At the same time, their polymeric analogs with the same substituents in the side groups, poly(n-biphenyl acrylate) (III) and poly(p-acyclohexylphenyl acrylate) (IV)

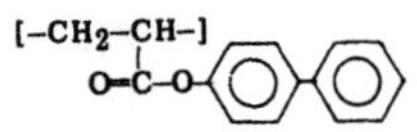

(III)

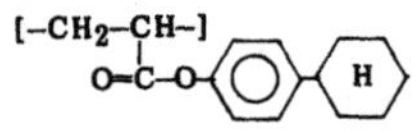

(IV)

can exist in the smectic LC state and have a clearing point in the 260-285°C range (polymer III) and at 205°C (polymer IV) [170-172]. The causes of the appearance of mesomorphism in these polymers have not been definitively determined; it is only possible to conjecture that bonding the rigid biphenyl fragments to the flexible main chain in one unit intensifies their dispersion interaction in a melt, which also results in the formation of a mesophase. At the same time, addition of the same biphenyl mesogenic groups to a main chain of the methacrylate type or a change in the "alignment" of the ester group for poly(n-diphenyl methcrylate) (V) and poly(vinyl-n-phenyl benzoate) (VI)

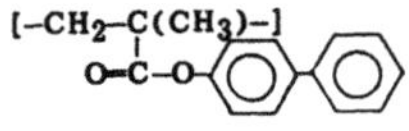

(V)

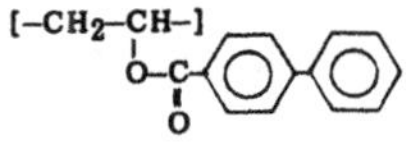

(VI)

totally excludes the possibility of mesophase formation and polymers V and VI are consequently amorphous.

The presence of rigid units of low-molecular-weight compounds in the macromolecules can result in the appearance of LC properties in the polymers even if the corresponding low-molecular-weight compounds do not form a LC phase. The possibility of inducing mesomorphism in organic compounds is very interesting with respect to increasing the group of LC compounds and "drawing" nonmesogenic low-molecular-weight substances into the class of polymeric LC compounds.

On the other hand, the flexibility of the main chain and bonding character of the mesogenic groups with the main chain of the macromolecule has been assigned an important role in the formation of the LC state. Direct attachment of the mesogenic groups to the macromolecular backbone of the chain usually hinders the ease of controlling the mesogenic groups with the formation of a mesophase; if a LC state is formed, the glass transition temperature and clearing points of these polymers are relatively high, frequently near their chemical decomposition temperatures. In some cases, no transition is observed from the glassy to the viscoelastic state in general. This significantly complicates the study of the structure and properties of polymers in the LC phase, i.e., in the state which determines the segmental mobility of the macromolecules and the anisotropy of the properties of the polymer.

Most of these polymers are essentially glassy liquid crystals for which the formation of LC orientational order is typical and the observed optical anisotropy is fixed in the glassy polymer matrix.

The term "liquid-crystalline polymer" in these cases thus only indicates the LC structure of the polymers, frequently formed during synthesis, although the region of the LC phase state can be absent in these polymers (especially if crosslinked polymers are involved). This structure of the macromolecules, where the bulky anisotropic mesogenic groups are directly bound to the main chain, is usually also responsible for the relatively high glass transition temperature, and this significantly limits (or totally excludes) the region of the LC state. However, this can be avoided by using comb-shaped polymers.

4.4. COMB-SHAPED LIQUID-CRYSTALLINE POLYMERS

One of the most highly recommended approaches to solving the problem of creating thermotropic LC polymers is the use of comb-shaped polymers, first proposed for this purpose by Shibaev, Freidzon, and Platé [1-4, 173-175], and subsequently successfully used by German investigators [176-177]. We have departed from our previously elaborated concepts on the relation between the structure and properties of comb-shaped polymers bearing long aliphatic side chains in each monomeric unit (see Chapter 1).

The autonomous behavior of the side chains of comb-shaped polymers, demonstrated by their capacity to form layered structures in melts and to crystallize regardless of the main chain configuration, makes it possible to create LC polymers from them. Actually, since chemical bonding of the asymmetric side chains in these polymers only takes place with respect to the end groups of the side chains, these groups can be considered to be a structurally organized set of long-chain molecules bound to the main chain, as indicated above, which is probably one of the necessary conditions for formation of the LC structure. In other words, the structure of comb-shaped polymers causes ordering in the arrangement of the side groups on one hand, and the independence of the side chain behavior with respect to the main chain on the other hand (Fig. 4.9). If mesogenic groups are added to the end groups of the side chains of comb-shaped polymers, i.e., if they are moved some distance from the main chain, it will be possible to decrease significantly the steric hindrance applied by the main chain to the packing of the mesogenic groups in comparison to polymers in which the mesogenic groups are directly "bound" to the main chain.

These polymers can be obtained using the same methods discussed in the preceding section, but with one important addition. A spacer in the form of aliphatic (or oxymethylene) units is introduced between the main chain and the mesogenic group. The spacer fulfills several functions. The first function, as noted, is ensuring the autonomous behavior of the mesogenic side groups. Although there is some correlation in the behavior of the main and side chains for

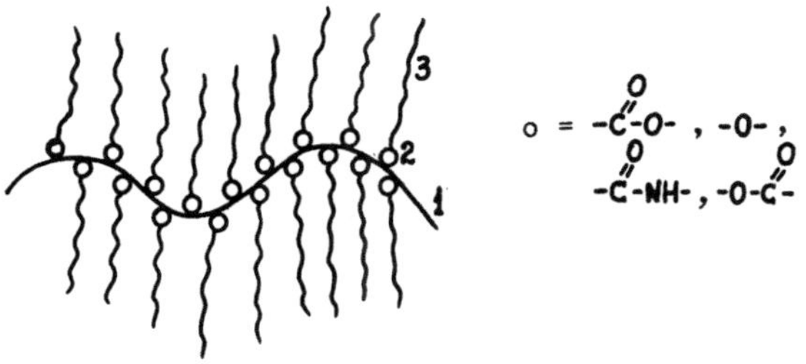

Fig. 4.9. Schematic representation of a comb-shaped macromolecule: 1) main chain; 2) linkage (chemical bond); 3) n-aliphatic side chain.

polymers with no spacer groups, it disappears to a certain extent at a certain length of the spacer group. The results from the studies of the solutions of comb-shaped polymers with mesogenic groups obtained by Tsvetkov et al. [178-180] (see Section 4.9, Table 4.32) are a clear illustration of this. As a comparison of the value of the optical anisotropy and the specific Kerr constant (which essentially characterize the degree of coherence of the individual elements) shows, separation of the mesogenic segment from the main chain by a "polymethylene" spacer of ten methylene units (see below, Table 4.32, polymer 5) results in a sharp decrease in the correlation of the orientations of the side groups and the main chain.

Another, no less important function of the spacer is related to its plasticizing effect on the polymer. As in the case of comb-shaped polymers with no mesogenic groups (cf. Chapter 1), an increase in the length of the aliphatic bridge also results in a shift in the glass transition temperature to the region of lower temperatures. By varying the length of the spacer, it is thus possible to significantly affect the T_g, i.e., to alter the lower temperature boundary of the LC state (Fig. 4.10). This is very important to emphasize, since polymers with no spacer groups usually have high glass transition temperatures (>100°C), which limits their use in normal conditions. It was possible to substantially decrease the glass transition temperature by synthesis of LC polysiloxanes with spacer groups [14]. The approximate temperature range of LC polymers with mesogenic side groups existing in the LC state is −100°C to +220°C.

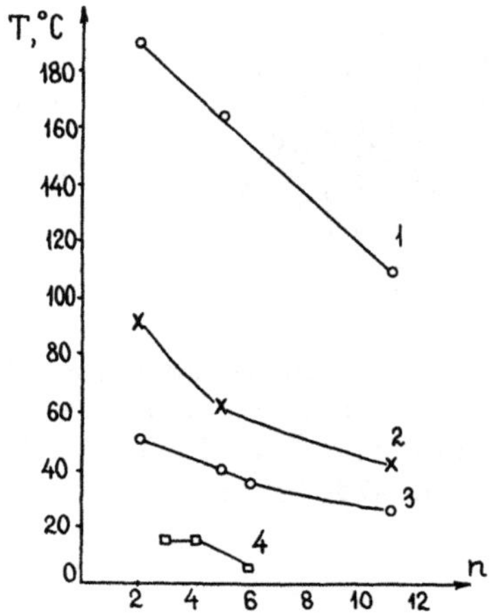

Fig. 4.10. Effect of the spacer length on the glass transition temperature of some comb-shaped LC polymers:

1) 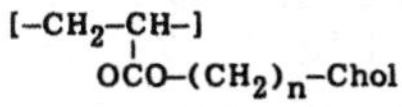 [12];

2) 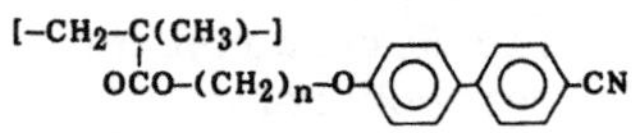 [182];

3) 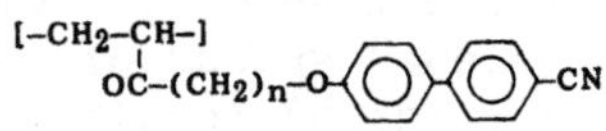 [181];

4) 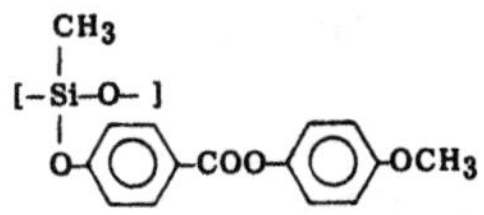 [14].

Finally, a third no less important function of the spacer group is related to its effect on the type of mesophase. By altering the number of methylene units in the linkage while the other parameters of the polymers remain the same, it is usually possible to go from the nematic to the smectic mesophase, which will be discussed in more detail below.

Thus, comb-shaped polymers have been convenient matrices for the synthesis of LC polymers with mesogenic side groups. For this reason, the overwhelming majority of subsequent studies were conducted with comb-shaped LC polymers, and the methods of their synthesis have been used in many different countries.

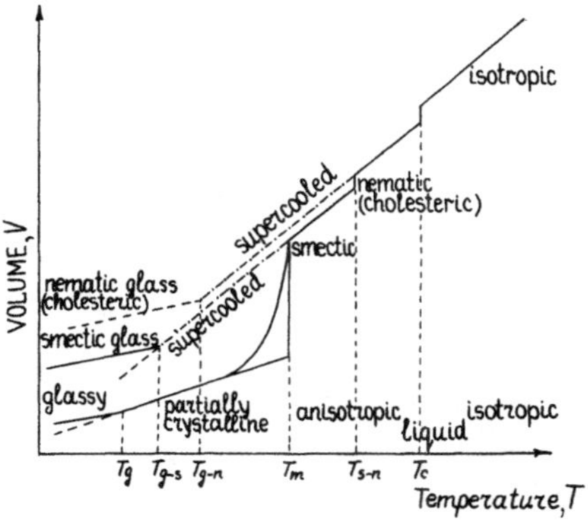

Fig. 4.11. Schematic volume—temperature diagram of a LC polymer [186].

4.5. FEATURES OF THE PROPERTIES OF THERMOTROPIC LIQUID-CRYSTALLINE POLYMERS RELATED TO THEIR MACROMOLECULAR NATURE

In turning to the examination of the properties of the LC state of polymers with mesogenic side groups, the dual nature of their macromolecules should be approached first. On the one hand, they are "bearers" of "polymeric" properties which are determined by the main chain. On the other hand, the mesogenic side groups are "bearers" of mesomorphic properties. The more complex structure of LC polymers in comparison to low-molecular-weight liquid crystals (which are not yet definitively known and are still the subject of intense investigation) complicates the study of their properties, especially the interpretation of their physicochemical behavior.

The question of whether the LC phase formed by such complex polymers is thermodynamically stable naturally arises. This question was answered in the studies by Frenzel and Rehage [183-186]. Using the well-known Ehrenfest equations, they described the behavior of a series of LC polymers in the form of P-V-T diagrams and showed that the phase transitions in LC polymers between the crystalline—smectic—nematic (cholesteric) and isotropic phases are first-order transitions, as in the case of low-molecular-weight liquid crystals. Figure 4.11 schematically shows the change in the specific volume as a function of the temperature for a hypothetical LC polymer which forms all possible types of mesophases, and

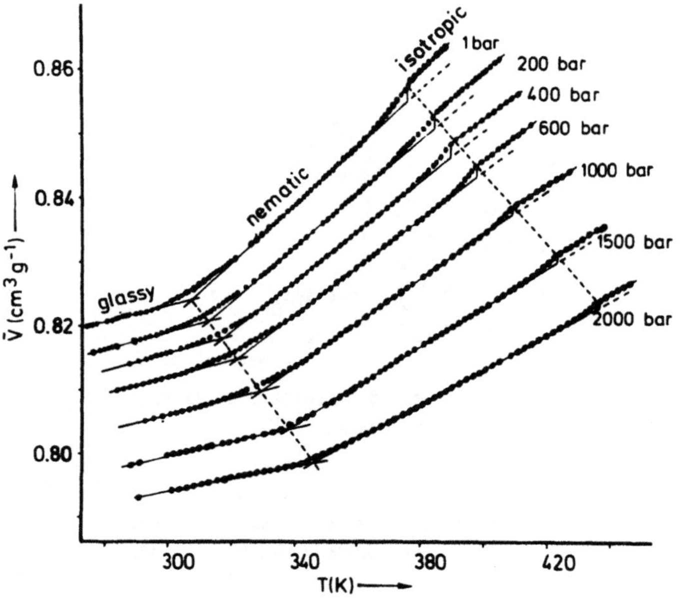

Fig. 4.12. Specific volume-temperature curves at different pressures for a LC nematic polymer [184]:

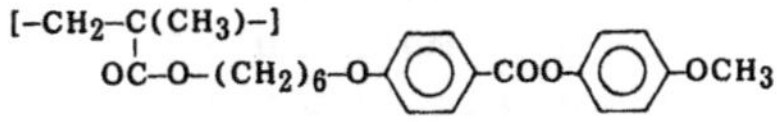

Fig. 4.12 shows a series of curves of the dependence of the specific volume on temperature for one of the LC nematic polymers at different pressures. The curves clearly establish the sudden change in the values of V at the LC phase–isotropic liquid transition point. In the thermodynamic sense, the LC phase of polymers behaves like the LC phase of low-molecular-weight liquid crystals and only differs with respect to the significantly higher viscosity, which can exceed the viscosity of low-molecular-weight liquid crystals by several orders of magnitude, attaining values of 10^5-10^6 P. This viscosity is essentially a function of the molecular weight (degree of polymerization) [187]. The elevated viscosity and slowness of all relaxation processes in LC polymers determine the specific polymeric features of their physicochemical behavior.

One of the features of LC polymers, as briefly indicated above, is related to the possible existence of a glassy LC state (Fig. 4.13) [10].

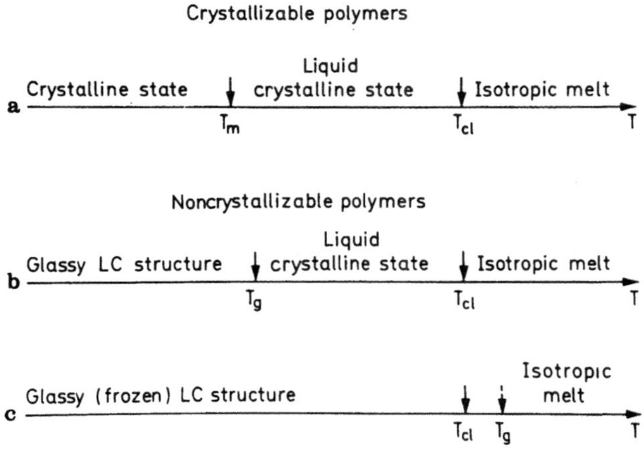

Fig. 4.13. Relationship between the glass transition temperature T_g, melting point T_m, and clearing point T_{cl} for LC polymers.

In the case of crystallizable polymers, primarily polymers with mesogenic groups in the main chains and some polymers with mesogenic side groups, the liquid-crystalline state exists above the melting point (T_m), and up to the clearing point (T_{cl}) the polymer has properties of anisotropy and flow, that is, it behaves like low-molecular-weight liquid crystals which have a significantly higher viscosity (Fig. 4.11 and Fig. 4.13a).

A different situation is observed for noncrystallizable polymers, which include the overwhelming majority of polymers with mesogenic side groups (Fig. 4.13b). The glass transition temperature T_g (and not the melting point as in crystallizable polymers), above which so-called segmental mobility appears due to the lability of the individual segments of the macromolecules, is the lower temperature boundary of their LC state. This temperature is usually below the clearing point, and in the T_g-T_{cl} region the polymer is in the LC state in the form of an elastomer or viscous melt.

In contrast to low-molecular-weight liquid crystals which primarily crystallize on cooling, in most cases polymers with mesogenic groups enter the glassy (solid) state on cooling with preservation of the LC structure corresponding to a certain mesophase: smectic, nematic, or cholesteric (Fig. 4.11, left part of the diagram). This is one of the

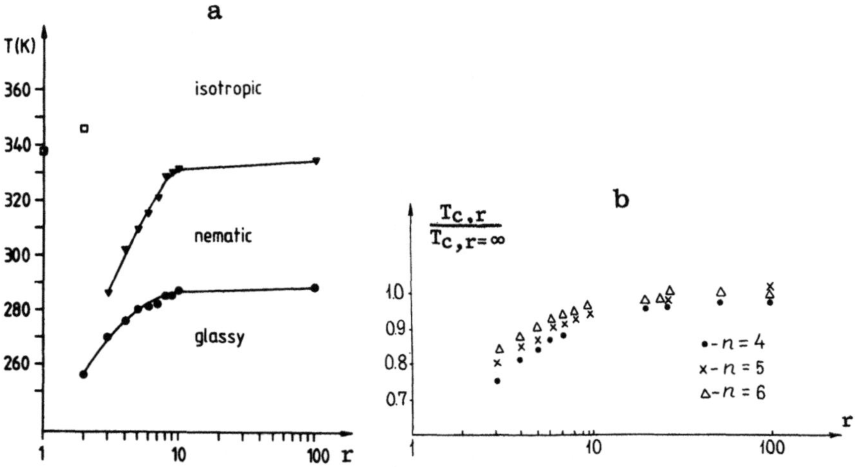

Fig. 4.14. a) Dependence of the phase transition temperatures T_m, T_g, and T_{cl} of monodisperse oligomers on the degree of polymerization r (□ : transition from the crystalline to the isotropic phase); b) normalized phase transition temperatures of an oligomer and a polymer versus r ($T_{c,r}$ is the clearing point of the oligomer; $T_{c,r=\infty}$ is the extrapolated T_{cl} of the polymer with $r = \infty$ [189].) n = 3 (a); 4, 5, 6 (b).

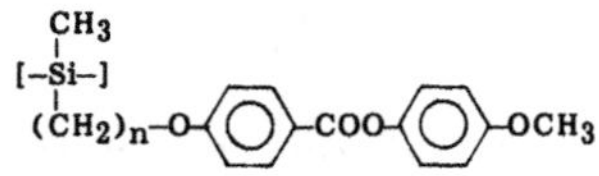

most interesting properties of liquid-crystalline polymers; by using the anisotropy of the "flowing" liquid crystalline phase and cooling the polymer below T_g, it is possible to vitrify and fix the liquid-crystalline structure in a solid (material) with its characteristic anisotropy of mechanical, optical, electrical, and other properties (see Section 4.8). If a polymer does not soften when heated below T_{cl}, then its hypothetical T_g is essentially higher than T_{cl} and the polymer is in the glassy ("frozen") state with a liquid-crystalline structure over the entire temperature range (Fig. 4.13c). Most LC polymers containing mesogenic groups directly attached to the main chain are glassy LC compounds.

The possibility of a mesophase existing in a highly elastic state [173-175] characteristic of macromolecular

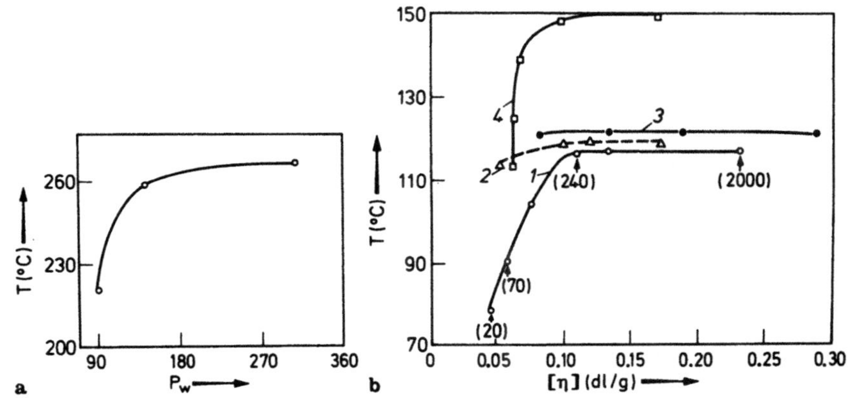

Fig. 4.15. Clearing points of poly(p-biphenyl acrylate) [171] (a) and cyanobiphenyl-containing polymers

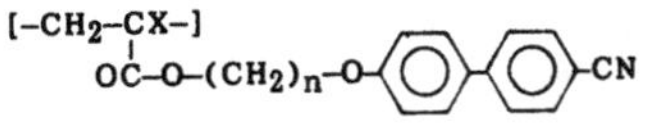

(b) as a function of the degree of polymerization P_w and intrinsic viscosity [η] of solutions in 1,2-dichloroethane [181]: 1) X = H, n = 5; 2) X = CH_3, n = 5; 3) X = H, n = 11; 4) X = CH_3, n = 11 (in parentheses: P_w for some fractions of polymer 1) [181].

compounds alone, along with their capacity to exhibit large reversible deformations, opens up interesting prospects both for creating new types of liquid-crystalline materials in the form of elastic films and developing a theory of the viscoelastic behavior of these unusual elastomers [188].

Despite the fact that the mesophase in LC polymers is formed by mesogenic groups and the type of mesophase is primarily (although not always!) determined by their chemical structure, the polymer chains also make a significant contribution to the physicochemical behavior of these polymers.

The effect of the degree of polymerization (DP) on the temperature range of the LC state in polymers should be mentioned first [181, 182, 189]. Although an increase in the glass transition temperature and clearing point is relatively well known for oligomers and "limitation" of the values of T_g and T_{cl} is observed regardless of the length of the spacer with low DP ≈ 10 (Fig. 4.14a, b), in the case of macromolecular samples of polymers (Fig. 4.15) this dependence requires detailed study. As a comparison of Figs. 4.14 and

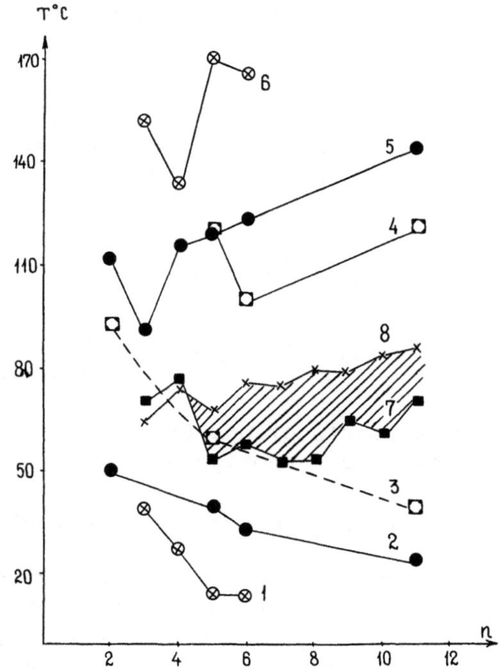

Fig. 4.16. Glass transition temperatures (1-3), clearing points (3-6, 8), and melting points (7) as a function of the number n of carbon atoms in the aliphatic substituent for [182, 191]: 1, 6) LC polysiloxanes 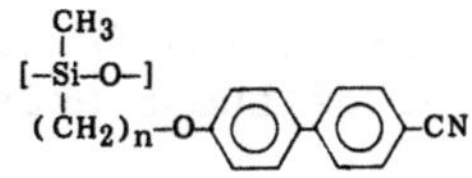 ; 2, 5) LC polyacrylates 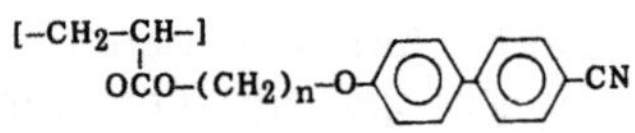 3, 4) LC polymethacrylates 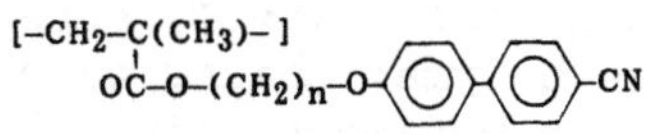 ; 7, 8) alkoxycyanobiphenyls 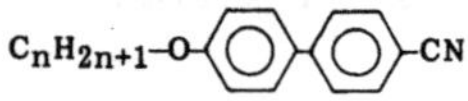

4.15 shows, the critical values of the DP (beginning at which T_{cl} is not dependent on DP) differ significantly. This is an extremely important feature of LC polymers with mesogenic side groups which essentially determines the upper temperature boundary of the LC state and which has not always been taken into consideration in studying LC polymers. At the same time, this molecular characteristic (and probably also the MWD) should always be considered in examining regularities in the properties of different LC polymers and in their comparative characteristics, and in comparing the temperature boundaries of the LC state. Ignoring the differences in the DP can result in serious errors and incorrect conclusions, since fractions of the same polymer at the same temperature will be in different phase states, as Fig. 4.14 clearly shows. The comparative analysis of LC polymers should be conducted using samples with a DP above the critical value, i.e., for samples in which T_{cl} is not dependent on DP.

As the above and a comparison of the phase transition temperatures for monomers and the corresponding polymers shows (Table 4.6), the presence of a polymer chain usually results in a sharply increased thermal stability of the mesophase of LC polymers in comparison to low-molecular-weight liquid crystals. This is clearly seen from a comparison of the transition temperatures of low-molecular-weight alkoxycyanobiphenyls and polymeric cyanobiphenyl derivatives of the acrylic, methacrylic, and siloxane series which have identical mesogenic groups [182] (Fig. 4.16). Figure 4.16 shows that the temperature range of the existence of a mesophase is significantly broader for LC polymers than for their low-molecular-weight LC analogs.

In examining the role of the main chain in determining the properties of LC polymers, the necessity of considering the slowness of all relaxation processes which take place in LC polymers should be especially noted. This "common" property of polymers (from the point of view of polymer chemists) is not always taken into consideration. Inhibition of all structural and relaxation processes results in the appearance of a large number of nonequilibrium states which can frequently be erroneously interpreted as equilibrium states. The transition of a LC polymer into an equilibrium state often requires very long annealing of the polymer in conditions which cause the rapid occurrence of relaxation processes (usually above T_g). The frequently observed, very uncharacteristic optical textures of a series of LC polymers, which produce a pretty optical picture similar to the textures of low-molecular-weight liquid crystals after appropriate heat treatment, can be an example of this (see Section 4.7.1.2). The DSC curves shown in Fig. 4.17a are a graphic illustration of the above [185]. The unannealed sample of

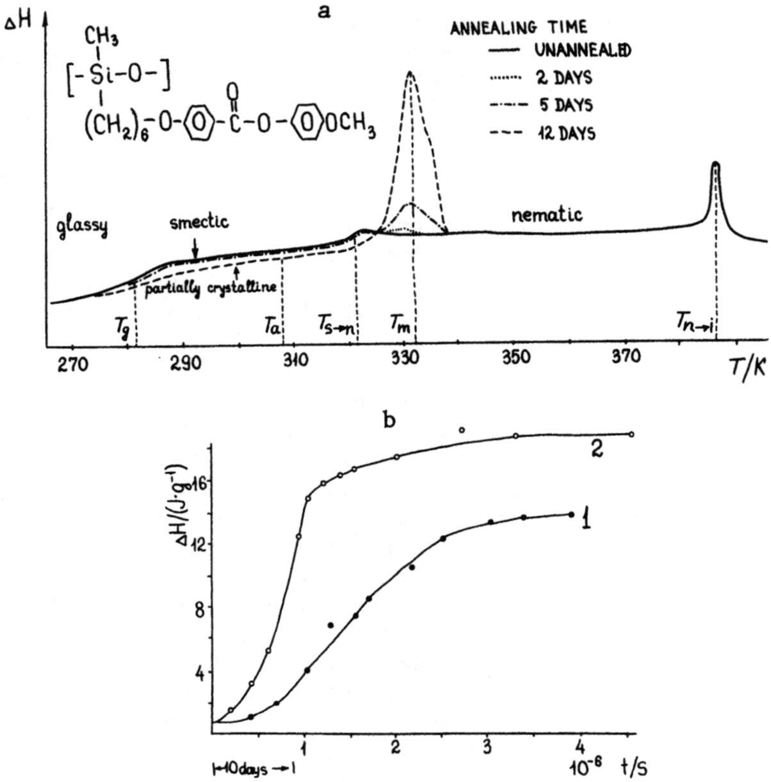

Fig. 4.17. a) DSC curves of a LC polymer before and after annealing at 35°C [185]; b) time dependence of the specific transition enthalpy ΔH at the melting point (T_m = 47°C) for the crystalline → nematic phase transition. Annealing temperatures, °C: 1) 22; 2) 353.

LC polysiloxane (curve 1) does not crystallize and there is a very small endothermic peak at 47°C, which could be interpreted as a S → N phase transition. However, only long annealing of the same sample above T_g (35°C) for several days "revealed" the crystalline character of this polymer, indicated by the data from an x-ray study and the significant value of the melting peak at 57°C. The increase in the heat of fusion ΔH as a function of the annealing time clearly demonstrates the effect of the annealing temperature (Fig. 4.17b) on attainment of equilibrium values of ΔH.

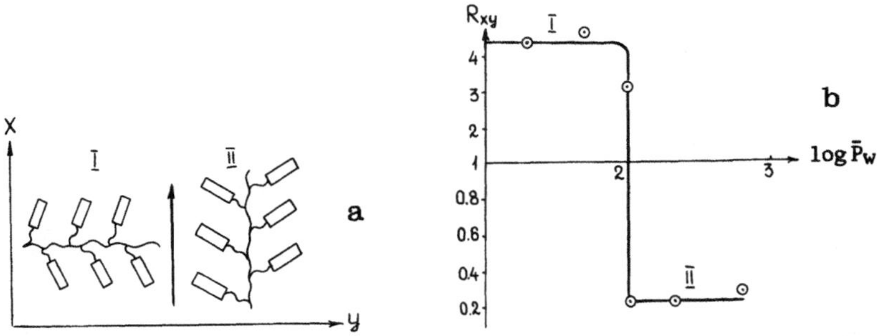

Fig. 4.18. Dependence of the dichroic ratio on the degree of polymerization (a) and schemes of mesogenic group arrangement (b) in oriented samples of a polymer: 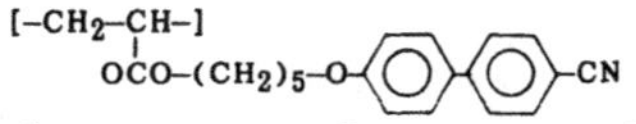 (The arrow shows the direction of orientation) [192].

The role of the length, i.e., the DP of the main chain, also appears in the study of the process of orientation of LC polymers. As established in [192], two opposite methods of orientation of the mesogenic groups as a function of the DP are possible. Using IR spectroscopy, the dichroic ratio R_{xy} for the characteristic absorption bands in the orientation of a series of fractions of a LC polymer in a mechanical field was calculated. As Fig. 4.18 shows, with degrees of polymerization below 200, the mesogenic groups are oriented in the direction of the effect of the force, i.e., like the molecules of low-molecular-weight liquid crystals. Exceeding some critical value of DP results in the opposite orientation of the mesogenic groups. However, there is also a fraction of the polymer (with DP $\approx$ 100) for which orientation of the first and second types can be realized. The different orientations of the mesogenic groups are apparently determined by the ratio of the rates of rearrangement of the mesogenic groups and main chains under the effect of a mechanical field which acts on the different structural elements of the polymer as a function of the DP; in the first case, the LC domains composed of mesogenic groups are oriented as a whole and, in the second case, the main chains of the macromolecules are oriented, while the mesogenic side groups are arranged perpendicular to the axis of the main chain. This feature is apparently inherent in not only the polymers examined here, but most LC polymers with different DP.

TABLE 4.7. Characteristics of Two Samples of Polymethacrylates Prepared in Conditions of Radical (Sample R) and Anionic (Sample A) Polymerization [190]

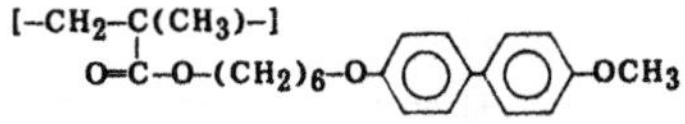

Properties	Sample R	Sample A
Tacticity, %		
iso-	2	68
syndio-	63	25
hetero-	35	7
Transition temperature, °C		
crystal—liquid crystal	117	131
liquid crystal—isotropic melt	127-131	135
Heat of transition		
liquid crystal—isotropic melt, J/g	7.4	8.6
Entropy of liquid crystal—isotropic melt transition, J/g	0.018	0.021

The role of the configuration of the main chain in the formation of the LC phase is similar to that of comb-shaped polymers. Since the LC phase is formed by mesogenic groups, the difference in the configurational structure of the macromolecules has little effect on the properties of the mesophase. The comparison of the properties of two samples of LC polymers with the general formula

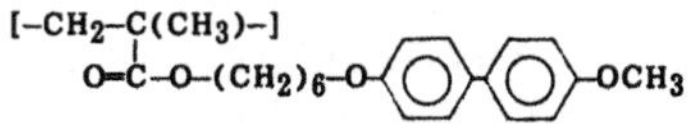

prepared under conditions of radical (sample R) and anionic polymerization (sample A), and which consequently have a different microstructure, did not reveal any significant difference in the phase transition temperatures and heats of transition (Table 4.7) [190].

In examining the role of the main chain in mesophase formation, it is impossible to ignore the significant effect of the flexibility of the macromolecule on the temperature range of the existence of the LC state. The glass transition temperatures and clearing points for two series of polymers with the same mesogenic groups, the same spacer group length, but different flexibility of the main chain are reported in Table 4.8. An increase in the clearing point is

TABLE 4.8. Effect of the Chain and Spacer Flexibility on the Phase State of Polymers with Identical Mesogenic Groups R = 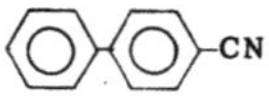

Main chain	Side group	Phase transition temperatures		Reference
$[-CH_2-C(CH_3)-]$, $O=C-O-$	$-(CH_2)_2-O-(CH_2)_2-OR$	T_g = 65°, no mesophase		[213]
	$-(CH_2)_5-OR$	T_g = 65°	S $\xrightarrow{120°}$ I	
$[-CH_2-CH-]$, $OC-O-$	$-(CH_2)_2-O-(CH_2)_2-OR$	T_g = 35°, no mesophase		[213]
	$-(CH_2)_5-OR$	T_g = 40°	N_{cyb}* $\xrightarrow{120°}$ I	
$[-Si(CH_3)-O-]_{36}$	$-(CH_2)_3-O-(CH_2)_2-OR$	T_g = 17°	N_{cyb} $\xrightarrow{60°}$ I	[214]
	$-(CH_2)_5-OR$	T_g = 14°	S $\xrightarrow{166°}$ I	
$[-Si(CH_3)-O-]_{50}$	$-(CH_2)_3-OR$	T_g = 40°	S $\xrightarrow{152°}$ I	[191]
	$-(CH_2)_4$	T_g = 12°	S $\xrightarrow{130°}$ I	

*N_{cyb}: Cybotactic nematic phase.

observed with an increase in the flexibility of the polymer backbone progressing from polymethacrylates to polysiloxanes (indicated by the decrease in the values of the glass transition temperatures) for the first group of polymers (with hydroxycyanobiphenyl mesogenic groups), while the directly opposite effect is observed for the second group of polymers. The interpretation of these data is unfortunately difficult, since the polymers in these two groups form different types of mesophases: polymers with cyanobiphenyl groups form a smectic mesophase, and polymers with methoxyphenylbenzoate groups form a nematic mesophase (although this difference in the mesophase could determine the character of the change in T_{cl}).

The role of the flexibility of the main chain in the formation of the LC state and its effect on the type of mesophase and clearing point have not yet been definitively determined. Systematic studies on the synthesis of LC polymers with varying main chain structures holding all other molecular units of the polymer chain constant, as well as a quantitative evaluation of the flexibility of the macromolecules, must be conducted. At this point, detailed studies have only been conducted on polymers of the acrylic, methacrylic, and siloxane series, which significantly complicates the determination of the general mechanisms.

The features of LC polymers examined above definitely do not exhaust their entire variety. We have only examined the most typical aspects of the macromolecular nature of LC polymers. The determination of mechanisms of their physicochemical behavior is complicated by the mutual effect on that behavior of the individual structural elements of the main chain, linkage, and mesogenic groups. At the same time, it is necessary to establish some correlation between the molecular structure of thermotropic liquid-crystalline polymers (both comb-shaped and linear) and the character of their orientational ordering so that a general theory of the LC state in polymer systems may be formulated. Some approaches to constructing a very simple statistical theory of nematic LC ordering in melts of linear and branched polymers and some results are examined in the next section.

4.6. THEORY OF THE LIQUID-CRYSTALLINE ORDERING OF MELTS OF LINEAR AND BRANCHED MACROMOLECULES WITH MESOGENIC GROUPS IN THE MAIN AND SIDE CHAINS

The problem of predicting the thermodynamic boundary conditions of the formation and structure of mesophases based on the structure of liquid-crystal molecules has not been solved for low-molecular-weight liquid crystals, despite its importance, and investigators only use some empirical rules for this purpose [27, 32, 67]. In the case of LC polymers, the situation is complicated by both the significantly smaller number of systems studied and the more complex structure of the objects themselves. However, approaches to solving such problems as predicting the thermodynamic boundary conditions of the formation of mesophases, the effect of external fields on LC ordering, the structural organization of polymeric thermotropic mesophases, and others are beginning

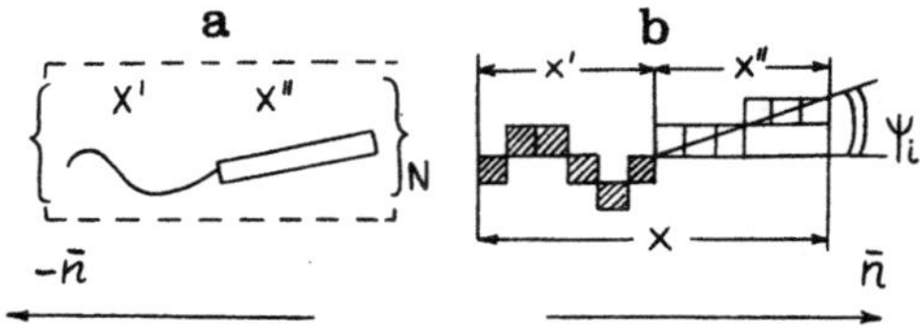

Fig. 4.19. Repeating unit of the macromolecules (a) and its arrangement in the lattice (b) [132, 197].

to be actively developed with respect to linear polymers [87, 88, 81, 93, 130-132, 193-196]. Since linear polymers are undoubtedly simpler objects than comb-shaped polymers, we will first examine the basic approach to constructing a theory of nematic ordering and the basic conclusions of this theory applied to linear macromolecules consisting of alternating mesogenic (rigid) and flexible segments, and we will then apply these approaches to comb-shaped polymers.

The model of a linear macromolecule constructed of N repeating units containing rigid (mesogenic) and flexible sections is shown in Fig. 4.19a. The rigid unit is approximated by a rod, whose size is given in terms of the ratio of its length L to its diameter d as x" = L/d, and the flexible spacer is composed of x' monomeric units connected so that each unit can be in several discrete states with respect to the preceding unit with probabilities of the same order of magnitude [130-131, 197].

When discussing the statistics of polymer chains arranged in a spatial lattice (Fig. 4.19b), the term "trans" is arbitrarily used for two sequentially bound units positioned in the same direction, and the term "gauche" is used for two sequentially bound units rotated at a right angle to each other. Since the energy difference of the trans and gauche states of a unit is equal to ε, the flexibility of the flexible chain in the absence of orientational ordering is characterized by the value f_0, i.e., the fraction of gauche units [86]:

$$f_0 = \frac{(z-2)\exp(-\varepsilon/kT)}{1+(z-2)\exp(-\varepsilon/kT)} \tag{4.1}$$

where z is the lattice coordination number, k is the Boltzmann constant, and T is the temperature. When ε = 0, f =

0.8 (maximally flexible chain). A decrease in the value of f_0 corresponds to an increase in ε, i.e., the rigidity of the chain.

The LC transition in polymers containing rigid rods in the main chain, free linkages between rods, and flexible segments was examined in [130, 198] based on the well-known lattice method [86]. One of the basic assumptions used in [130, 198] was the hypothesis that a change only takes place in the properties of the rigid segments in an isotropic medium formed by the flexible component in the formation of the LC phase. However, as indicated in [131, 132], this hypothesis is not the only possible one. The free energy of an anisotropic athermal melt ΔF_0 of macromolecules of this type is composed of the following factors: the contribution of the rods due to the entropy of the orientational ordering of an ideal gas F_{or}; the conformational entropy of the flexible segments F_{conf}; the translational entropy F_{trans} of the macromolecules formed by the linkage of the individual flexible and rigid units; and the contribution from the steric interaction of oriented particles F_{st}:

$$\Delta F_0 = F_{or} + F_{conf} + F_{trans} + F_{st} \tag{4.2}$$

It is clear that the self-consistent orienting field F_{st} which arises when orientational order is established not only acts on the rigid segments through F_{or} but also on the flexible segments by straightening them, i.e., by altering their configurational properties and, subsequently, the form of F_{conf}. As a consequence, the description of the state of these segments should not only include the difference between the "trans" and "gauche" form of each unit but also the difference in the orientations of the unit with respect to the axis of ordering. A number of experimental data from [122, 198] also indicate ordering of the flexible segments in the formation of the LC phase [122, 198].

Vasilenko, Kokhlov, and Shibaev [131-132, 197] developed a method which is a generalization of the lattice method [86] and accounts for the additional partial ordering of the flexible units of the polymer chain in the formation of the LC phase.

The phase diagrams calculated in [132] for the LC transition in a melt of chains of the type examined for different flexibility values $f_0 = 0.8$ (Fig. 4.20a), $f_0 = 0.7$ (Fig.

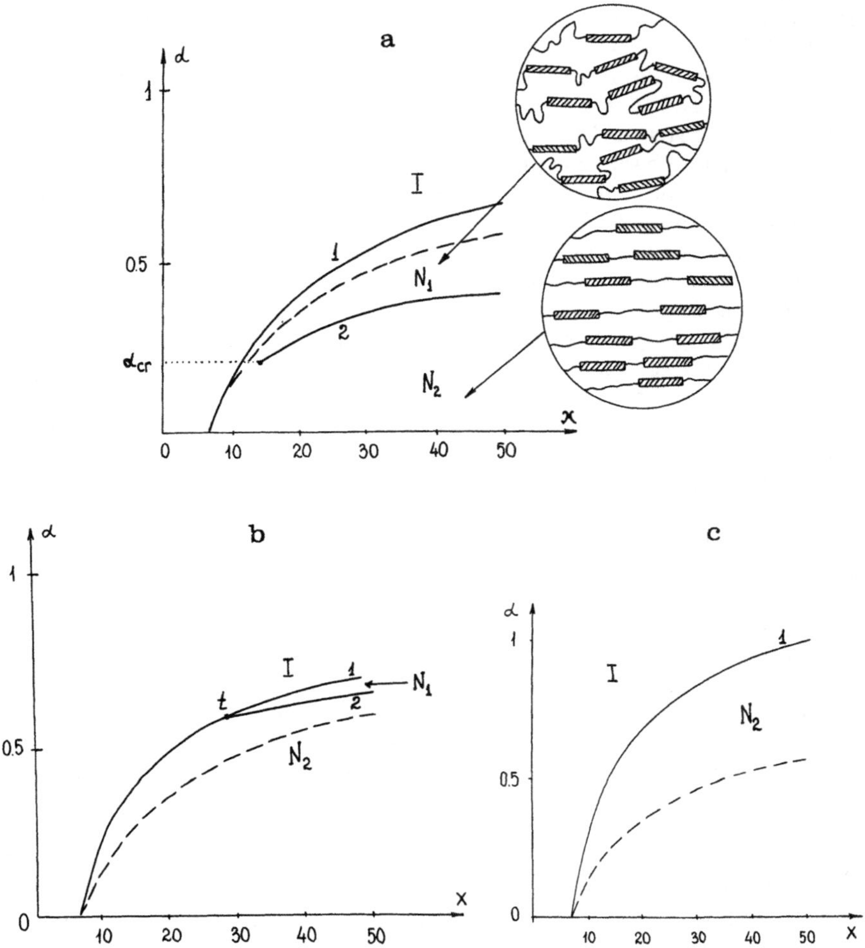

Fig. 4.20. Phase diagrams for an athermal melt of macromolecules with rigid and flexible sections in the main chain and induced ordering of the flexible segments [132] with f_0 = 0.8 (a), 0.7 (b), and 0.6 (c). The curve for the isotropic–nematic transition derived in the Matheson–Flory approximation without consideration of the induced ordering of the flexible segments is indicated by the dashed line. I) Isotropic phase; N_1 and N_2 are, respectively, weakly and highly ordered phases.

4.20b), and f_0 = 0.6 (Fig. 4.20c) are shown in Fig. 4.20. The molecular parameters of the model α and X are used as the variables for the phase diagrams, where d = X'/(X' + X") characterizes the relative concentration of the flexible component,* and X = X' + X" corresponds to the total length of the repeating unit. These results are compared with the corresponding phase diagram plotted on the assumption of interaction of the rigid sections alone [198]. This diagram is indicated by the broken line in Fig. 4.20. Let us characterize its basic features: the nematic LC phase corresponds to the absolute minimum of the free energy in the region to the right of the curve; within the limit of $X \to \infty$, $d \to 1$ and is proportional to 1/X", i.e., within this limit, the LC transition takes place in approximately the same way as in a dilute athermal solution of long rods; parameter α plays the role of the concentration of the effective solvent. In the case examined, the phase diagram is not dependent on either N or f_0.

An examination of Fig. 4.20 primarily shows that the LC transition takes place under conditions which are essentially dependent on the value of parameter f_0, as this should correspond to the physical sense of the problem. If the ordering of the flexible fragments is considered, the region of stability of the anisotropic phase in the phase diagram can be significantly extended (equilibrium curves 1 of the isotropic and anisotropic phases are shifted to the left with respect to the dashed curve), which becomes more important for lower values of f_0.

Generally speaking, the presence of a flexible component in the chain in macromolecules of this type can produce two opposite tendencies. With small α, the flexible sections are so strongly oriented in the LC transition that they play the role of a chain "stiffener" instead of an effective solvent, which increases the degree of asymmetry of the macromolecules in the anisotropic phase. On the other hand, with large α, the flexible component of the chain is only weakly ordered in the LC transition and plays the more natural role

*We note that, according to the model of the macromolecule used in [132], when α = 0, the macromolecule is not transformed into a stiff rod of length N·X" but into a sequence of N flexibly bound rods of length X". At the same time, when α = 1, this is a single flexible chain of length N·X'.

of an effective solvent, that of a plasticizer, in the anisotropic phase. The first tendency causes the formation of a strongly anisotropic phase, and the second tendency makes the existence of a weakly anisotropic phase more advantageous. The presence of both tendencies results in the possible existence of an additional transition between these two phases (curve 2 in Fig. 4.20 corresponds to this phase transition). A sufficient flexibility in the flexible component f_0 is a condition for realization of this possibility. With large f_0 ($f_0 > 0.745$), regions of weakly anisotropic and strongly anisotropic phases are always present in the phase diagram, and curve 2 originates at the critical point (Fig. 4.20a). With a decrease in f_0, the region of stability of the weakly anisotropic phase is reduced and is shifted toward larger α, and beginning with $f_0 = 0.745$, the critical point in the phase diagram is transformed into the triple point t (Fig. 4.20b). Finally, when the value of f_0 becomes less than the critical value of $(f_0)_{cr} = 0.63$, the region of the weakly anisotropic phase disappears (Fig. 4.20c). The value of $(f_0)_{cr}$ corresponds to the value of the flexibility at which LC ordering should be observed in a melt of homogeneous flexible chains (in our case, when $\alpha = 1$). With values of $f_0 < 0.63$, both rigid and flexible segments are very strongly ordered in the entire range of changes in α. Finally, with small f_0 (for example, $f_0 \sim 0.4$), the phase diagram almost becomes independent of α. In other words, when $f_0 > 0.63$, the flexible component of the chain always plays the role of a stiffener. At the same time, when $f_0 > 0.63$, additional improvement in the orientational order due to the phase transition between two anisotropic phases is possible within the LC phase.

The physical meaning of these somewhat unexpected results, which could not be obtained within the framework of the Matheson–Flory approximation [198], is illustrated in Fig. 4.21, which shows the dependence of the order parameters of rigid S^r and flexible S^f segments* on α when $f_0 = 0.8$ [132]. Note that the order parameter of the rigid component remains relatively high in a wide range of changes in α, while the order parameter of the flexible component sharply increases above α_{cr} due to the induced increase in the rigidity. The

*Order parameter S^r is determined by the ratio $S^r = \frac{1}{2}\{3(x'')^2 \times [(x'')^2 + 2(x'' \langle \sin\varphi_i \rangle)^2]^{-1} - 1\}$ (4.3) and $S^f = (3h - 1)/2$ (4.4), where φ_i is the angle of the i-th rod with the axis of anisotropy; h is the fraction of units parallel to the axis of ordering.

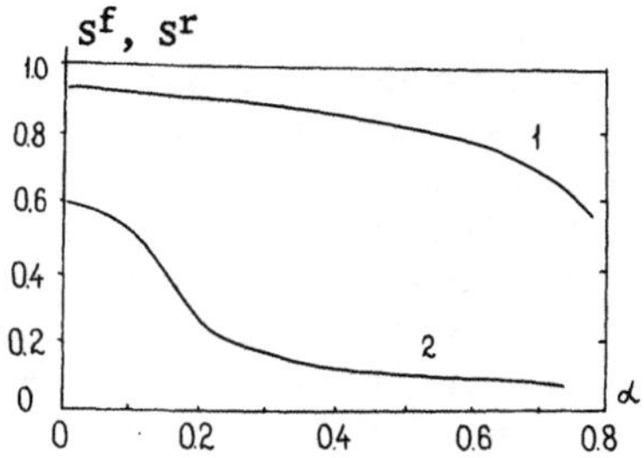

Fig. 4.21. Order parameters S^r (1) and S^f (2) in the liquid-crystalline phase calculated using the curve of the coexistence of the isotropic and anisotropic phases with f_0 = 0.8 [132].

theoretical estimation of the order parameters of flexible and rigid segments for a series of LC polyethers are in good agreement with experimental results [119, 120, 190]. In the development of theoretical studies, the study [200] should be examined in which the observed alternation of the melting points and heats of fusion of a series of liquid-crystalline polymers with even and odd numbers of methylene units in the flexible component [201-202] is substantiated.

In generalizing the results of the theoretical study of LC ordering in linear systems, the approach proposed by Vasilenko, Khokhlov, and Shibaev [203-204] for the description of polymers with mesogenic side groups was applied to comb-shaped LC polymers. A model of this macromolecule is shown in Fig. 4.22. They studied the case of nematic ordering for both a system of rods directly joined to the main chain (x^0 = 0) and with a spacer group of finite length. For both cases, it was shown that the basic difference in the properties of a nematic melt formed by comb-shaped macromolecules in comparison to linear polymers containing mesogenic units is the greater independence of the behavior of the flexible polymeric component from the ordering imposed by the rigid sections. This is primarily demonstrated by the fact that the order parameters of the flexible segments are smaller for comb-shaped chains and the region of stability of the anisotropic phases in the phase diagram is narrower than for the linear analogs. In the case of a flexible main chain (f_0 = 0.8) with directly joined mesogenic groups (x^0 = 0), a new region (hatched in Fig. 4.23a) appears in the phase diagram (Fig. 4.23) corresponding to ordering of

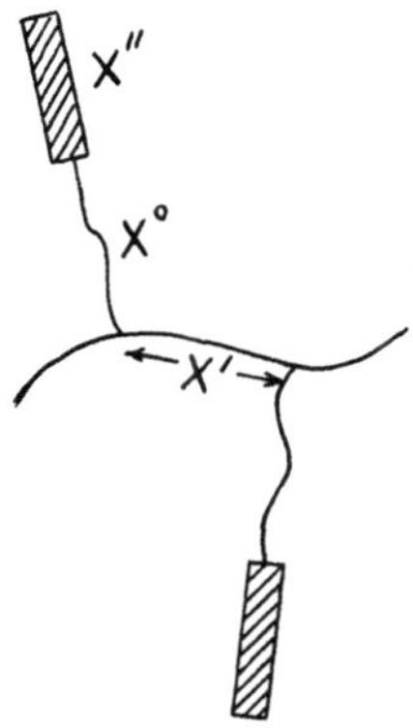

Fig. 4.22. Schematic representation of a comb-shaped macromolecule with mesogenic side groups. x' is the distance (number of monomeric units) between the side branches along the main chain; x'' is the length of the mesogenic group; x^0 is the spacer length.

the mesogenic groups alone (with small α), and the phase transition from a weakly ordered (phase N_1) to a strongly ordered state (phase N_2) takes place with higher values of α and x than in the case of linear polymers (dashed line). In addition, as the main chain is being stiffened, i.e., in the transition from $f_0 = 0.8$ to $f_0 < 0.63$ (Fig. 4.23b), a new phase N_3 appears in the phase diagram; however, curve 1, corresponding to the transition between the isotropic and strongly anisotropic phases, remains, as in the case of linear systems. The new phase N_3 appears with relatively high values of α, i.e., when the mesogenic side groups are significantly distant from each other; in real systems, this case corresponds to copolymers consisting of mesogenic and nonmesogenic comonomers and possessing a rigid main chain. The physical meaning of the formation of an anisotropic phase of this type consists of the fact that when the flexibility of the main chain is low, the effective asymmetry parameter of the main chain is of the same order of magnitude as that of the mesogenic side groups and, in many cases, can even exceed the value of the latter. In this case, the main chain exhibits a more pronounced tendency toward LC ordering than the side chains, and a phase in which the order parameter of the main chain is close to unity and the short rigid side

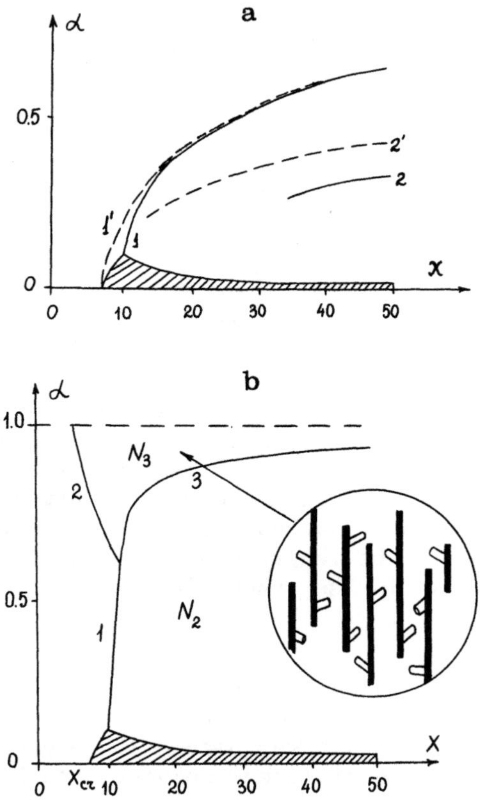

Fig. 4.23. Phase diagrams for a melt of comb-shaped polymers with $f_0 = 0.8$ (a) and $f_0 = 0.4$ (b); $x^0 = 0$. The dashed line corresponds to a linear polymer with alternating rigid and flexible sections in the main chain [204].

chains are disordered appears as a result. The theoretical approach for describing LC ordering in polymer systems proposed in [130-132, 197, 203-204] thus allows for the prediction of the conditions of formation of the LC phase and conclusions on the character and degree of structural ordering of the anisotropic phases based on the structure of linear and comb-shaped polymers (taking into consideration the flexibility of the main chains and spacers and the length and spacing of the mesogenic groups). In the further development of this research, it is necessary to consider the extension of this approach to cases of more complex ordering in the mesogenic segments — in the smectic and cholesteric mesophases — and the forces of interaction between these segments, as well as the study of dynamic processes in LC polymers under the effect of external forces and other questions, some of which are now beginning to be elaborated.

4.7. MECHANISMS OF THE FORMATION AND PROPERTIES OF THE SMECTIC, NEMATIC, AND CHOLESTERIC MESOPHASES OF LC POLYMERS WITH MESOGENIC SIDE GROUPS

The current classification of low-molecular-weight liquid crystals is based on the results of optical and x-ray studies and, to a significant degree, on the study of the miscibility of known (reference) liquid crystals with unknown compounds [206]. In the case of LC polymers, the situation is more complex. Due to the high viscosity of polymeric mesophases, it is not always possible to obtain a characteristic type of optical texture, and for high-molecular-weight LC samples, a very "unexpressive" fine-grained texture (see below) is frequently the most characteristic optical picture. In these cases, a typical "polymeric method" is usually used; the samples of polymers are annealed in the glass transition temperature—clearing point range, which produces more specific optical textures for use in comparison with the optical textures of classic liquid crystals and for classification. Studies involving the use of mixtures of polymeric liquid crystals to identify the polymeric mesophases have only begun and concentrate on polymers with mesogenic groups in both the main and side chains [207-210]. Despite the importance of this type of research, it would be desirable to see the principle of miscibility developed in detail for low-molecular-weight LC substances with similarly sized small molecules. The use of this principle for mixtures of polymers with low-molecular-weight liquid crystals whose molecules differ strongly in size does not yet seem valid. The possible strong difference in the temperature regions of the mesophases of low-molecular-weight and polymeric substances, resulting in a significant narrowing of the concentration range in which these compounds are miscible together, should also be noted. For this reason, the x-ray method of investigating oriented and unoriented samples of LC polymers is the basic method for studying the structure of LC polymers. Certain additional information can be obtained with calorimetry, since the sequence of events in phase transitions during

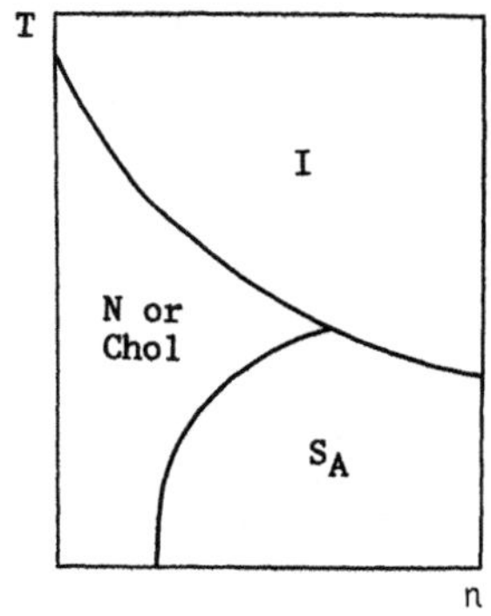

Fig. 4.24. Phase diagram for a homologous series of LC compounds with mesogenic groups terminated by alkyl chains [212].

heating (or cooling) can also provide the necessary information on the proposed structure of LC polymers. Based on the Demus—Sackmann classification of low-molecular-weight liquid crystals [48, 206, 211], we will examine the structure and properties of polymeric smectic, nematic, and cholesteric liquid crystals in consideration of the above comments.

4.7.1. Smectic Mesophases

4.7.1.1. Chemical Structure and Thermal Properties. With respect to structure, smectic mesophases are the most highly ordered type of liquid crystals, similar to solid crystalline compounds. According to the McMillan theory [212], the smectic phase first occurs at a certain length of the alkyl chain as shown in the phase diagram (Fig. 4.24) describing the phase transitions between the isotropic, nematic, and smectic modifications for a homologous series of compounds consisting of rigid mesogenic units with flexible alkyl tails.

In the case of comb-shaped LC polymers, the presence of a long spacer group or an increase in the length of the end alkyl (or alkoxy) group in the mesogenic segment of the side chain also allows for the formation of a smectic mesophase. The smectic phase is also characteristic of LC polymers in which there is generally no spacer and the side groups are directly bound to the main chain, which clearly distinguishes these LC polymers from low-molecular-weight liquid crystals.

TABLE 4.9. Examples of Polysiloxane Copolymers with Different Types of Mesophases:

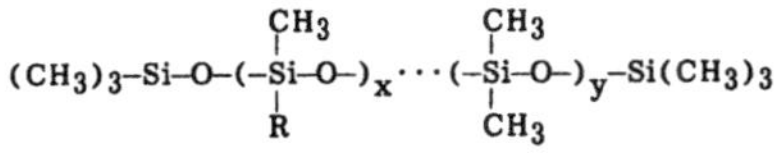

No.	R	x	y	T, °C	Phase transitions, °C	Reference
1	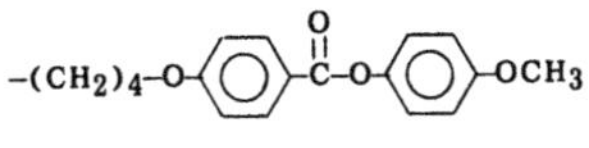	120	0	15	$N \xrightarrow{95} I$	[138]
2		60	60	−6	$S \xrightarrow{50} I$	
3	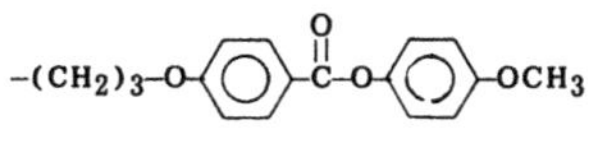	120	0	15	$N \xrightarrow{61} I$	[138]
4		60	60	3	$N \xrightarrow{21} I$	
5	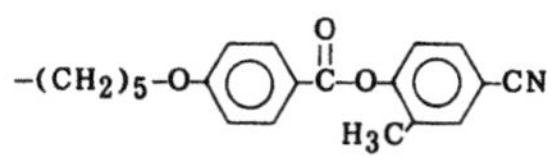	50	0	19	$N \xrightarrow{65} I$	[191]
6		29	21	15	$N \xrightarrow{56.5} I$	

Examples of these LC polymers are reported in Table 4.6 and in Section 4.7.1.2 below. In some cases, the smectic phase is also formed in copolymers with an increased distance between mesogenic groups, although the nematic modification can also be the equilibrium phase for the starting homopolymer (Table 4.9, polymers 1 and 2). In this case, an increase in the distance between mesogenic groups would be equivalent to lengthening the flexible tail according to [138], which causes the formation of the smectic phase (cf. Fig. 4.24). However, it is difficult to say definitely that this mechanism will be valid for any copolymer, since there are cases where preservation of the nematic mesophase has been observed for copolymers of polysiloxanes (Table 4.9, polymers 3-6).

The flexibility of the polymer main chain and the nature and flexibility of the spacer connecting the mesogenic groups with the main chain also play an important role in the formation of any type of mesophase [213-214] (Table 4.8). In comparing polymers 2, 4, and 6, for example, we find that despite the identical mesogenic groups, only polymer 4 forms a

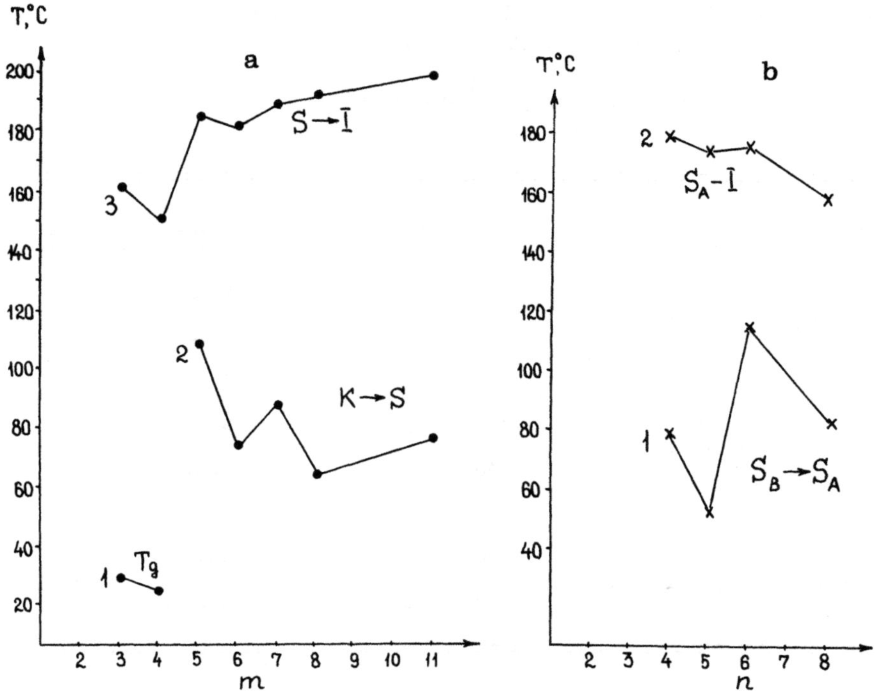

Fig. 4.25. Transition temperatures as a function of m (a) and n (b) for smectic LC polysiloxanes [191]:

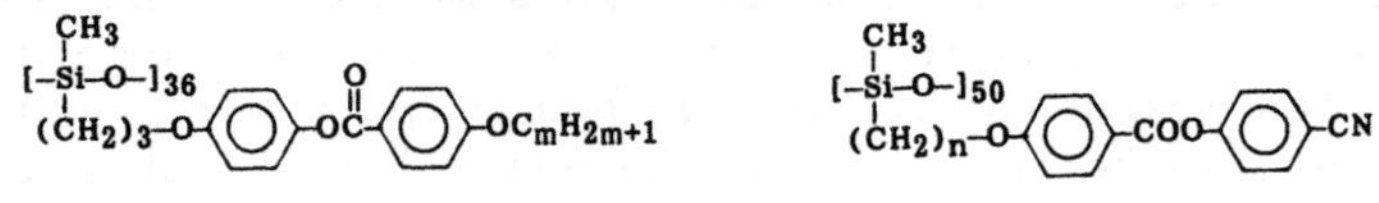

(a) (b)

cybotactic nematic phase,* while polymers with more flexible (polymer 6) and more rigid (polymer 2) main chains form smectic phases. Based on a comparison of polymers 1, 2, 3, and 4, it is possible to conclude that the oxyethylene spacer (polymers 1 and 3) does not favor the formation of a mesophase. A difference in T_g and T_{cl} is also characteristic of polymers 7 and 8, which differ only by the chemical nature of one group in the spacer; this also indicates the important effect of the spacer on the formation of the mesophase, as

*The type of mesophase of polymer 4 is a function of the DP, and an increase in the DP increases the tendency toward layered packing.

in the above example. As Table 4.8 shows, it is still relatively difficult to establish a direct correlation between the type of mesophase and the indicated molecular parameters. Only the accumulation of experimental data will be of use in this respect, and further systematic studies of homologous series of polymers through successive changes in the individual structural units of the macromolecules are required, as in the study of the structure and properties of comb-shaped polymers and copolymers (Chapters 1 and 2).

The dependence of the melting points on the length of the aliphatic linkages (Fig. 4.25a) and end alkyl substituents for two types of LC polysiloxanes (Fig. 4.25b) are shown in Fig. 4.25. Both series of polymers form only the smectic mesophase, even with extremely short spacer groups (m = 3, 4) and alkoxy groups (OCH_3). A comparison of these graphs with each other and with Fig. 4.16 allows one to draw the following conclusions. First, note the increase in the clearing points of the LC phase with an increase in the length of the spacer group (Fig. 4.16, curves 4-6; Fig. 4.25a, curve 3), which differs from the dependence shown in Fig. 4.24 for low-molecular-weight liquid crystals. At the same time, the decrease in T_{cl} with an increase in the length of the end alkoxy substituent in the mesogenic group in the polysiloxane series (Fig. 4.25b, curve 2) apparently correlates well with the right branch of the curve in Fig. 4.24. Another interesting feature of the thermal properties of smectic polymers is the alternation of the melting points and clearing points (Fig. 4.16, curves 5, 6; Fig. 4.24a, b). These so-called "odd—even" effects which are well known for low-molecular-weight liquid crystals are due to a change in the ratio of the trans and gauche isomers in progressing from uneven to even homologs [30, 51], and they are also observed for polymers with mesogenic groups in the main chains. The same mechanisms also apparently act in the case of comb-shaped LC polymers.

4.7.1.2. Structure of Smectic Mesophases. Smectic mesophases are most characteristic of LC polymers with mesogenic side groups. The optical textures of smectic polymers are similar to the confocal texture of low-molecular-weight smectic liquid crystals and consist of a set of birefringent regions with a size of the order of 2-10 μm (so-called fine-grained texture) (Fig. 4.26a). The clearest characteristic textures can be obtained by using prolonged annealing of polymer films near T_{cl} (Fig. 4.26b, c). Figure 4.26a, b

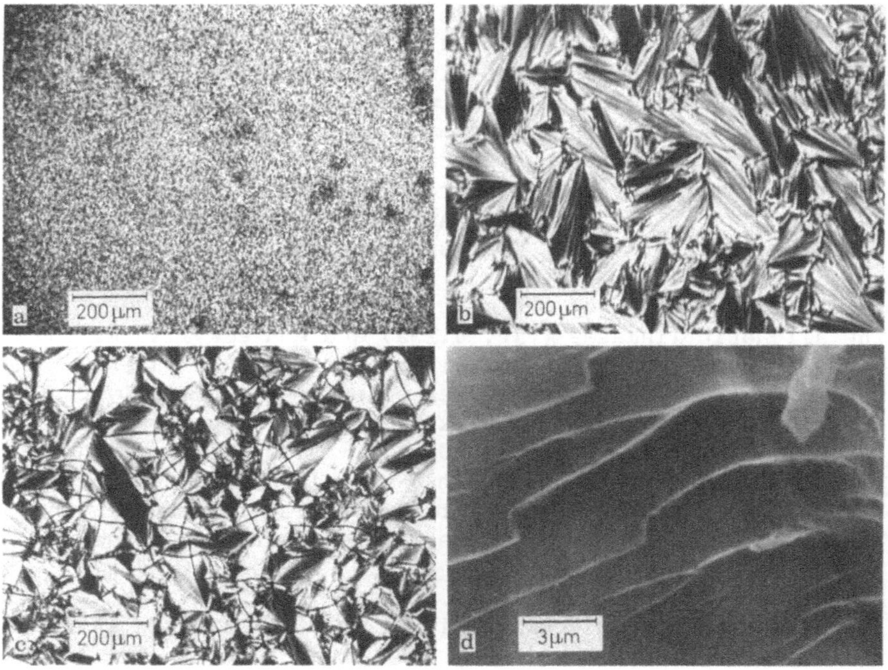

Fig. 4.26. Optical (a-c) and electron microscopic photographs (d) of smectic LC polymers before (a) and after annealing (b-d): 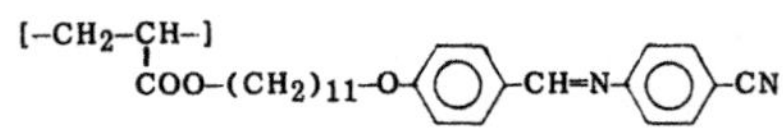 (a, b); 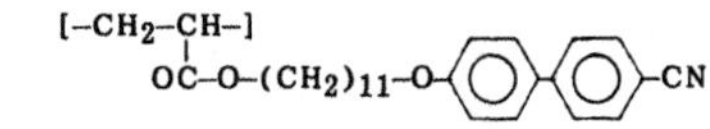 (c); 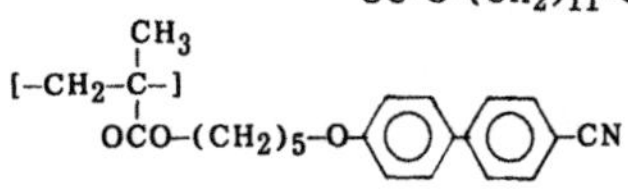 (d).

shows the optical textures of the same smectic polymer before and after annealing; the fan-shaped texture shown in the photographs is most characteristic of smectic liquid-crystalline polymers. The existence of smectic layers in polymer films in the form of distinctly delimited lamellae (sheets) is clearly visible in the photomicrograph made with the scanning electron microscope (Fig. 4.25d) [182].

The number of known smectic modifications is still basically limited to the A, B, and C phases (S_A, S_B, and S_C, respectively), although there is isolated information on the existence of other types of polymeric smectic phases such as

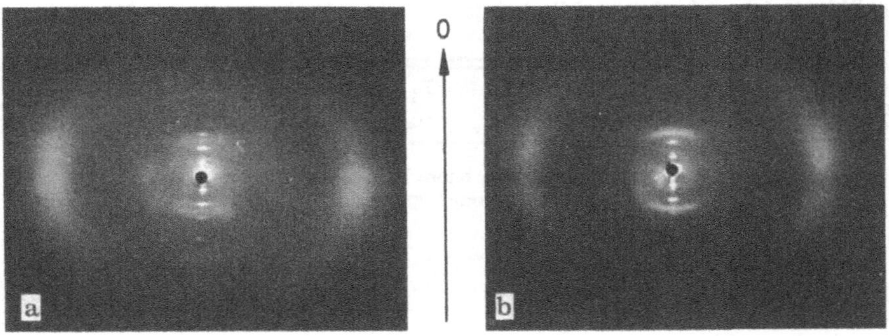

Fig. 4.27. X-ray diffraction patterns of oriented smectic LC polymers in the S_A mesophase [218, 219]: a) polymer 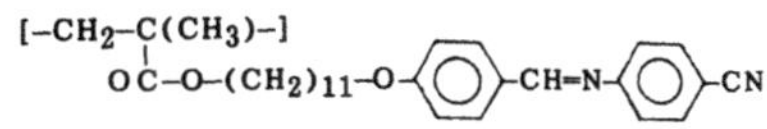 ; b) polymer 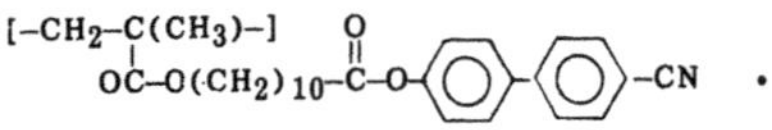 .

E, F, G, and I [163, 217]; at the same time, more than ten polymorphous modifications have already been counted for low-molecular-weight liquid crystals. However, this does not mean that LC polymers are inferior to low-molecular-weight liquid crystals in this respect. This is due to the fact that the identification of polymeric smectic phases based on optical textures is almost impossible except in rare cases due to their similarity (see above for the causes), and the detailed x-ray study of LC polymers has only just begun.

The x-ray patterns of smectic polymers are usually characterized by the presence of a number of distinct layered meridional reflections with small x-ray scattering angles, as clearly seen in the x-ray diffraction patterns in Fig. 4.27, for example [218, 219]. The presence of small-angle reflections indicates the existence of smectic order in the packing of the side chains, primarily arranged normal to the plane of the smectic layer in which the main chains of the macromolecules lie.

Either amorphous scattering corresponding to the presence of short-range order between the side groups of the macromolecules alone ("unordered smectic phase") or one to three relatively narrow reflections indicating the ordered

TABLE 4.10. Smectic LC Polymers with Undetermined Smectic Modifications

No.	Structure of the polymer unit		T_g, °C	T_{cl}, °C	ΔH_{cl}, J/g	Reference
	A. Polymers containing phenyl, diphenylbenzoate, and diphenyl groups					
1.	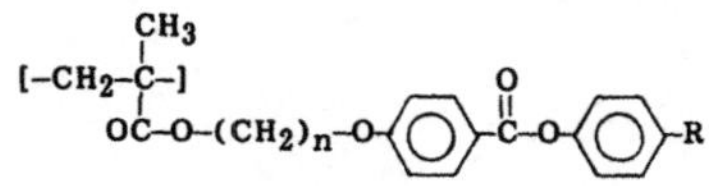					
1.1	n = 2	$R = -C_6H_{13}$	100	156	–	[224]
1.2	n = 2	$R = -OC_3H_7$	120	129	9.2	[224-225]
1.3	n = 2	$R = -OC_6H_{13}$	100	140	11.3	[224-225]
1.4	n = 3	$R = -OC_6H_{13}$	100	120	6.8	[226]
1.5	n = 3	$R = -C_6H_{13}$	100	120	–	[226]
1.6	n = 6	$R = -C_6H_{13}$	60	90	–	[226]
1.7	n = 6	$R = -OC_6H_{13}$	60	115	15.5	[224]
2.	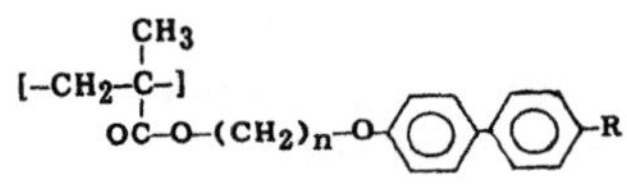					
2.1	n = 6	$R = -OC_5H_{11}$	80	129	16.8	[227]
2.2	n = 6	$R = -OC_6H_{13}$	–	159	22.7	[227]
3.*	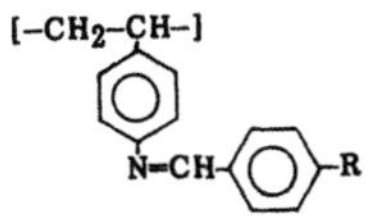					
3.1*		$R = -OC_4H_9$	55	–	–	[161, 228-229]
3.2*		$R = -OC_6H_{13}$	–	–	–	[228]
	B. Cholesterol-containing polymers					
3.	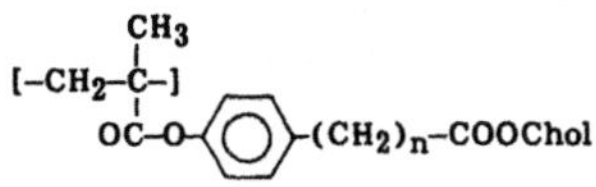					
3.1	n = 2		–	–	–	[225]
3.2	n = 6		100	182	6.7	[225]
3.3	n = 12		100	168	4.2	[225]

TABLE 4.10 (continued)

No.	Structure of the polymer unit	T_g, °C	T_{cl}, °C	ΔH_{cl}, J/g	Reference
4.	$[-CH_2-C(CH_3)-]$ with $OC-O-(CH_2)_n-COOChol$				
4.1	X = CH_3 n = 5	85	190, 210 †	2.1	[230, 231]
4.2	n = 10	60	124, 158 †	0.4, 1.4	[230, 231]
4.3	n = 14	50	54, 153‡	4.2, 6.3	[230, 231]
4.4	X = H n = 5	55	–, 218 †	– 4.2	[230, 231]
4.5	n = 10	35	–, 148 †	– 3.2	[230, 231]
	C. Siloxane polymers				
5.	$[-O-Si(CH_3)-]$ with $(CH_2)_3-R$				
5.1	R = COOChol	45	115	2.7	[136]
5.2	R = 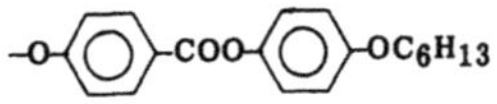	15	85	11.6	[136]
5.3	R = 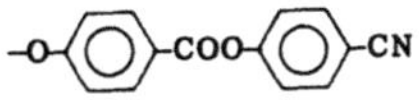	20	61	1.9	[136]

*This type of mesophase was determined as "intermediate" between the S and N phases in [228-229].
†In addition to the smectic phase, the polymer forms a cholesteric mesophase, and T_m** corresponds to melting of the cholesteric mesophase.
‡Two smectic modifications are observed in the polymers.

arrangement of the side chains ("ordered" smectic phase) is observed in the large-angle region.

Unfortunately, despite the relatively large number of LC polymers with mesogenic side groups, the number of purely structural x-ray studies of such systems is still very limited. The smectic type of packing has been attributed to a number of polymers in various studies without indicating to which of the "ordered" or "unordered" smectic phases they be-

TABLE 4.11. LC Polymers with a Hypothetical S_A Structure

No.	Structure of the polymer unit	T_g, °C	T_{cl}, °C	ΔH_{cl}, J/g	Reference
	A. Polymers containing derivatives of Schiff bases				
1.	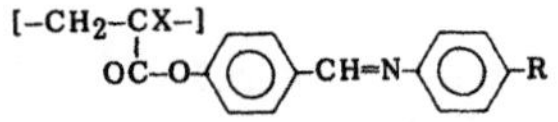				
1.1	X = H R = $-OC_2H_5$	180	310	–	[146]
1.2	X = $-CH_3$ $-OC_2H_5$	–	308	–	[224, 225]
1.3	X = $-CH_3$ R = $-COOH$	124	–	–	[229]
2.	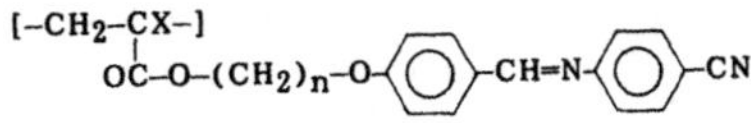				
2.1	X = H n = 11	10	169	8.6	[216]
2.2	X = $-CH_3$ n = 6	35	125	4.2	[216]
2.3	X = $-CH_3$ n = 11	25	155	7.3	[216]
	B. Polymers containing diphenyl groups				
1.	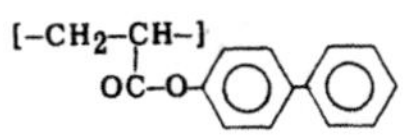	110	285	21.4	[170-172]
2.	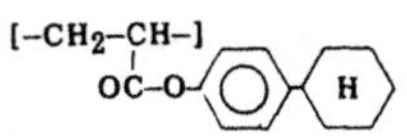	101	205	17.9	[170-172]
3.	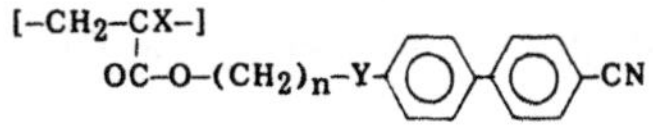				
3.1	X = CH_3, y = $-O-$ n = 5	60	121	6.7	
3.2	X = CH_3, y = $-O-$ n = 11	40	121	8.8	[182, 215]
3.3	X = CH_3, y = $-CH_2-$ n = 10	30	81	7.1	
3.4	X = CH_3, y = $-C(=O)-O-$ n = 10	45	93	4.2	
3.5	X = H, y = $-O-$ n = 11	25	145	6.3	

TABLE 4.11 (continued)

No.	Structure of the polymer unit		Tg, °C	Tcl, °C	ΔHcl, J/g	Reference
4.	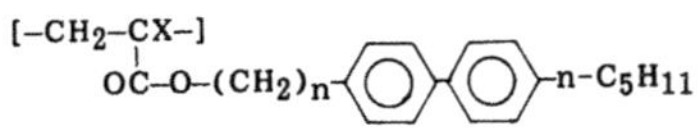					
4.1	X = H	n = 1	67	130		
4.2		n = 2	65	192		[199, 233]
4.3*		n = 3	69	90*, 124		
4.4		n = 4	–	129		
4.5		n = 5	66	93		[191, 233]
4.6*		n = 6	–	102*, 110		
4.7	X = CH_3	n = 2	–	97		
	C. Polymers with substituted phenyl- and diphenylbenzoate groups					
1.	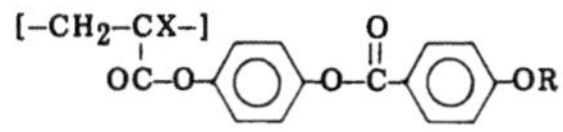					
1.1	X = CH_3	R = $-C_4H_9$	180	270	27.8	
1.2	X = CH_3	$-C_9H_{19}$	140	220	7.2	
1.3	X = CH_3	$-C_{12}H_{25}$	130	215	7.7	[222, 234, 235]
1.4	X = CH_3	$-C_{16}H_{33}$	160	220	8.5	
1.5	X = H	$-C_{16}H_{33}$	100	200	–	
	D. Cholesterol-containing polymers					
1.	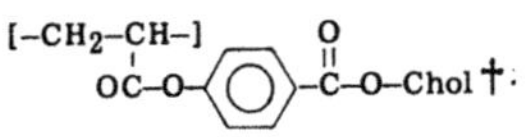		–	–	–	[229]
2.	[$-CH_2-C(CH_3)-$] with side group OC–OChol		160	–	–	[229]
3.	[$-CH_2-CX-$] with side group OC–NH–$(CH_2)_n$–COOChol					
3.1	X = CH_3	n = 2	180	–		
3.2	X = CH_3	n = 5	130	220		
3.3	X = CH_3	n = 6	130	215		[175, 236]
3.4	X = CH_3	n = 8	130	200		
3.5	X = CH_3	n = 10	125	185		

TABLE 4.11 (continued)

No.	Structure of the polymer unit		Tg, °C	Tcl, °C	ΔHcl, J/g	Reference
3.6	X = CH_3	n = 11	120	180	3.3	
3.7	X = H	n = 2	190	250	–	
3.8	X = H	n = 5	165	220	–	[175, 236]
3.9	X = H	n = 11	110	185	–	[237]
	E. Polysiloxane polymers					
1.	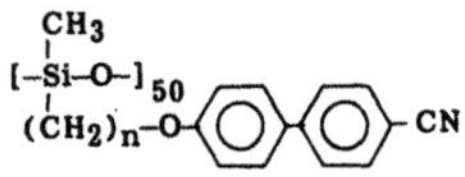					
1.1		n = 3	40	152	4.2	
1.2		n = 4	28	132	3.4	
1.3		n = 5	14	170	5.4	[191]
1.4		n = 6	13	165	5.4	

*Two smectic modifications exist; the second is S_A.
†Chol: Cholesteryl unit.

long. In addition, the results of calorimetric studies alone and data on the miscibility with low-molecular-weight liquid crystals have sometimes been used as the criteria for designating polymers as smectic (and nematic). Some types of smectic polymers for which the actual type of smectic structure has not been determined are reported in Table 4.10 to demonstrate the variety of the chemical structure of LC polymers which form the smectic mesophase. It was found that "unordered" (S_A, S_C) or "ordered" (S_B, S_F, S_E, etc.) smectic phases can be designated, as in the case of low-molecular-weight liquid crystals. The packing of the macromolecules in each of these types of mesophases is examined below.

4.7.1.2a. S_A Mesophases. The overwhelming majority of LC polymers with mesogenic side groups which have been synthesized up to now have an S_A structure (Table 4.11) [1, 220-223].

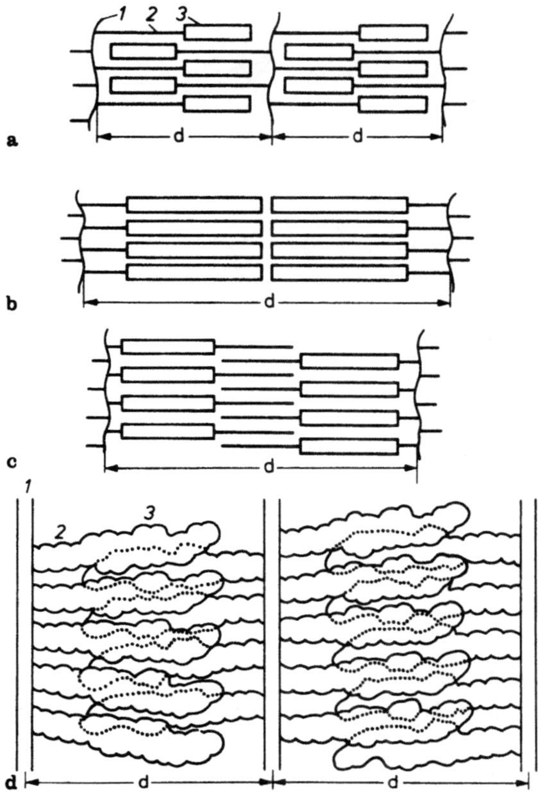

Fig. 4.28. Schematic representation of the side chain packing of macromolecules of oriented LC polymers in the S_A mesophase: a) single-layer packing; b) two-layer packing; c) packing with overlapping alkyl "tails"; d) packing with partial overlapping of mesogenic units (the side chains are indicated by the van der Waals radii). 1) Main chain; 2) spacer; 3) mesogenic group [the side groups of neighboring macromolecules (in cases a, c, and d) are arranged in different planes parallel to the plane of the figure].

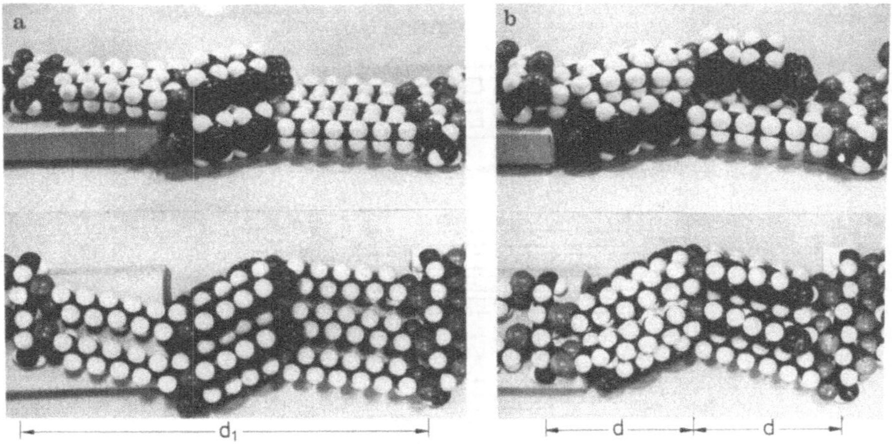

Fig. 4.29. Molecular models of layer packings of macromolecules in the S_A mesophase with d_1 (a) and d_2 (b) spacing for the polymer [182, 219]:

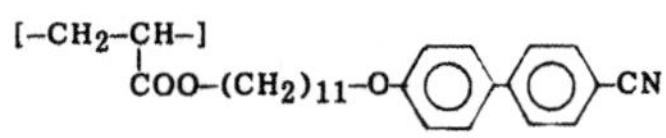

In the S_A modification of the smectic polymers, the side groups of the macromolecules are arranged in layers parallel or antiparallel to each other so that the axes of the side chains are perpendicular to the plane of the smectic layers in which the main chains lie.

The arrangement of the side chains in layers is unordered (diffuse large-angle reflection on the x-ray diagrams) but translational order is retained normal to the plane of the layer. Typical x-ray patterns of S_A smectic polymers are shown in Fig. 4.27. The analysis of the data in the literature shows that the side chains can be arranged in both single-layer and two-layer packing of the mesogenic groups, depending upon the chemical structure of the LC polymers (Fig. 4.28a, b), and with partial overlapping of the side groups (Fig. 4.28c, d) [10, 221-223, 232]. The selection of a packing variant is based on a comparison of the experimental data on estimation of the value of the d spacing (corresponding to the thickness of the smectic layer) with the calculated value of the length of the side chain.

Of the packing variants in the S_A phase for LC polymers with no spacer group or with a short aliphatic bridge, two-layer packing is preferred, where the thickness of the layer is defined by twice the length of the side chain (cf. polymers A.1-A.1.3, B.1, B.2, C.1, etc., in Table 4.11). The single-layer antiparallel packing common in low-molecular-weight smectic liquid crystals is less frequently encountered in polymeric liquid crystals (cf. polymers D.3.1-3.6 in Table 4.11). There are cases of the coexistence of two independent types of packing with d_1 and d_2 spacings, as for some CN-containing polymers, for example (polymers A.2.1, A.2.2, B.3.2-B.3.5 in Table 4.11). Two-layer packing with overlapping aromatic groups is the basic type of structure in these polymers (Fig. 4.29a). The thickness of the smectic layer is approximately one and a half times the length of the side chain d_1. This arrangement of the mesogenic groups is apparently stabilized by the interaction of strong constant dipoles caused by the presence of nitrile groups. Single-layer packing with the complete overlapping of the antiparallel side groups, where they are more closely packed, corresponds to the second type of structure with d_2 spacing (Fig. 4.29b).

4.7.1.2b. S_B, S_E, and S_F Mesophases and Other Types of Structures with Translationally Ordered Groups in Layers. The structure of the S_B mesophase is characterized by arrangement of the side chains in layers where the axes of the side chains are perpendicular to the plane of the layers. A distinctive feature of the S_B mesophase is the greater ordering of the side chains in layers in comparison to the S_A mesophase. The side chains usually form packing of the hexagonal type, which brings this type of packing closest to that of a crystalline compound.

For this reason, we combined two groups of polymers in this section and in Table 4.12: polymers forming an "ordered" smectic phase (S_B, S_E, S_F, etc.) and polymers with mesogenic groups for which the formation of a crystal structure has been proposed. This is due to the fact that it is frequently very difficult to draw a boundary between the crystalline and LC state of ordered smectic phases. The interpretation of the data on the crystalline phases of the series of polymers shown in Table 4.12 is thus also possible for smectic polymorphism.

TABLE 4.12. LC Polymers with Translationally Ordered Side Groups in Layers

No.	Polymer	Tg, °C	Phase transitions	ΔH, J/g	Number and type of large-angle reflections	Hypothetical type of structure †	Reference
1.	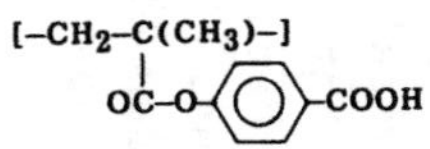	110	–	–	1 narrow	S_B C	[163] [165, 229]
2.	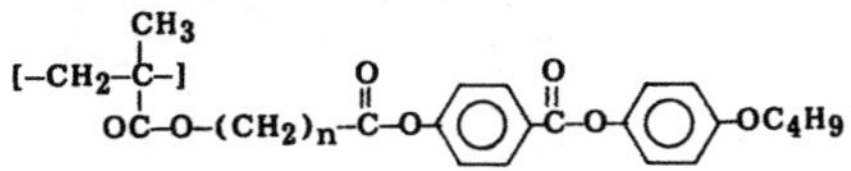	55	$M_1^* \xrightarrow{130} I$	11.3	1 narrow	S_B	[235]
3.	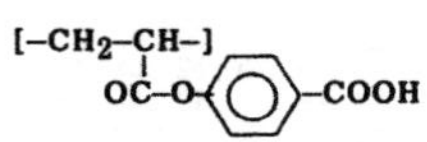	125	$M_1 \xrightarrow{252} I$	–	3 narrow	S_E C	[163] 229]
4.	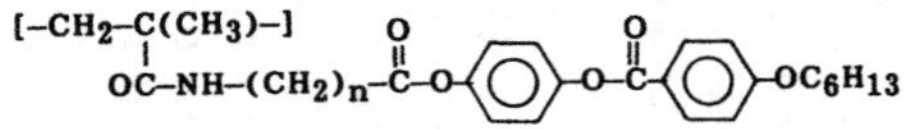						
4.1	n = 5	–	$M_1 \xrightarrow{139} M_2$ $\xrightarrow{183} I$	13.0 1.3	1 sharp 1 diffuse	C S	[238-240]
4.2	n = 11	–	$M_1 \xrightarrow{73} M_2$ $\xrightarrow{160} I$	2.5 8.4	1 sharp 1 diffuse	C S	
5.	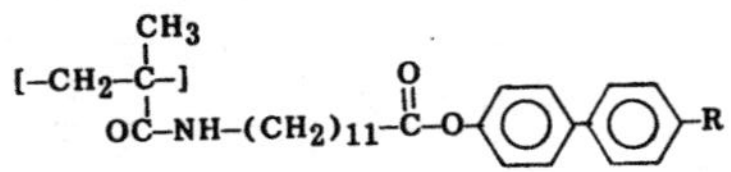						
5.1	R = H	–	$M_1 \xrightarrow{75} I$	6.3	1 sharp	C	[241]
5.2	R = $-OC_6H_{13}$	–	$M_1 \xrightarrow{143} M_2$ $\xrightarrow{160} I$	9.2 4.9	1 narrow	C S	[241]

6.	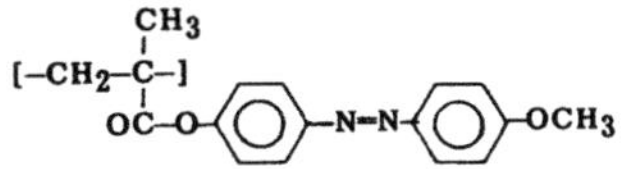	100	$M_1 \xrightarrow{270°} I$	–	4	sharp	C	[242]
7.	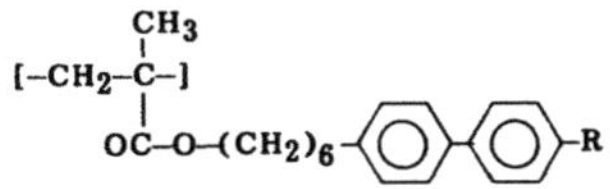		$M_1 \xrightarrow{119°} M_2 \xrightarrow{136°} I$	12.6; 7.1		–	C S	[227]
7.1	R = $-OCH_3$		$M_1 \xrightarrow{117°} N(?) \xrightarrow{131°} I$	3.2 7.4	3	sharp	C N(?)	[190]
7.2	R = $-OC_2H_5$		$M_1 \xrightarrow{134°} M_2 \xrightarrow{162°} I$	10.5; 10.9		– –	C S	[227]
8.	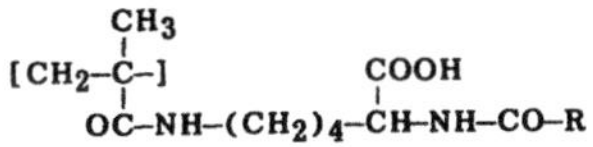							
8.1	R = $H-C_{15}H_{31}$	–	$M_1 \xrightarrow{32°} M_2$	–	3	narrow	C	
8.2	R = $H-C_{17}H_{35}$	–	$M_1 \xrightarrow{34°} M_2$	20.9	3	narrow	C	[167-168]
8.3	R = $H-C_{21}H_{43}$	–	$M_1 \xrightarrow{56°} M_2$	49.0	3	narrow	C	
9.	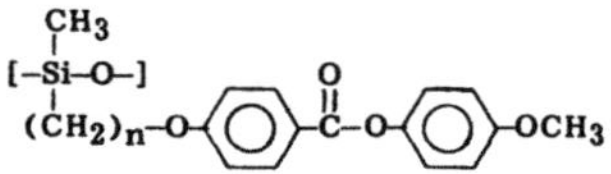							
9.1	n = 5	–	$M_1 \xrightarrow{87°} M_2 \xrightarrow{115°} I$	–		–	C	[139, 243]
9.2	n = 6	–	$M_1 \xrightarrow{52°} M_2 \xrightarrow{112} I$	–		–	C	

TABLE 4.12 (continued)

No.	Polymer	T_g, °C	Phase transitions	ΔH, J/g	Number and type of large-angle reflections		Hypothetical type of structure	Reference
10.	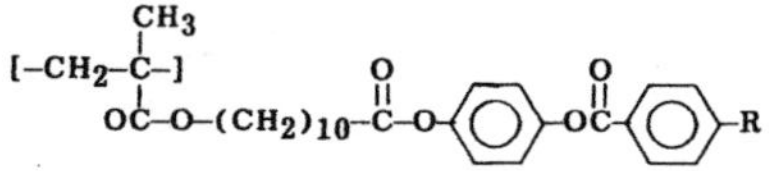							
10.1	R = $-OC_4H_9$	55	$M_1 \xrightarrow{66°} M_2 \xrightarrow{122°} I$	10.6 3.6		sharp diffuse	S_B S_A	[244-245]
10.2	R = $-OC_7H_{15}$	60	$M_1 \xrightarrow{79°} M_2 \xrightarrow{133°} I$	9.7 7.3	1 1	sharp diffuse	S_B S_A	[244-245

*M_1 and M_2 are mesophases with translationally ordered and unordered arrangement of the side chains in layers; I, isotropic phase.
†C = crystalline structure.

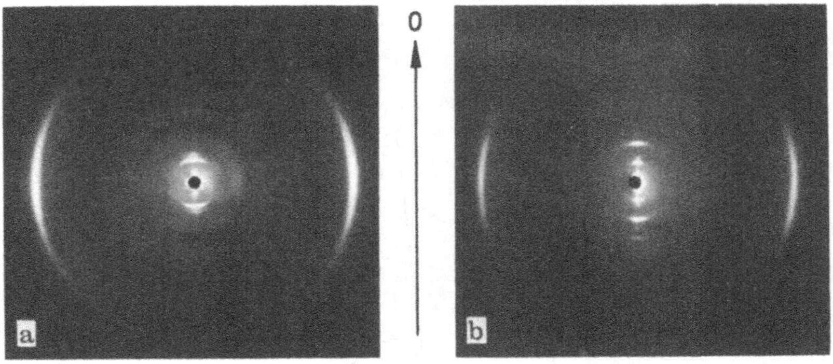

Fig. 4.30. X-ray patterns of oriented smectic LC polymers in the S_B mesophase [216, 218]: a) polymer

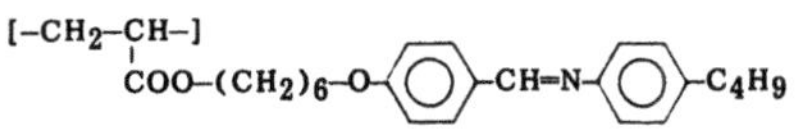

; b) polymer

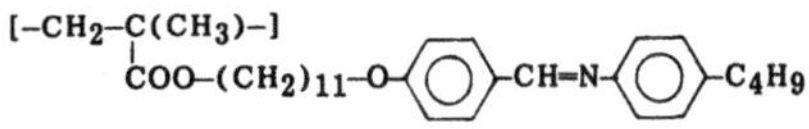

(0 is the texture axis).

As Table 4.12 shows, the number of polymers with S_B mesophases is very limited. In addition to a number of small-angle reflections, the x-ray patterns of polymers in this modification usually exhibit a narrow and relatively intense large-angle reflection (Figs. 4.30 and 4.31). Two packing variants are possible for LC polymers of this type: the perpendicular arrangement of the side groups to the plane of the layer, the true S_B phase (polymer 2 in Table 4.12), or the tilted S_{B_t} phase [163]. According to the classification of low-molecular-weight liquid crystals, the S_I mesophase should be assigned to the latter [48].

The second group includes polymers with three-dimensional ordering of the side chains (for example, forming the M_1 phase) (Table 4.12). The large-angle x-ray diffraction patterns of these polymers are characterized by the presence of 3-4 narrow reflections. This type of structure is usually determined as crystalline or as an ordered smectic liquid crystal structure in the S_E or S_H phases.

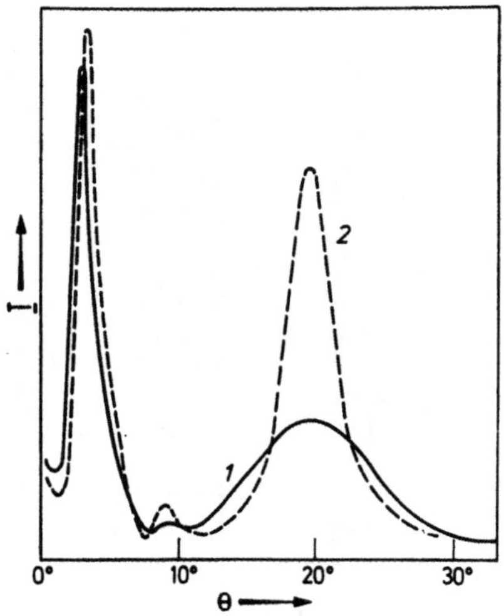

Fig. 4.31. X-ray diffraction curves of LC polymers: 1) polymer 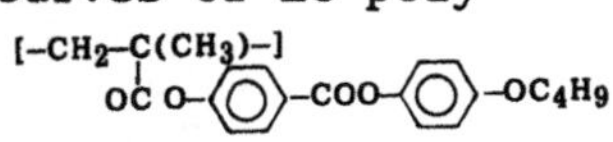 in the S_C mesophase at 160°C [222]; 2) polymer 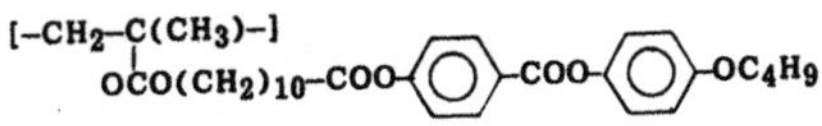 in the S_B mesophase at 100°C [221, 223].

The S_E mesophase found for a polysiloxane polymer possesses the most ordered arrangement of the mesogenic groups [214]:

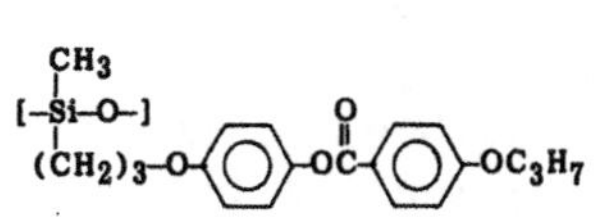

(VII)

$$S_E \xrightarrow[10.7\ J/g]{46°C} I$$

As Fig. 4.32 shows, the appearance of two to three sharp large angle-reflections and a number of orders of reflections of the value of the d spacing is characteristic of this type of packing. Polymer VII is characterized by an orthorhombic cell with an orthogonal arrangement of the mesogenic groups.

The tilted arrangement of the side chains is characteristic of the highly ordered S_G mesophase of polymer VIII pre-

Fig. 4.32. X-ray pattern of oriented polymer VII in the S_E mesophase.

pared by Ya. S. Freidzon (MSU) and studied by us together with V. V. Tsukruk, V. V. Shilov, and Yu. S. Lipatov (Kiev Institute of Macromolecular Compounds, Academy of Sciences of the Ukrainian SSR).

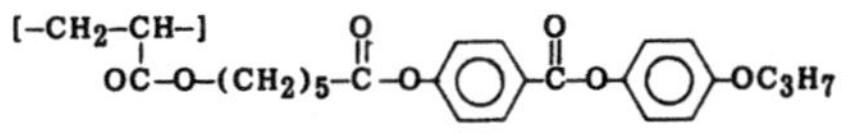

(VIII)

This polymer forms a number of polymorphic modifications, and the transitions between them are shown below:

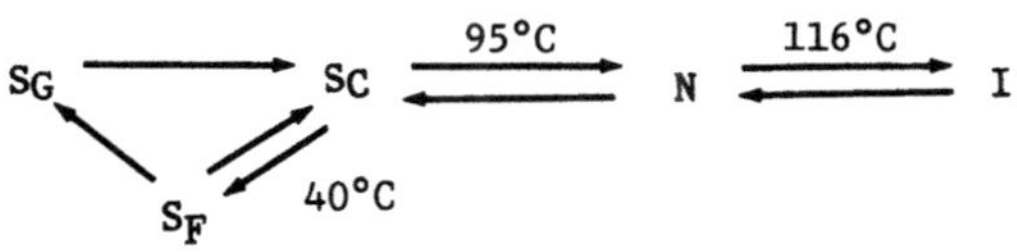

It is important to note that the formation of the S_G phase takes place through the formation of the nematic mesophase and two types of smectic mesophase, S_C and S_F, on cooling from an isotropic melt. Prolonged annealing of the polymer (approximately 100 days) is required for formation of the S_G mesophase, resulting in the appearance of a number of sharp additional maxima corresponding to a monoclinic cell in its large-angle x-ray pattern. The x-ray diffraction

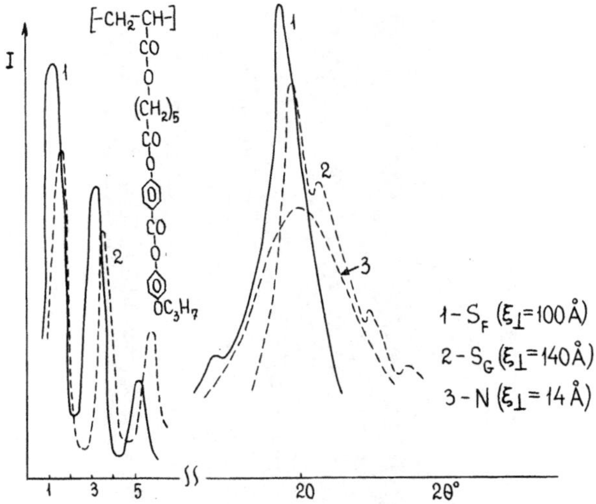

Fig. 4.33. X-ray diffraction patterns of the polymer 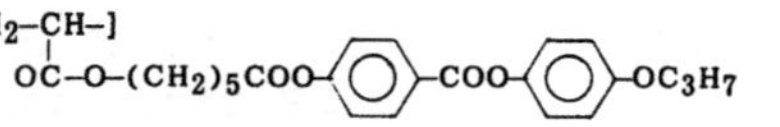 in the S_G (1), S_F (2), and N (3) mesophases.

patterns of polymer VIII in different modifications and the values of the correlation lengths $\xi_\perp$ in the direction perpendicular to the side chains (see Section 4.7.1.3 for the calculation) are shown in Fig. 4.33. Heating the polymer in the S_G phase causes its transition into the tilted unordered S_C phase, and subsequent cooling of the polymer results in the formation of an ordered but metastable S_F phase. Schematic representations of the packing of the macromolecules in the tilted smectic ordered S_F and unordered S_C phases are shown in Fig. 4.34.

The stable tilted S_F phase is characteristic of polymers similar in chemical structure, such as polymer VIII, but in this case differs with respect to the direction of the arrangement of the end ester bond.

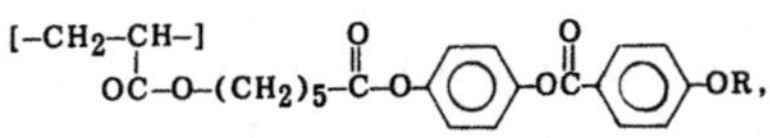

$R = C_3H_7$ $S_F \xrightarrow{60°} N \xrightarrow{120°} I$ (IXa)

$R = C_4H_9$ $S_F \xrightarrow{70°} N \xrightarrow{135°} I$ (IXb)

In this case, the S_F phase changes directly into the nematic mesophase when the polymer is heated.

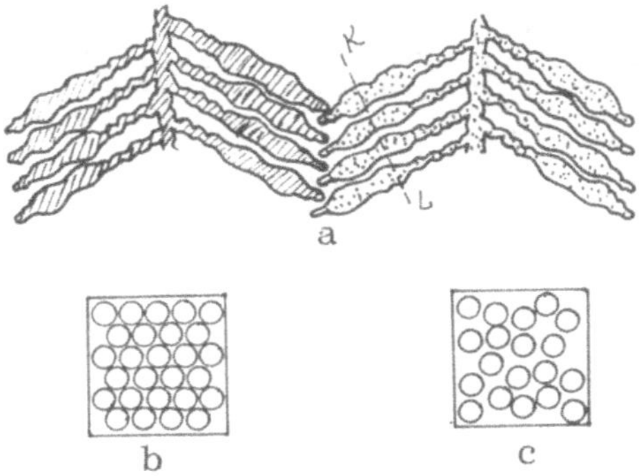

Fig. 4.34. Schematic representations of the side chain packing of macromolecules in the tilted smectic phases. a) The arrangement of macromolecules with mesogenic groups are indicated by van der Waals radii; b, c) arrangement of the mesogenic groups in the S_F (b) and S_C (c) mesophases in plane KL perpendicular to the long axis of the side chains [217].

A comparison of the behavior in orientation of two polymers with a similar structure – VIII and IX – revealed their significant difference. The smectic layers of polymer VIII are arranged parallel to the axis of stretching on orientation, and the mesogenic groups are at some angle φ to the plane of the layer in which the main chains lie.

The heats of transition from the anisotropic state to an isotropic melt are usually small and do not exceed the heats of transition characteristic of low-molecular-weight liquid crystals for the corresponding modifications. The similar structural parameters of ordered smectic and crystalline polymers significantly complicate their precise identification.

As for low-molecular-weight smectic materials, polymorphism is also possible in polymeric smectic mesophases. Examples of polymorphism of the ordered–unordered smectic type $S_B \rightleftarrows S_A$ for polymeric smectic phases of the acrylic and polysiloxane series are shown in Table 4.13.

TABLE 4.13. $S_B \rightleftarrows S_A$ Polymorphism in Liquid-Crystalline Polyacrylates and Polysiloxanes [214, 232]

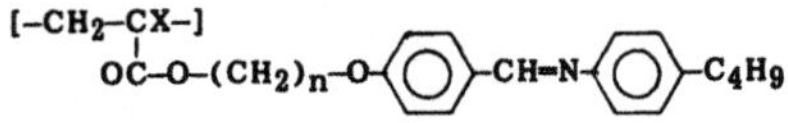

X	n	$S_B \rightleftarrows S_A$ transition: T_{cl}, °C	$S_B \rightleftarrows S_A$ transition: ΔH, J/g	$S_A \rightleftarrows$ I transition: T_{cl}, °C	$S_A \rightleftarrows$ I transition: ΔH, J/g
H	3	70	–	165	2.5
H	6	76	4.2	115	8.4
H	11	90	6.7	149	11.3
CH_3	6	77	3.3	115	7.9
CH_3	11	86	5.0	140	12.5

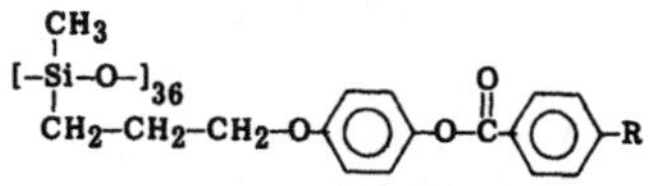

R	T_g, °C	T_{cl}, °C	ΔH, J/g	T_{cl}, °C	ΔH, J/g
$-OC_4H_9$	65	79	7.7	179	5.2
$-OC_5H_{11}$	43	52	1.0	177	6.0
$-OC_6H_{13}$	75	115	2.7	176	10.8
$-OC_8H_{17}$	67	82	3.1	156	10.7

4.7.1.2c. S_C Mesophases. This type of structure with a tilted arrangement of the side groups to the smectic layers is rarely encountered among LC polymers. The existence of the S_C phase was postulated by Hungarian investigators in [209], although no detailed structural data on the packing of the macromolecules were given. The polymers were assigned to this type of mesophase based on the miscibility of cholesterol-containing polymers (polymers 1.1-1.3, Table 4.14) with the S_C phase of a low-molecular-weight liquid crystal, terephthalyl-bis-butylamine. At the same time, we had previously noted that the principle of identification of the LC phases based on the miscibility, which works well for low-molecular-weight liquid crystals, is not always valid for LC polymer—low-molecular-weight liquid-crystal systems. The existence of a tilted phase was proposed [242, 246, 247] for polymers 2.1-2.2 (Table 4.14). X-ray data on the thickness of the smectic layer, the value of which was smaller than the calculated value based on a two-layer S_A structure, were the basis for this hypothesis. However (since oriented samples were not studied), the formation of the S_A mesophase with packing characteristic of poly-

TABLE 4.14. LC Polymers with a Hypothetical S_C Structure

No.	Polymer	T_g, °C	T_{cl}, °C	Type of packing, angle of mesogenic groups in the layer	Reference
1.	[$-CH_2-C(CH_3)-$] with side group $OC-O-CH_2-R$				
1.1	R = $-OOC(CH_2)_2C(=O)-OChol$		–	S_C	
1.2	R = $-COOChol$	–	–	S_C	[209]
1.3	R = $-OCOChol$	–	–	S_C	
2.	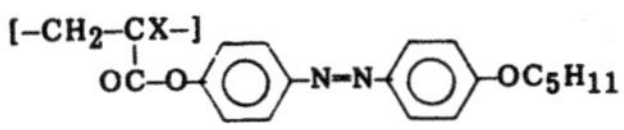				
2.1	X = H	75	>270	Two-layer, 36°	[242, 246, 247]
2.2	X = CH_3	135	>270	Two-layer, 50°	
3.	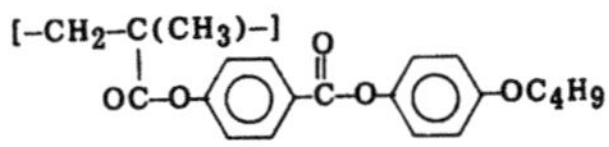	140	260	Herringbone structure	[222, 248]
4.	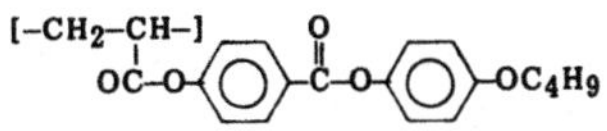	100	240	Herringbone structure	

meric liquid crystals with overlapping aliphatic "tails" could also be proposed as an alternative.

The x-ray data obtained from oriented samples could be the most convincing and unambiguous criterion for the existence of the S_C phase. The four-point splitting of the small-angle reflections, also found for polymers similar to 3.1 and 3.2 (Table 4.14), is a distinctive feature of x-ray patterns of this type. The formation of the S_C phase as a result of the phase transition from the S_A and S_F form was observed in [215, 217, 219] for polymers 1 and 2 in Table 4.15 and for polymer VIII, examined above. The presence of a $S_C \rightarrow S_A$ transition, first discovered for polymeric liquid crystals in [215, 217], is seen in four-point splitting of the small-angle x-ray reflection and a change in the value

TABLE 4.15. Polymorphism of LC Polymers with the S_C Mesophase*

$$[-CH_2-CH-] \text{ with side group } O{=}C-O-(CH_2)_{11}-OR$$

	R		Reference
1.	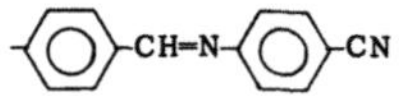	$S_C \xrightarrow[0.9]{31} S_A \xrightarrow[8.3]{169} I$	[215, 219, 232]
2.	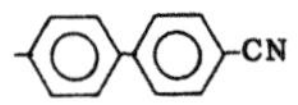	$S_C \xrightarrow[-]{30} S_A \xrightarrow[6.3]{145} I$	
	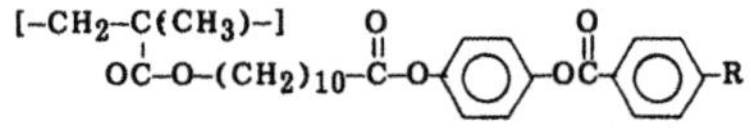		[244, 245]
3.	$-OC_4H_9$	$S_F \xrightarrow[2.2]{57} S(?)^{\dagger} \xrightarrow[2.0]{66} S_C \xrightarrow[3.6]{120} I$	
4.	$-OC_7H_{15}$	$S_F \xrightarrow[4.1]{80} S_C \xrightarrow[5.3]{139} I$	
5.	$-\overset{O}{\overset{\Vert}{C}}-O-(CH_2)_2-CH-(CH_3)_2$	$S_C \xrightarrow[1.7]{64} S_A \xrightarrow[5.3]{76} I$	

*The number above the arrow is the transition temperature in °C; the number below the arrow is the heat of transition in J/g.
†S(?): type of smectic mesophase not determined.

of spacing d_1. This polymorphic transition is also accompanied by a change in the ratio of trans and gauche isomers in the spacer groups [192], as IR spectra analysis at different temperatures shows. As the calculations indicate, the number of gauche isomers increases by approximately 20% in the transition of the S_C to the S_A mesophase. Schemes of the packing of the macromolecules in the S_A and S_C phases are shown in Fig. 4.35, and the heats of transitions between different types of mesophases are reported in Table 4.14. The $S_C \rightarrow S_A$ transition is not only demonstrated by the presence of an endothermic peak on the thermograms (cf. transition enthalpy values in Table 4.15) but also by a significant change in the optical texture (Fig. 4.36). The fan-shaped texture

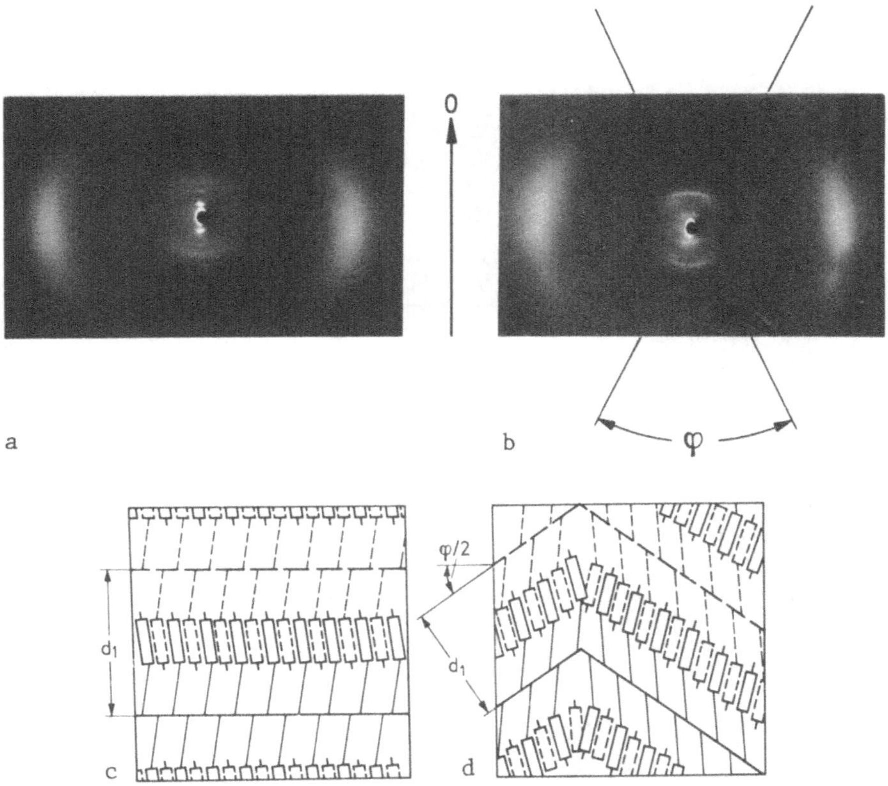

Fig. 4.35. X-ray patterns of oriented samples (a, b) and schematic representations of the molecular packing of mesogenic groups in the S_A (c) and S_C (d) mesophases of the polymer:

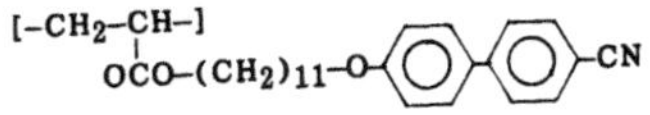

of the S_A mesophase is characteristic of both polymers above the $S_C \rightarrow S_A$ transition temperature (Fig. 4.36a). Ring-shaped bands initially appear in the fans at the temperature corresponding to the beginning of the transition (Fig. 4.36b), and the fans are then totally destroyed (Fig. 4.36c). It is important to note that the close single-layer packing (Fig. 4.28a) which simultaneously exists with two-layer packing with overlapping mesogenic groups (Fig. 4.35) remains unchanged during the transition from S_C to S_A, i.e., the two smectic types of packing with a single-layer arrangement of the side groups (S_A) and a two-layer arrangement of the side

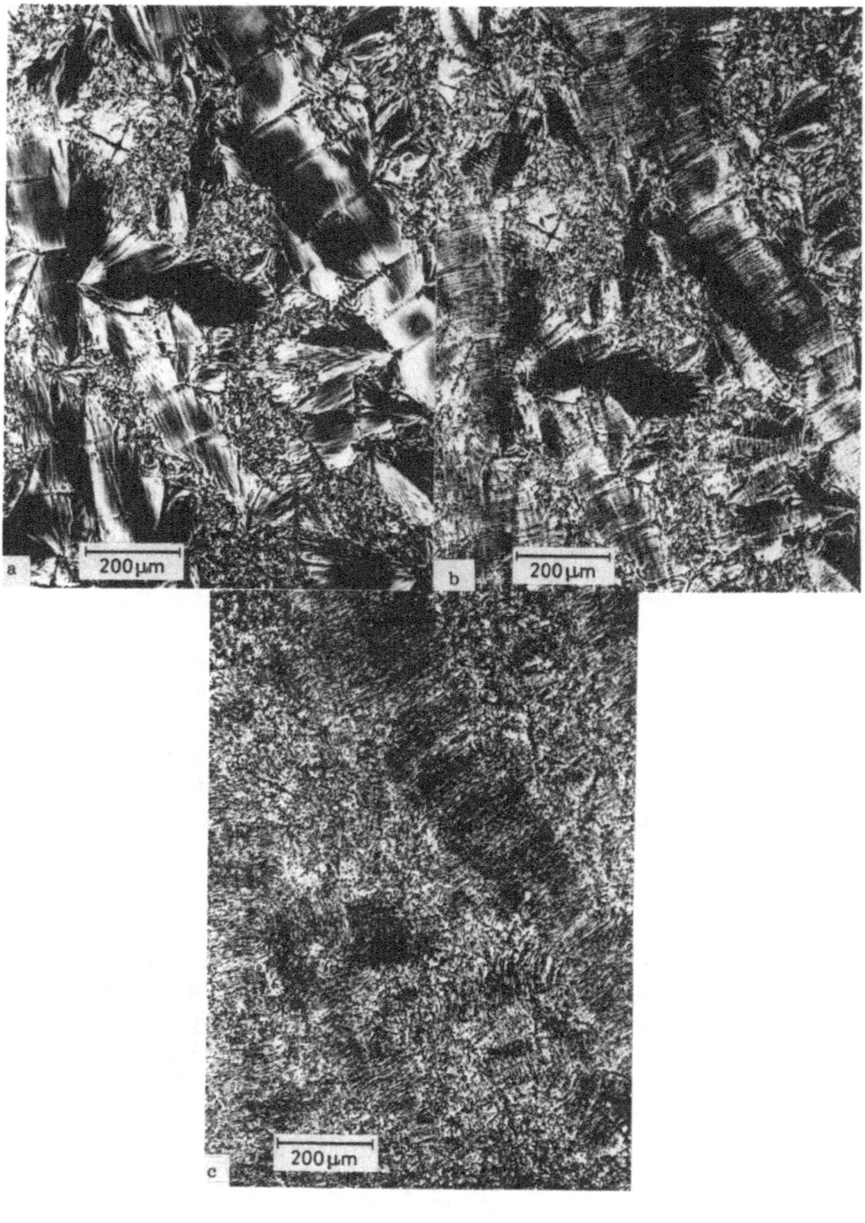
a
200 µm
b
200 µm
c
200 µm

groups tilted toward the plane of the layer essentially coexist. The coexistence of two types of packing was also observed in [222, 248] for polymers 3.1 and 3.2 (Table 4.14).

The coexistence of at least two (and perhaps more) types of packing is apparently one of the features of the structure of LC polymers with mesogenic side groups. This feature is probably characteristic of polymers, since such events have not been observed in low-molecular-weight liquid crystals.

An interesting observation was made in [249], where the formation of the S_C mesophase was observed on substitution of the methyl group in the main chain by a chlorine atom while retaining the same mesogenic groups. The properties of these polymers are compared below:

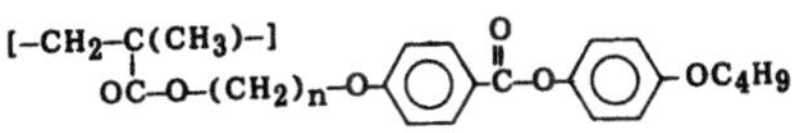

$n = 2 \quad S_A \xrightarrow{163°} I$

$n = 6 \quad S_A \xrightarrow{109°} I$

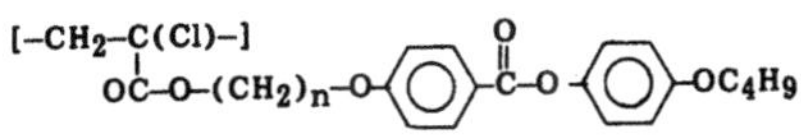

$n = 2 \quad S_C \xrightarrow{175°} I$

$n = 6 \quad S_C \xrightarrow{110°} N \xrightarrow{117°} I$

4.7.1.3. Some General Comments on the Structure of Smectic Polymers. In the preceding sections on the description of the structure of different types of smectic polymers, most of the attention was focused on an examination of the nature of the arrangement of the mesogenic units within the boundaries of the smectic layers, i.e., on a scale of several

Fig. 4.36. Change in the optical texture of a LC polymer

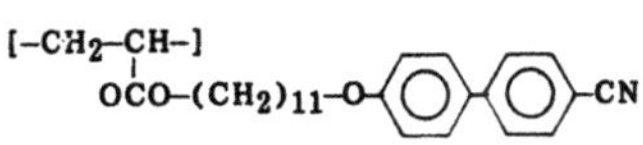

during the $S_A \rightarrow S_C$ transition: a) S_A phase at 35°C; b) during the transition at 30°C; c) S_C phase at 25°C [219].

TABLE 4.16. Some Structural Characteristics of Smectic Polymers [222, 250-251]

No.	Structural formula of the polymer	Type of mesophase	d, Å	L, Å	ξ, Å	g, %
	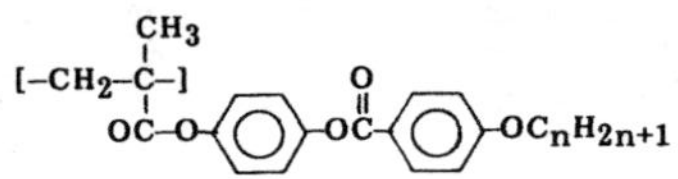					
1.	n = 4	S_A	44.0	340	90	7
2.	n = 9	S_C	38.0	230	40	16
3.	n = 12	S_A	41.0	200	42	17
4.	n = 16	S_A	46.0	550	130	11
5.	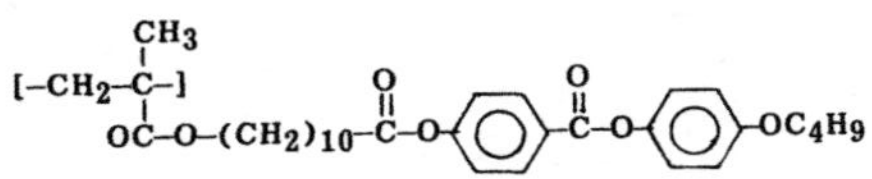	S_B	32.0	680	160	6
	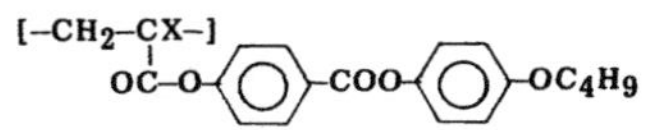					
6.	X = H	S_C	36.0	240	50	6
7.	X = CH_3	S_C	35.0	200	45	6

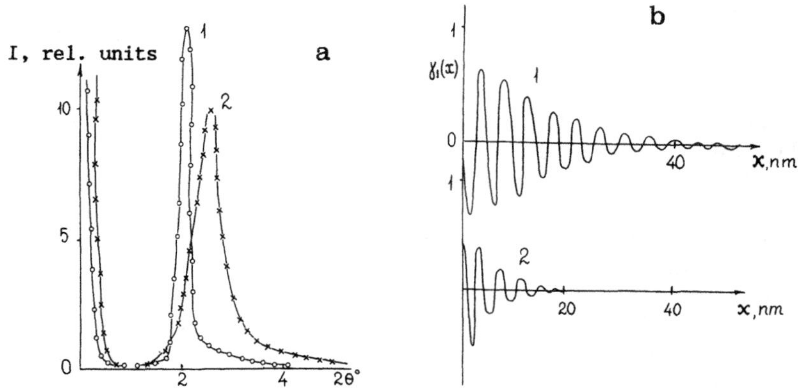

Fig. 4.37. X-ray diffraction curves (a) and one-dimensional correlation functions $\gamma_1(x)$ (b): 1) polymer 4 (Table 4.16); 2) polymer 3 (Table 4.16) [250].

tens of angstroms. Since the supermolecular organization of LC polymers covers a scale of several hundred thousand angstroms, it is still relatively difficult to represent the general picture of the structure of LC polymers. However, due to the recent intense research by Lipatov, Shilov, and Tsukruk on the x-ray analysis of smectic LC polymers, some mechanisms have already been found [221-223].

The limited size of the regions of LC ordering, in contrast to the macroscopic order which exists in low-molecular-weight liquid crystals, is a specific feature of smectic polymers. This follows from the analysis of small-angle x-ray scattering and the calculation of the one-dimensional correlation functions $\gamma(x)$ for some smectic polymers (Table 4.16).

The main feature of all of the small-angle x-ray scattering curves, in addition to the existence of distinct reflections in the 2-5° range, is the presence of strong zeroth-order scattering (Fig. 4.37a). If the small-angle maxima are due to the formation of layered structures of the smectic type, the intense zeroth-order scattering indicates the presence ot regions with a different electron density several hundred thousand anstroms in size [221].

The one-dimensional correlation functions for a series of smectic polymers (Fig. 4.37b) obtained by Fourier transformation of the small-angle scattering curves (Fig. 4.37a) have a characteristic oscillating shape which reflects the distribution of the electron density in a system of closely packed layers along the normal to the plane of the layer, with the spacing of the oscillations equal to the thickness of the smectic layer d. This type of $\gamma_1(x)$ is characteristic of lamellar structures of a high degree of perfection (Fig. 4.37b). At the same time, the rapid attenuation of $\gamma_1(x)$ to zero with an increase in x indicates the existence of certain perturbations in the packing of the layers. Based on an analysis of the correlation functions and calculation of the width of the diffraction maxima of the different orders of reflections, the size of the regions of layered order L and the value of the paracrystalline perturbations g were estimated for some smectic polymers in [250] (Table 4.16). The value of L, the so-called layer stack, is determined by the value of x at which function γ_1 decreases to zero; in other words, parameter L characterizes the size of the regions of smectic order within which the positional order in the arrangement of the smectic layers is preserved (Fig. 4.37b). The value of g characterizes the fluctuations in the widths of the smectic layers with respect to their average value $\bar{d}$ caused by the presence of paracrystalline perturbations (bending of the layers, shifts of the mesogenic groups, etc.). The correlation between these values is determined by the relation

$$g^2 = d/\pi^2(\beta_{ex}{}^2 - 1/L^2)^{1/2} \qquad (4.5)$$

where $\beta_{ex}{}^2$ is the integral width of the diffraction maximum. Attenuation of the correlation function $\gamma_1(x)$ with an increase in x, related to the accumulation of paracrystalline perturbations, takes place according to an exponential law

$$\gamma_1(x) = \gamma_0 \exp(-2x/\xi) \qquad (4.6)$$

where ξ is the correlation length characterizing the distance at which the correlation in the arrangement of the layers remains almost constant [250]. The calculated structural parameters d, L, ξ, and g for some smectic polymers are reported in Table 4.16. As Table 4.16 shows, the values of L are within the limits of 200-500 Å for smectic A and C. This means that the number of smectic layers in these smectic phases does not exceed 5-10, which is clearly insufficient for systems with true long-range one-dimensional order. The values of the correlation lengths are within the limits of

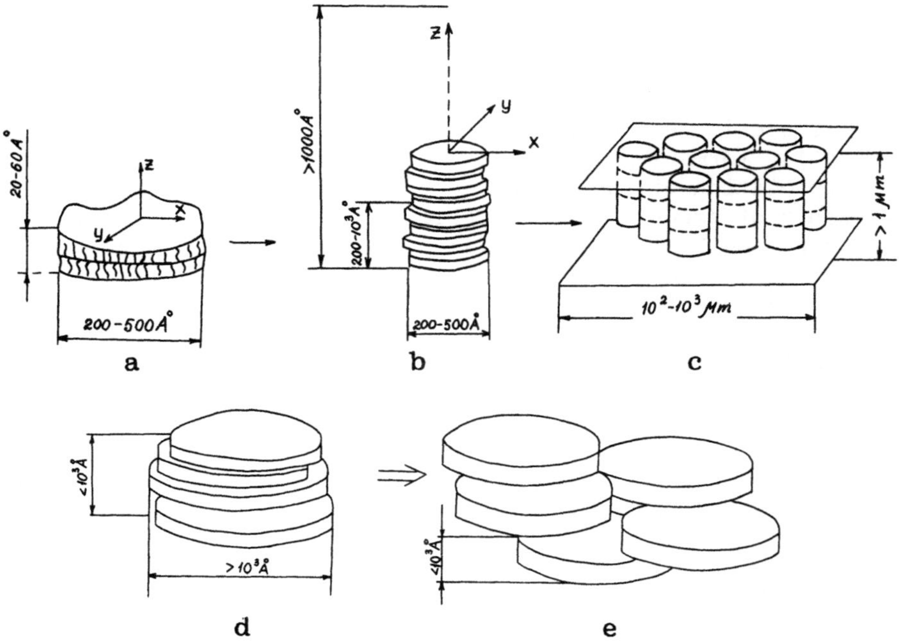

Fig. 4.38. Models of supermolecular structures in smectic polymers [221] (see text).

40-130 Å, which is characteristic of systems with short-range order. The values of the paracrystalline perturbations are within the limits of 6-17%. However, in highly ordered smectic materials, the number of closely packed layers is significantly higher and exceeds 20 (Table 4.16).

As a consequence, the presence of positional order in the packing of only a small number of smectic layers with a high degree of defects in the packing indicates the absence of true long-range order. Positional order extends over no more than ten neighboring layers forming the individual layer stack (Fig. 4.38a, b). Neighboring layer stacks are not positioned in correlation with each other. The order in the arrangement of the smectic layers is perturbed both due to the accumulation of paracrystalline perturbations and due to random shifts and rotations of the layers in the plane parallel to the plane of the smectic layer (in plane xy in Fig. 4.38b). The latter perturbations result in loss of the positional order in the packing of the layers (interruption of coherence), but the translational order along the normal to the plane of the layer is preserved.

However, in the opinion of Tsukruk et al. [223], layer stacks are in turn structural units from which larger supermolecular formations are constructed. The determination and interpretation of the presence of strong zero radiation for a number of smectic polymers (Fig. 4.37a) permitted Lipatov, Tsukruk, and Shilov to postulate the formation of large cylindrical supermolecular formations 10^2-10^3 Å in diameter and 10^3-10^4 Å long from the layer stacks (Fig. 4.38c). These cylindrical formations should be arranged according to the principle of close packing with the preservation of some parallel orientation. Since the heights of the cylinders differ insignificantly (since there is one-dimensional order along the normal to the plane of the layer), macrolayers of the order of several microns thick can be formed as a result of the packing of these cylinders (Fig. 4.38d, e).

The model of the supermolecular organization of smectic polymers proposed in [223] is based on both the results of x-ray studies and on the few light-scattering and electron-microscopic data [223]. The idea of the microheterogeneity of smectic LC polymers as the basis of the structural model [246, 247] is well correlated with the concepts on microphase separation in branched, graft, and comb-shaped polymers [253].

However, in the structural models examined above, the basic attention was concentrated on the packing of the mesogenic groups, and the role of the main chain and its conformation in the formation of the supermolecular structure was hardly considered. Different opinions have been advanced concerning the arrangement and conformational structure of the main chains in mesomorphic polymers. Some investigators attribute a random coil conformation to the main chains [254], while others propose the existence of conformationally ordered main chains in the mesophase [255]. However, it is very clear that unambiguous information on the conformational state of the main chains and the methods of their arrangement on all levels of structural organization is necessary for constructing a complete, universal structural model of a smectic LC polymer. Constructing such a structural model for both smectic and nematic (cholesteric) mesophases is one of the most important problems for investigators.

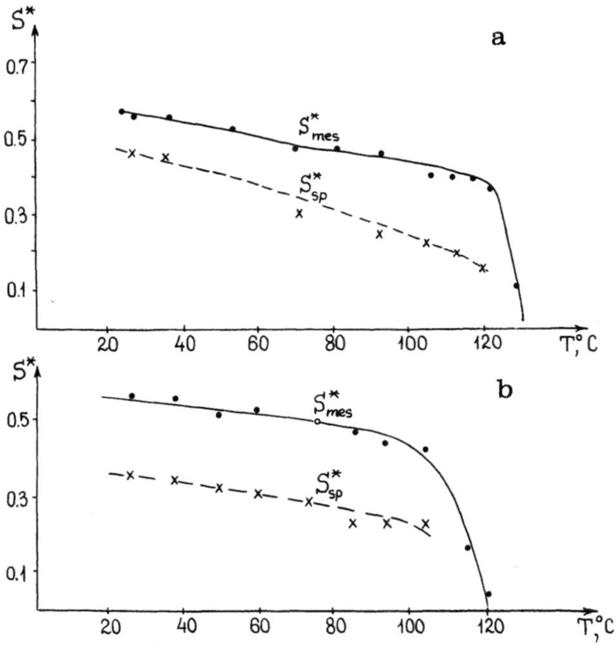

Fig. 4.39. Temperature dependence of the orientational parameters of mesogenic S_{mes}* and spacer S_{sp}* groups for polymers X (a) and XI (b) [192].

As indicated above, an important role in the formation of the mesophase has been assigned to spacer groups. A differentiated quantitative evaluation of the contribution of the mesogenic units and spacer groups to the formation of the smectic order under the effect of mechanical shear stresses in a series of LC polymers, some of which are indicated below, was conducted in [192, 256]:

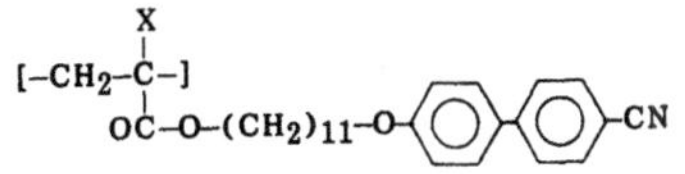

X = H (X)

X = CH_3 (XI)

Using the polarized IR spectra of polymers X and XI, the degree of orientation of the mesogenic groups S_{mes}* and spacer group S_{sp}* and their temperature dependence was studied (Fig. 4.39).† Figure 4.39 shows that the degree of orientation of the mesogenic units is greater than the values of S_{sp}^* for both polymers throughout the entire temperature range. The comparative lack of orientation of the hydrocarbon segments is due to the fact that the methylene chains are not in the form of stretched trans conformations but are partially in the form of coiled gauche isomers. Using polarized IR spectroscopy, it is possible to follow the molecular rearrangements which accompany the structural rearrangements and phase transitions under the effect of external fields in such complex systems as LC polymers.

A number of features in the LC structure of polymers were discovered in studying the effect of the temperature on their behavior.

It is known that an increase in the temperature usually causes perturbation of the structural ordering in low-molecular-weight liquid crystals, as shown by an increase in the average intermolecular distances and a decrease in the intensity of the small-angle diffraction maxima. In the case of LC polymers, the temperature dependence of the intensity of the small-angle maximum on the x-ray scattering curves as a function of the degree of binding of the mesogenic groups to the main chain can differ significantly, as Fig. 4.40 shows for a comparison of LC polymers XII-XIV [248]:

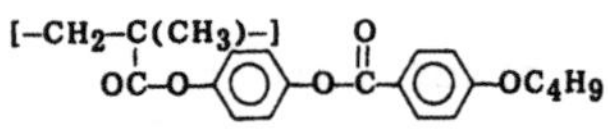

(XII)

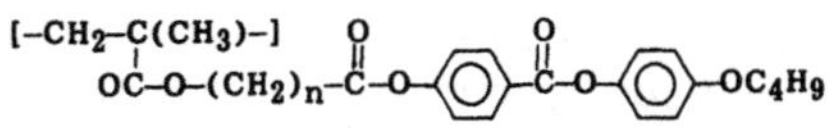

n = 3 (XIII)

n = 10 (XIV)

The changes in the intensity of the small-angle maxima I and their widths β are due to an increase or decrease in the concentration and size of the regions of LC ordering (layer stacks). In polymer XII, the values of I_{max} and the integral intensity I_{int} increase monotonically and the width of maximum β remains constant due to an increase in the vol-

†The methods of calculation used for low-molecular-weight liquid crystals were used for finding S_{mes}^* and S_{sp}^*. In the absence of an orientational effect $S_{mes}^* = 0$ and $S_{sp}^* = 0$, and with total orientation $S_{mes}^* \to S$ and $S_{sp}^* \to S'$, where S is the order parameter; $S = \frac{1}{2}(3\cos^2\theta - 1)$.

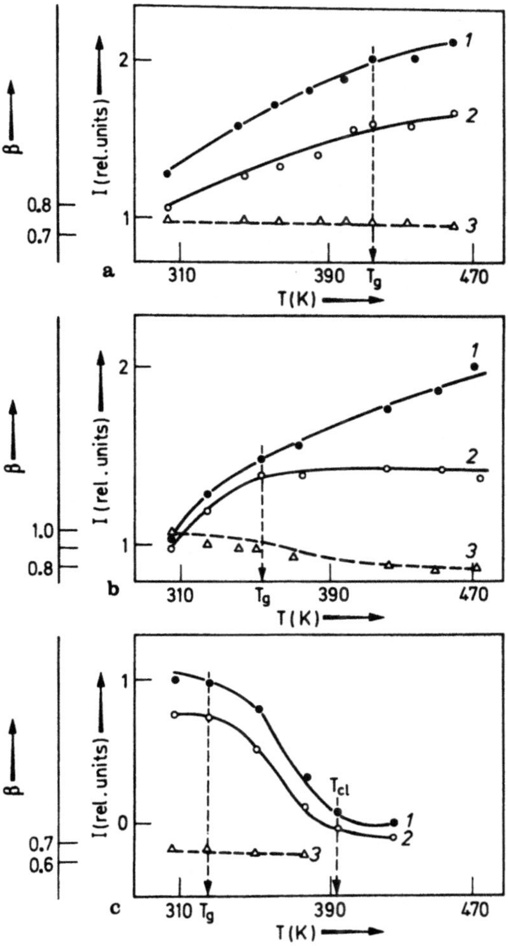

Fig. 4.40. Temperature dependence of the intensity of the first diffraction maximum I_{max} (1), integral intensity I_{int} (2), and integral width β (3) of the small-angle peaks for polymers XII (a), XIII (b), and XIV (c) [248].

ume fraction of regions of LC ordering with no significant change in the perfection of the LC structure. An increase in the perfection with no change in the concentration of LC regions basically takes place in polymer XIII. For polymers XII and XIII, the unusual temperature dependence of the intensity of x-ray scattering observed is thus due to an increase in the mobility of the main chains of the macromolecules, which either results in the formation of new regions of LC ordering (polymer XII) or perfection of the LC structure (polymer XIII) (Fig. 4.40a, b). In the case of polymer XII where the most perfect packing is attained due to the "remoteness" of the mesogenic groups from the main chain, an increase in the temperature has a disordering, i.e., opposite, effect on the LC structure (Fig. 4.40c). The examples given clearly demonstrate the difference in the structural changes which take place in LC polymers with a different spacer group length with changing temperature.

Polymers with mesogenic groups in the side chains thus form smectic phases of the same structural types as low-molecular-weight liquid crystals. This allows the traditional classification of mesophases to be used in the description of LC polymer structure, especially for the best-studied types of mesophases: S_A, S_B, and S_C. At the same time, it is possible that the structure of a number of comb-shaped polymers (cf. Chapters 1 and 4 and Table 4.12) which are considered crystalline could be treated as one of the varieties of highly ordered smectic mesophases (S_H or S_I), which have only begun to be studied.

Recently synthesized polymethacrylates form chiral tilted smectic phases called smectic S* phases [244-245], which have ferroelectric properties and thus are of special interest. Based on the concepts of symmetry of ferroelectric liquid crystals elaborated in [257, 244], comb-shaped LC polymethacrylates containing optically active groups in the side chains were synthesized in [244]:

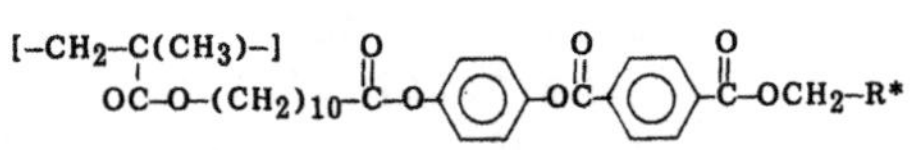

where R* is a chiral group [for example, $-\overset{*}{C}H(CH_3)-C_2H_5$ (XV)].

In polymers with this structure, a second-order polar axis S_2, directed along the layers and perpendicular to the plane of inclination of the chiral segments, is the only element of symmetry. Dipole ordering caused by the hindered rotation of the chiral fragments around their long axes arises along this axis. The character of this ordering and,

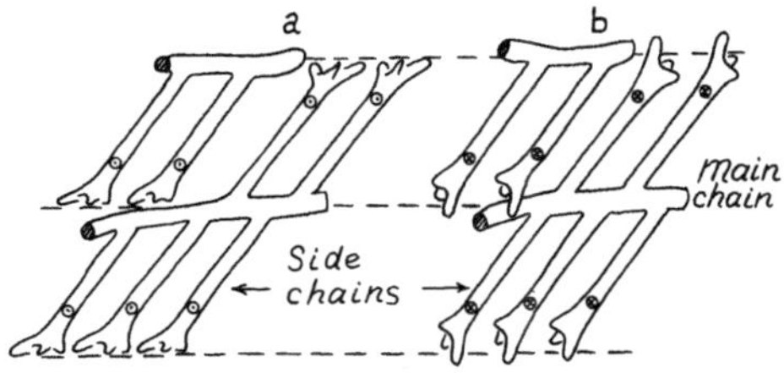

Fig. 4.41. Schematic representation of arrangement of the side chains in comb-shaped macromolecules reflecting the dipole ordering in the chiral smectic S_C* phase [245].

consequently, the appearance of spontaneous polarization P_S is shown in Fig. 4.41, where the position of the side chains of the macromolecules is illustrated.

The end chiral groups of the macromolecules are in the form of tilted tripods which form the layered packing. In this structure, the transverse dipole moments m of the chiral groups are not compensated and are added in a direction perpendicular to the inclined plane, since position "a" will be preferred to position "b" for tripods in the layer in the tilted positions (Fig. 4.41).

Some characteristics of a S* chiral smectic polymer are shown below:

Characteristics	T_g, °C	$S_C \rightarrow S_A$, °C	ΔH_{C-A}, kJ/mole	S_A–I, °C
Polymer IV	43	73	1.6	85

To give this polymer ferroelectric properties, it is heated above the glass transition temperature and after application of an electric field for several seconds, it is cooled for several seconds in the field (or without it) below the glass transition temperature. It is important to note that the strength of the electric field needed is approximately 10^3 times less than for such a well-known ferroelectric as polyvinylidene fluoride.

The temperature dependence of the spontaneous polarization P_S and pyroelectric coefficient γ for the polymer is shown in Fig. 4.42 [245]. The maximum value of P_S for the S* smectic phase is $(0.8\text{-}0.9)\cdot 10^{-5}$ $C\cdot m^{-2}$, which is close to

TABLE 4.17. LC Polymers with a Hypothetical Nematic Structure

No.	Polymers	T_g, °C	T_{cl}*	ΔH_{cl}, J/g	Method of investigation†	Reference
	A. Polymers containing derivatives of Schiff bases and an azoxy group					
1.	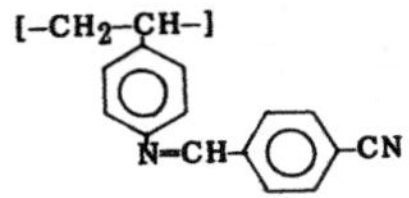	90	190	–	X	[228, 258]
2.	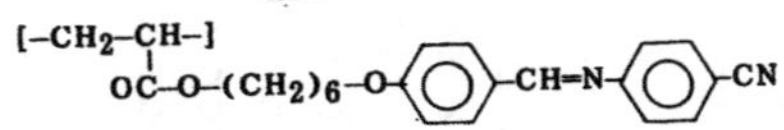	20	154	1.8	X, DSC	[216]
	B. Polymers containing substituted phenyl, phenylbenzoate, and diphenyl groups					
1.	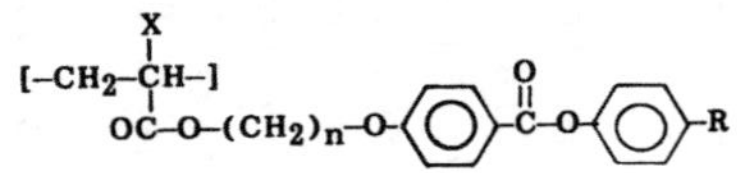					
1.1	$X = CH_3$, $n = 2$, $R = -OCH_3$	101	121	2.3	DSC, X, opt	[225]
1.2	$X = CH_3$, $n = 2$, R = 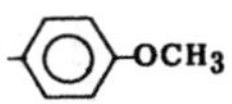	120	174	3.1	DSC	[177]
1.3	$X = CH_3$, $n = 2$, R = 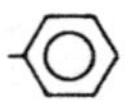	–	S 124 N 187	1.0-0.8	DSC	[21]
1.4	$X = CH_3$, $n = 3$, R = 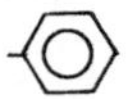	–	S 170 N 187	–	DSC	[21]
1.5	$X = CH_3$, $n = 3$ R = 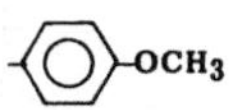	–	S 170 N 197	–	DSC	[225]

1.6 X = CH_3, n = 3, R = 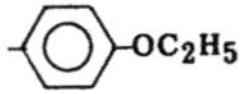	120	300	–	DSC	[225]
1.7 X = CH_3, n = 6, R = $-OCH_3$	95	105	2.1	DSC, M,† opt	[224]
1.8 X = CH_3, n = 6, R = 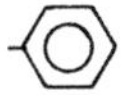	132	S 164 N 184	2.3	DSC, opt	[225]
1.9 X = CH_3, n = 6, R = $-CH_3$	70	84	1.3	DSC, opt	[226]
1.10 X = H, n = 2, R = $-OCH_3$	60	115	–	DSC	[14]
1.11 X = H, n = 2, R = $-CN$	62	93	–	DSC	[259]
1.12 X = H, n = 2, R = 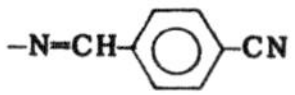	72	267	–	DSC	[259]
1.13 X = H, n = 6, R = $-OCH_3$	60	S 98 N 125	–	DSC	[14]
1.14 X = H, n = 6, R = $-CN$	60	109		DSC	[259]
1.15 X = H, n = 6, R = 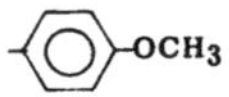	100	S 121 N 271			[259]
1.16 X = H, n = 6, R = 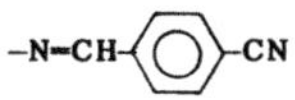	56	C 110 S 138 N 257			[260]
2. 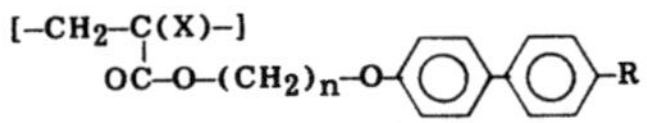					
2.1 X = CH_3, n = 2, R = $-OCH_3$	120	152	2.8	DSC	[227]
2.2‡ X = CH_3, n = 6, R = $-OCH_3$	–	C 117 N(?)	32, 7.4	DSC, X, opt	[190]
	–	C 119 S 136 I	12.6, 7.1	DSC	[227]
2.3 X = H, n = 2, R = CN	50	112	0.4	DSC, X, opt	[261, 215]

TABLE 4.17 (continued)

No.	Polymers	T_g, °C	T_{cl}*	ΔH_{cl}, J/g	Method of investigation†	Reference
3.	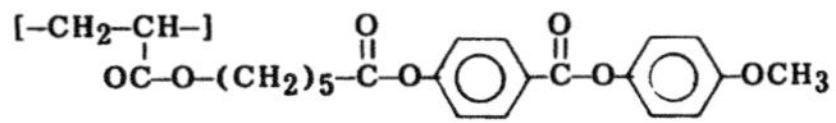	15	105	2.1	DSC, X, opt	[262]
	C. Siloxane polymers					
4.	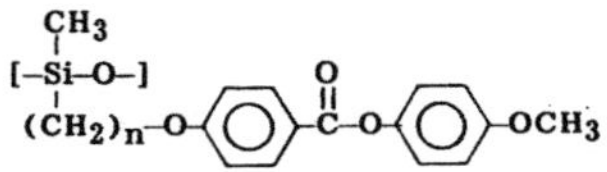					
4.1	n = 3	15	61	2.2	DSC	[136, 243]
4.2	n = 4	15	95	–	DSC	[139]
4.3	n = 5	–	C 87 N 115	–	DSC	[243]
4.4	n = 6	–	C 52 N 112	–	DSC	[139, 243]

*T_{cl} and ΔH_{cl} correspond to the melting point and heat of fusion of the nematic mesophase; if there is another structure, it is indicated by the corresponding symbol.
†X) x-ray; DSC) differential scanning calorimetry; opt) optical microscopy; M) study of the miscibility of LC polymers with low-molecular-weight liquid crystals.
‡The polymer forms a fan-shaped texture.

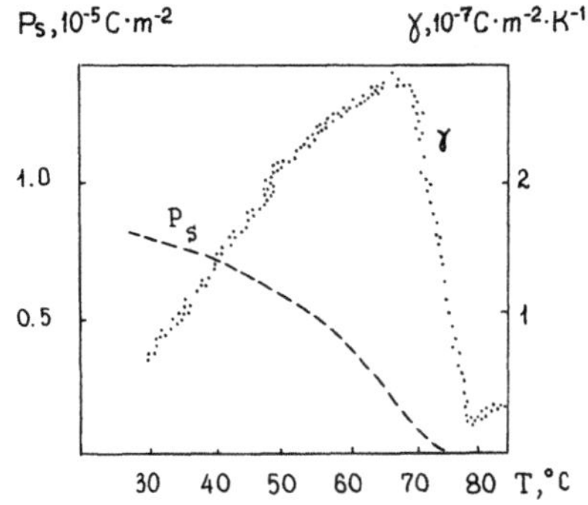

Fig. 4.42. Temperature dependence of the spontaneous polarization P_s and pyroelectric coefficient γ for ferroelectric LC polymer XV [245].

the values for low-molecular-weight LC ferroelectrics [257]. However, the preservation of both the spontaneous polarization and the pyroelectric coefficient in the glassy state without application of an external polarizing field is a characteristic difference between polymeric LC ferroelectrics and their low-molecular-weight analogs, and this is a definite advantage in comparison to low-molecular-weight LC compounds for possible practical purposes.

4.7.2. Nematic Mesophase

The nematic type of ordering is encountered less frequently than smectic ordering among LC polymers with mesogenic side groups. The basic representatives of the different derivatives of nematic LC polymers are shown in Table 4.17. A nematic type of structure is usually obtained for homopolymers with short end substituents in the mesogenic groups (CH_3, OCH_3, CN) and containing from two to six methylene groups in the aliphatic spacer group (Table 4.17). As an analysis of the data in the literature and in Table 4.17 shows, a nematic type of structure is not characteristic of polymers with no spacer group and is more frequently encountered among comb-shaped LC polymers. The introduction of side substituents in the mesogenic groups, clearly seen from a comparison of the properties of the homologous series of polysiloxanes, causes the formation of the nematic phase [191]:

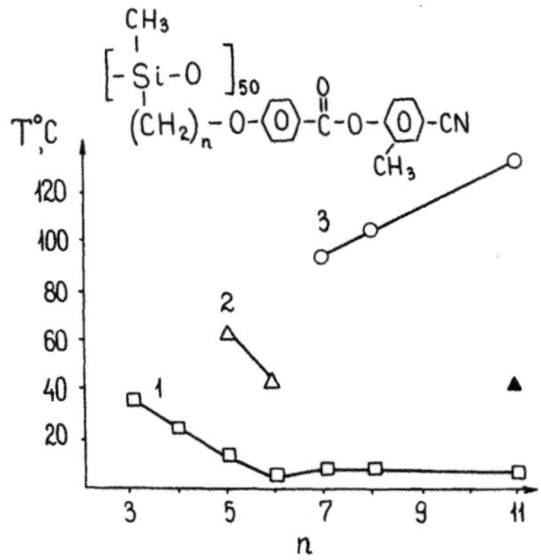

Fig. 4.43. Glass transition temperatures (1) and N → I (2) and S_A → I (3) phase transitions for polysiloxanes XVI (the dark triangle corresponds to the melting point of a polysiloxane with n = 11) [191].

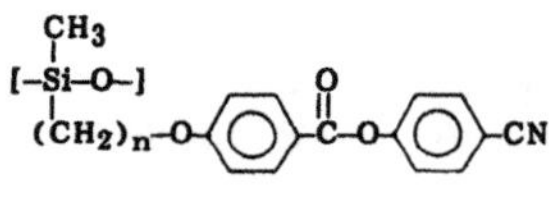

(XVI)

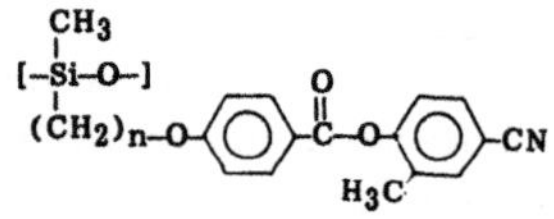

(XVII)

All polymers XVI beginning with m = 3 only form a smectic mesophase (Fig. 4.25b), while in the series of polymers XVII, the formation of a nematic mesophase takes place with n = 5, 6 (Fig. 4.43). A further increase in the length of the spacer group (n = 7-11) again induces layered ordering, i.e., the formation of the smectic mesophase.

Copolymerization of nematogenic and smectogenic monomers is one of the methods of obtaining nematic polymers; in this case, it is possible to go from smectic polymers to nematic copolymers by varying the ratio of monomeric units in the copolymer (see the section on copolymers below).

It is important to note that the identification of polymeric nematic mesophases is more complex than that of polymeric smectic mesophases. This is due to the fact that assignment of the type of mesophase to the nematic structure is based on the results of calorimetric studies in most cases (see Table 4.17, methods of investigation). The structural data are usually limited to finding no small-angle reflections in the x-ray patterns of unoriented samples. The low enthalpy of the transition from the anisotropic to the isotropic phase (Table 4.17), similar to the corresponding values characteristic of low-molecular-weight nematic liquid crystals, and the absence of layered reflections would seem to indicate a one-dimensional type of ordering. However, these data are apparently insufficient for the complete description of the structure of nematic polymers which can simultaneously exhibit both similarities and differences with respect to low-molecular-weight nematic materials.

The similarity with low-molecular-weight nematic materials is based on the fact that nematic polymers can form schlieren and marbled textures characteristic of nematic materials (Fig. 4.44a); the enthalpy of the transition from the LC state to an isotropic melt is also close to the values for low-molecular-weight nematic liquid crystals and usually does not exceed 2-3 J/g.

At the same time, there are also certain structural differences. Some elements of structural ordering in the arrangement of the side chains (a weak diffuse small-angle reflection) which could indicate a cybotactic nematic type of ordering are observed in the x-ray patterns of nematic polymers even in the unoriented state.

In addition, the existence of a new type of nematic phase for a comb-shaped polymer with phenylbenzoate groups was found in [217]:

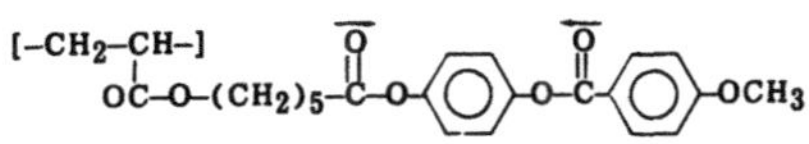

(XVIIIa)

The x-ray pattern of this polymer (Fig. 4.45a) is unusual for nematic liquid crystals: there is only one narrow reflection corresponding to an interplanar distance of 4.4 Å with no small-angle reflections. The presence of only one narrow reflection indicates that a nematic structure with an ordered (apparently hexagonal) arrangement of the mesogenic groups in the absence of spacing in the direction of their long axes is realized in this polymer. A similar type of structure has not been observed up to now in either low-molecular-weight or polymeric liquid crystals. This struc-

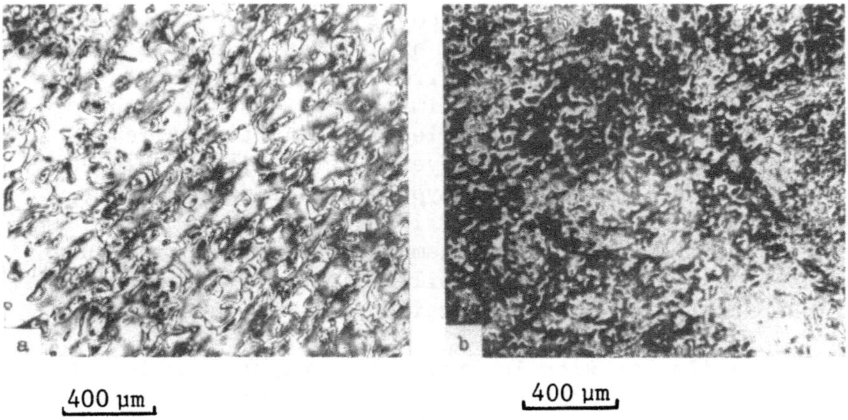

Fig. 4.44. Schlieren (a) and marbled (b) textures of nematic polymers.

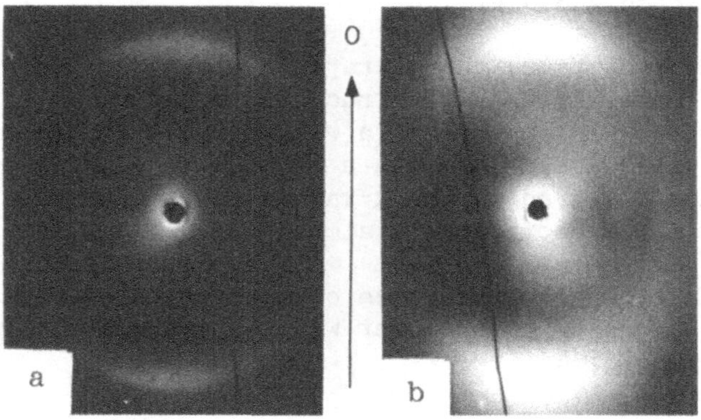

Fig. 4.45. X ray of polymer XVIIIa in the N_B (a) and N_A (b) mesophases [217]. (0 is the texture axis.)

ture was called nematic B (N_B), by analogy with smectic B, in which a hexagonal type of packing of the mesogenic groups also takes place. When polymer XVII is heated above 60°C, the narrow reflection disappears, and the usual nematic structure is formed (Fig. 4.45b) with orientational order in the arrangement of the mesogenic groups, i.e., there is a polymorphous transition: ordered nematic (N_B) → ordinary nematic (called N_A), which changes into an isotropic melt above

120°C: $N_B \xrightarrow{60°C} N_A \xrightarrow{120°C}$ I. It is interesting to note the structurally similar LC polymer with the same mesogenic group (XVIIIb) as polymer XVIIIa which differs only with respect to the direction of the second ester group:

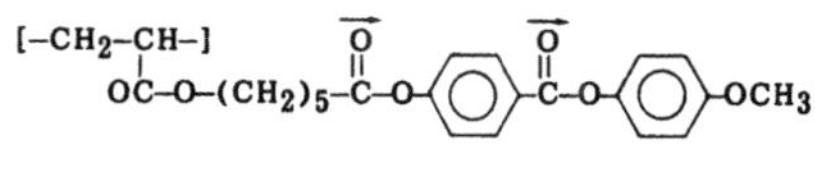

(XVIIIb)

This does not form an ordered nematic phase, but only the usual nematic mesophase N_A, and changes into an isotropic melt at the same temperature as polymer XVIIIa: $N_A \xrightarrow{120°C}$ I.

The orientation of the ester group in polymer XVIIIa (in comparison with the opposite orientation in polymer XVIIIb) apparently causes the more ordered arrangement of the mesogenic groups, which follows from both the previously examined behavior of polymers XIII and IXa and from a comparison of the series of polysiloxanes with identical mesogenic groups [214]:

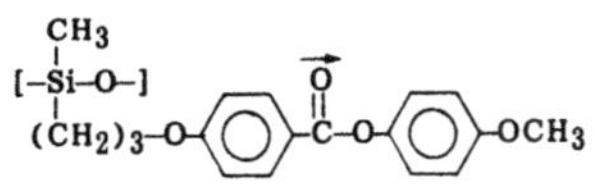

$N \xrightarrow{60°} I$ (XIX)

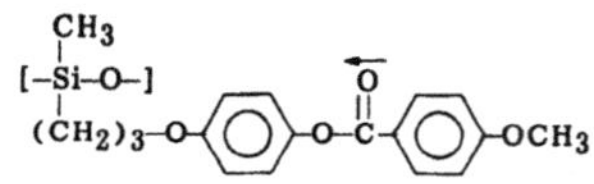

$S \xrightarrow{112°} I$ (XX)

Polymer XIX only forms the nematic phase, while polymer XX forms the smectic mesophase. The reason for this difference should be sought in the conformational state of the mesogenic groups, which causes different anisotropy of polarizability of the side groups.

The formation of fan-shaped textures typical of low-molecular-weight and polymeric smectic materials (for example polymer B.3.2 in Table 4.17) is characteristic of some nematic polymers. It can be postulated that these polymers have some elements of layered ordering which differ from the known elements for low-molecular-weight cybotactic nematic materials.

The data reported for some polymers thus show that there is significantly greater structural similarity in the

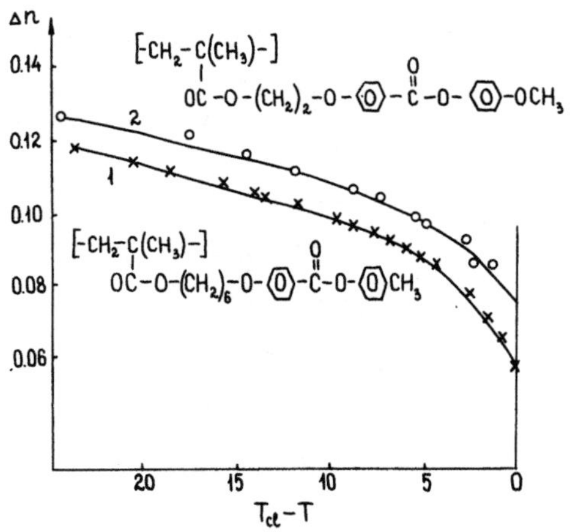

Fig. 4.46. Temperature dependence of the birefringence Δn for LC polymers: 1) "normal" nematic polymer (Δn > 0); 2) "anomalous" nematic polymer (Δn < 0).

nematic and smectic phases for polymeric liquid crystals than for low-molecular-weight liquid crystals.

Are these features only characteristic of the polymers indicated here or are they characteristic of the entire class of polymeric nematic materials? It is not yet possible to unambiguously answer this question due to the limited experimental data available in the literature. Nevertheless, it was previously observed (Chapter 1) that a certain tendency toward the formation of layered structures even in isotropic melts is observed for comb-shaped polymers, even if they contain no mesogenic groups. For this reason, it can be postulated that the polymeric nature of comb-shaped polymers with mesogenic groups can result in some ordering of the nematic structure formed by these groups. This is also indicated by the data from structural studies conducted for a fraction of polymer A.2 with a different degree of polymerization (Table 4.17) [216]. It was found that the formation of a "schlieren" texture typical of low-molecular-weight nematic liquid crystals is characteristic of the oligomeric fractions of this polymer. However, an increase in the degree of polymerization results in "degeneration" of this texture and the formation of structures which resemble the fan-shaped textures of smectic phases to a greater degree. A similar picture is also observed for polymers of the acrylic series containing cyanophenyl groups.

Another feature of polymeric nematic materials was revealed in studying oriented films of these materials. It was found in [226, 264] that most of the polymers indicated in Table 4.17 form an optically positive, uniaxial, homeotropic structure where the mesogenic side groups are arranged perpendicular to the axis of the fiber under the effect of a mechanical field; these polymers have a positive birefringence Δn (cf. polymers B.1.2, B.1.7, and B.1.8 in Table 4.17, for example). At the same time, a number of "anomalous" polymers have been found which are characterized by negative birefringence despite having a structure very similar to the structure of the polymers indicated above (cf. polymers B.1.1, B.1.9, and B.3.1 in Table 4.17). The temperature dependence of the birefringence is shown in Fig. 4.46 for two nematic polymers with $\Delta n > 0$ and $\Delta n < 0$. Despite the similar chemical structure of these polymers and the similar type of temperature dependences of Δn, their birefringence is of opposite sign; in [264], this difference was attributed to the stronger "perturbing" effect of the main chain on the packing of the mesogenic groups in the case of a polymer with a short spacer. It should be emphasized that polymers with $\Delta n < 0$ ("anomalous" polymers in Finkelmann's terminology) also form a somewhat unusual type of optical texture which differs from the texture of the usual nematic materials. The mesogenic side groups are arranged parallel to the axis of the fiber in the orientation of these "anomalous" polymers and polymers VII and XVIII according to the x-ray data. The causes of this type of anomaly are apparently related to the structure of LC polymers, primarily to the flexibility and length of the main chain and spacer group and the value of the external force, which determines the character of the orientation of the mesogenic groups (see above, Fig. 4.18).

The more "difference" there is in the behavior of the mesogenic side groups and main chains, the smaller the effect will be of the main chain on the orientation of the liquid-crystalline domains. Here, too, in orientation of the polymers, the mesogenic groups are oriented along the direction of the effect of the force (a low DP also causes this). The more "dependent" the behavior of the main and side chains (i.e., a short spacer), the stronger their interaction: the mesogenic side groups are arranged perpendicular to the orientation of the main chain upon orientation.

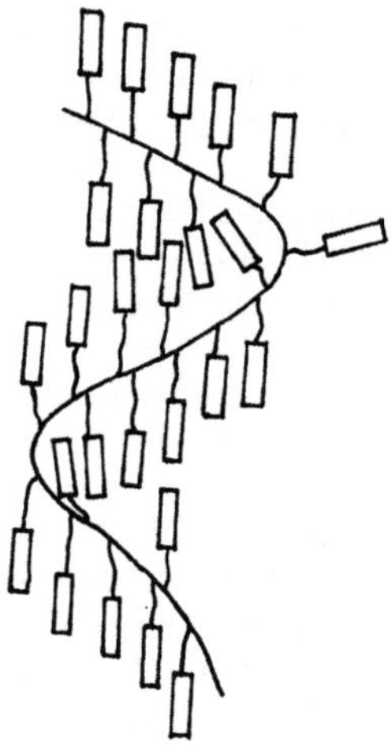

Fig. 4.47. Schematic representation of the macromolecule of a nematic polymer.

Concerning the packing models of macromolecules in the nematic mesophase, since (as mentioned above) nematic LC polymers have very "poor" x-ray patterns with some elements of layered order, most investigators are still limited to finding the presence of one-dimensional orientational order alone in these polymers (Fig. 4.47). Only the studies by Hungarian investigators, Cser in particular [21], are an exception: different concepts of the structure of nematic LC polymers were elaborated in these studies. In their proposed model [21] (Fig. 4.48), the LC order is formed due to the parallel packing of the main chains, while the side groups lie perpendicular to the directrix. The capacity to orient under the effect of an electromagnetic field is proposed in [21] as a criterion for assignment of a polymer to the smectic type. The structure of the polymer should be called smectic when the orientation of the side groups is parallel to the orienting field and nematic when the orientation of the side groups is perpendicular to the field. However, as recent intense research on the structure and behavior of LC polymers in electrical fields has shown (cf. Section 4.8), this type of packing of the macromolecules in nematic LC polymers is not sufficiently substantiated.

Most nematic LC homopolymers probably have elements of layered order in the arrangement of the mesogenic groups which approaches the cybotactic or smectic to a greater or

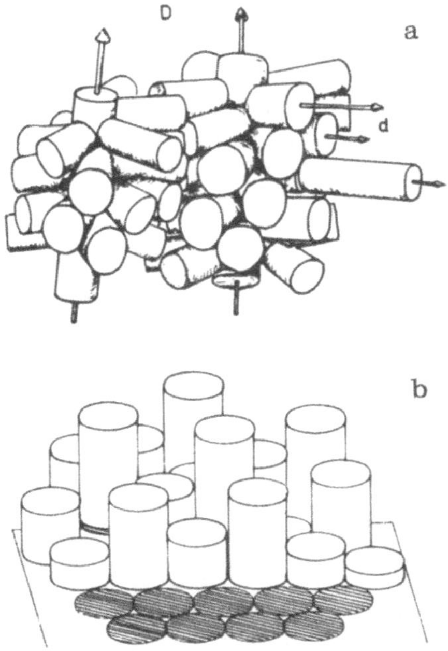

Fig. 4.48. Schematic model of cylindrical macromolecules (a) and their semihexagonal packing (b) for nematic LC polymers [21].

lesser degree; further structural studies are required to develop a special classification for this group of LC polymers. It is clearly necessary to use the methods of investigation successfully recommended for low-molecular-weight liquid crystals such as plotting of the radial distribution function, cylindrical function of interatomic distances, and the method of optical masks, which has only recently begun to be applied to LC polymers.

4.7.3. Comparison of Some Properties of Smectic and Nematic Liquid-Crystalline Polymers

4.7.3.1. Homopolymers. In the overwhelming majority of studies of LC polymers, most of the attention has been focused on the synthesis and study of structure. There is much less information on the physical properties of LC polymers, especially in comparison to low-molecular-weight liquid crystals. Some comparative data on such important characteristics of nematic and smectic LC polymers as the order parameter and the rheological and dielectric properties are examined in the present section. In addition, the mechanisms in the properties and structure of polymers with a sequentially altered macromolecular structure in going from the nematic to the smectic mesophase are examined using the example of copolymers of smectic and nematic monomers.

4.7.3.1a. The Order Parameter. The concept of the order parameter S, first introduced by Tsvetkov [265] for low-molecular-weight liquid crystals, can be used for describing the degree of orientational order in LC polymers

$$S = \tfrac{1}{2}\langle 3\cos^2\theta - 1\rangle \qquad (4.7)$$

where θ is the angle between the long axis of the molecule and the directrix of the system corresponding to their primary orientation, and the angular brackets indicate the average of all orientations of the molecules. With an ideal parallel orientation $S = 1$, and with an unordered orientation $S = 0$. In real systems, S assumes intermediate values which are strongly dependent on the temperature. In the case of LC polymers, the order parameter is usually considered with respect to the mesogenic groups, although the same parameter can be used for evaluating the orientational order of the main chains and spacer groups (see above, Section 4.6).

The results of studying the optical properties (birefringence, IR and UV dichroism), x-ray, NMR, and EPR methods (with specially introduced paramagnetic probes) are usually used for estimating the order parameter. The so-called "guest–host" effect (see below, Section 4.8.1.2c), where molecules of dichroic dyes in a low concentration of 10-15% are introduced in the LC polymer, is a widely used method. The order parameter (see below on the guest–host effect) is calculated from the data on the linear dichroism of the extraneous dye molecules which are isomorphous to the liquid crystal.

The values of the order parameter S for some polyacrylic derivatives of LC polymers are reported in Table 4.18. The values of S are for the temperature region below the glass transition temperature, i.e., in conditions where the LC phase is frozen in a glassy matrix. Note the difference in the values of S for nematic and smectic polymers: the values of S for nematic LC polymers are within the range of 0.45-0.65, and for smectic polymers, S = 0.85-0.92.

TABLE 4.18. Values of the Order Parameters for Some Polyacrylic LC Polymers Determined by Different Methods

No.	Polymer	Type of mesophase	Order parameter, 20°C	Method of determination	Reference
1.	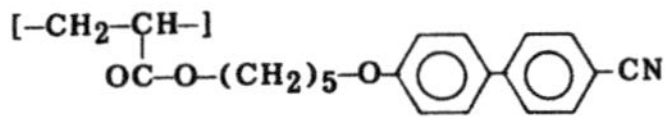	N	0.45 0.52	NMR "Guest–host" effect	[269]
2.	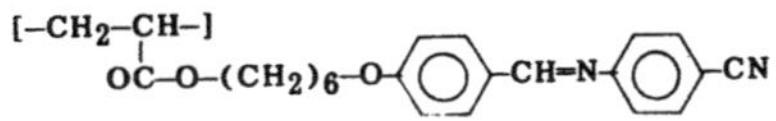	N	0.5	"Guest–host" effect	[267]
3.	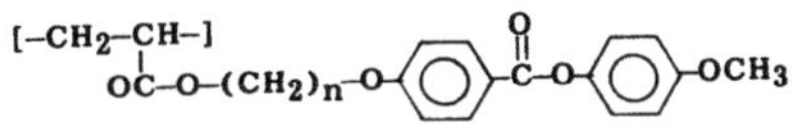	n = 2N n = 6S	0.65 0.92	EPR (label)	[266]
4.	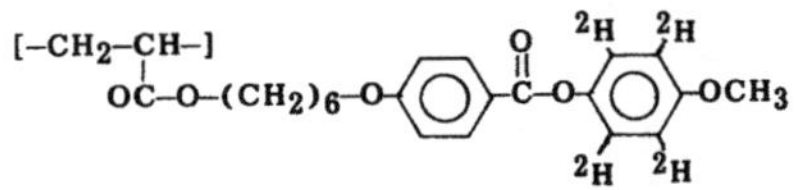	S	0.85 ± 0.05	NMR	[270]
5.	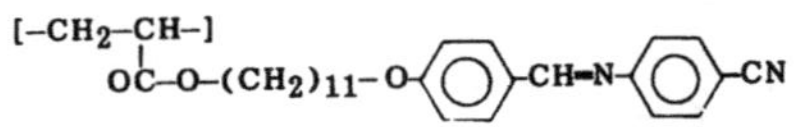	S	0.91	X-ray	[267]

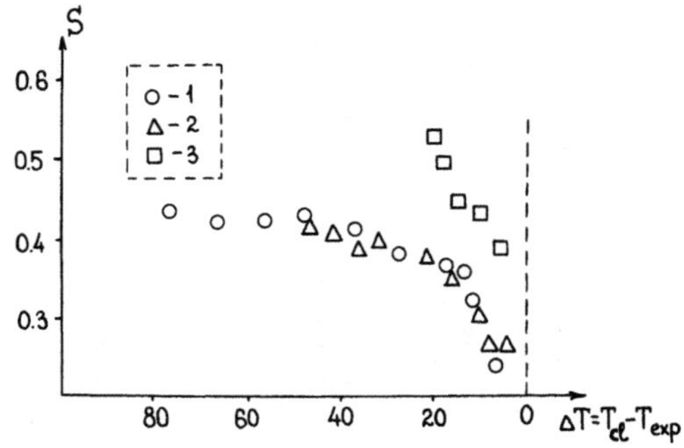

Fig. 4.49. Temperature dependence of order parameter S obtained by NMR [269]: 1) nematic homopolymer 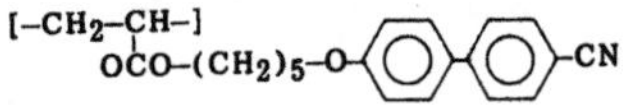 ; 2) nematic copolymer (1: 1) 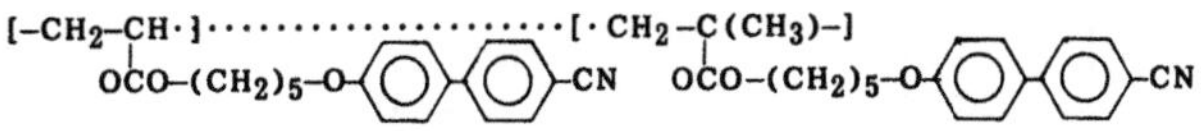 ; 3) 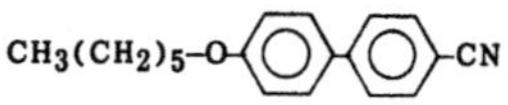 [269] (T_{exp} is the experimental temperature).

At the same time, the maximum value of S for low-molecular-weight nematic materials is usually 0.6-0.8; it reaches 0.9 for smectic materials. A comparison of these values shows that the ordering of smectic polymers is extremely close to the degree of order of low-molecular-weight smectic liquid crystals. At the same time, nematic polymers are characterized by a slightly lower degree of order than their low-molecular-weight analogs. This is clearly seen from a comparison of the temperature dependences of the order parameter S for polyacrylic polymers with the behavior of low-molecular-weight analogs of mesogenic groups (Fig. 4.49). The lower value of the degree of order for polymeric compounds containing a mechanically induced, covalently bound dye was also observed in [270-271].

However, in examining the results of these studies, it is important to emphasize that in determining the absolute values of S, attention should be paid to the selection of the extraneous dye introduced in the polymer matrix. In [271], using a monomeric compound 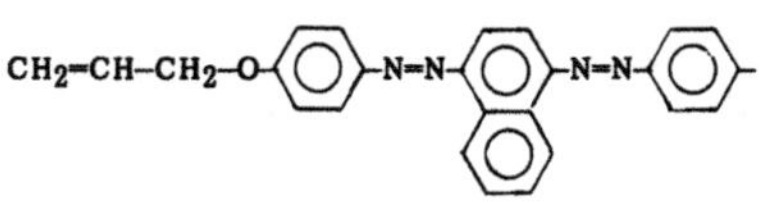 as

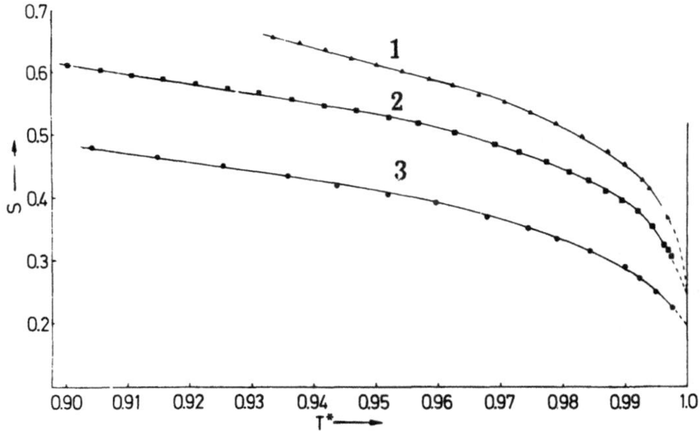

Fig. 4.50. Temperature dependence of the order parameter S of dye probes: 1) mixture A; 2) mixture B; and 3) copolymer XXI [271] ($T^* = T_{exp}/T_{cl}$, where T_{exp} is the experimental temperature and T_{cl} is the clearing point).

the dye, mixtures of a low-molecular-weight nematic liquid crystal 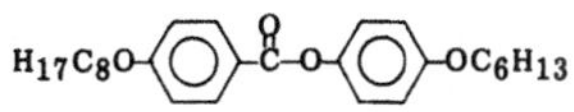 (mixture A) with a polysiloxane nematic polymer 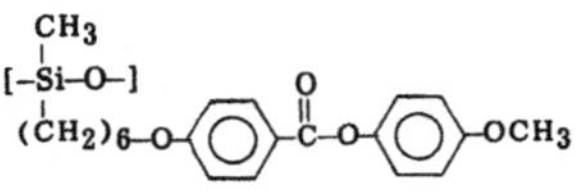 (mixture B) were prepared. As Fig. 4.50 shows (curves 1 and 2), the value of S for the low-molecular-weight liquid crystal is higher than for the polymer. However, if the same dye is chemically bound to the polymer, i.e., copolymer (XXI) is obtained

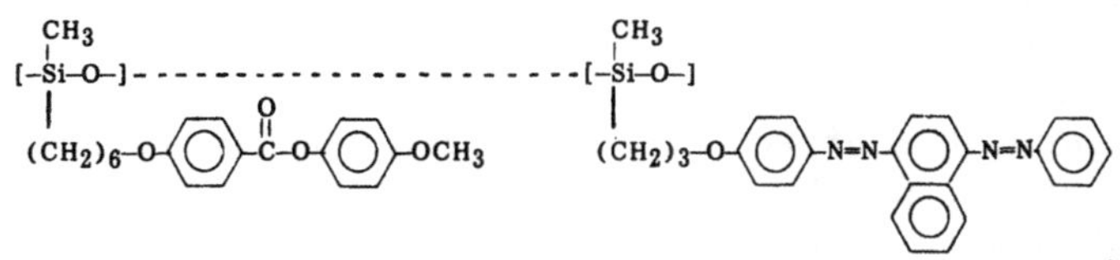 (XXI)

the temperature dependence of its order parameter S (Fig. 4.50, curve 3) lies below curves 1 and 2. This is apparently the result of perturbation of the LC order due to the effect of the dye units.

However, since the chemical structure of the dye molecules differs from the mesogenic groups and the mesogenic groups in turn differ slightly from the molecules of the low-molecular-weight liquid crystal, serious attention should probably not be given to the comparison of the differences in the absolute values of S in this case. It is only useful to examine the character of the temperature dependence of S.

In estimating the order parameter with a "marker" (dye, spin label), it is thus necessary to consider the fact that the value of S obtained can correspond to the parameter of a "perturbed region of the LC matrix" and not to the true value of S of an unperturbed LC phase. Lower values of S for LC polymers than for their low-molecular-weight analogs can be the consequence of the above.

An increase in the length of the spacer group in a series of nematic polymers has almost no effect on the value of the order parameter, according to investigations on polysiloxane polymers with spacer groups of three to six methylene groups [186]. The studies on deuterated polyacrylic LC polymers with "labeled" mesogenic groups and an aliphatic spacer are of great interest in this respect [272]. A comparison of the order parameters and their temperature dependence for the polymers and their low-molecular-weight analogs (including oligomers) showed that the values of S decrease in the order: low-molecular-weight liquid crystal → LC oligomer [degree of polymerization (DP) = 10] → LC polymer (DP = 100) for both the nematic and for the smectic mesophase. The value of the order parameter for the spacer group is less than for the mesogenic group, but the absolute difference in the value of S is essentially dependent on the position of the "label" methylene groups in the spacer. The farther they are from the mesogenic group (i.e., closer to the main chain), the smaller the value of the order parameter; both the conformation of the main chain and the type of mesophase in the polymer apparently play an important role in the disordering of the aliphatic spacer groups. Nevertheless, the effect of the nature and flexibility of the main chain and the length of the spacer groups on the order parameter for LC polymers cannot be considered resolved; for this, more precise and finer approaches to its evaluation are required.

The relatively high values of the order parameter of smectic polymers are apparently a reflection of the greater tendency of comb-shaped polymers toward layered ordering, the reason for their similarity to the order parameter of low-molecular-weight liquid crystals. In view of the tendency of nematic LC polymers toward layered ordering, it would seem that their order parameter should even exceed the value of S for low-molecular-weight analogs. However, as shown above, the opposite mechanism usually occurs; the causes of this are not yet completely understood.

4.7.3.1b. Rheological Properties. The study of the rheological properties of LC polymers is very interesting due to the capacity of these polymers to become oriented in electric and magnetic fields (cf. Section 4.8). A detailed study of the rheological behavior of LC polymers is necessary for understanding the mechanism and studying the kinetics of the structural transformations ultimately caused by the movement of the individual segments of the macromolecules. The study of the rheological properties of LC polymers with mesogenic side groups is also of independent interest due to the special comb-shaped structure of their macromolecules, which include, as two types of structural units, polymer chains and mesogenic groups, which simulate the structure of molecules of low-molecular-weight liquid crystals. The first information on the rheological behavior of these polymers was obtained in studying the shear flow of LC polymers XXII and XXIII, which have side groups of identical structure and similar degrees of polymerization, but a different main chain structure [187]. Polymer XIII has a smectic structure S_A and polymer XXIII is also characterized by layered packing of the side chains, and after prolonged annealing its structure is similar to the structure of polymer XXIIa:

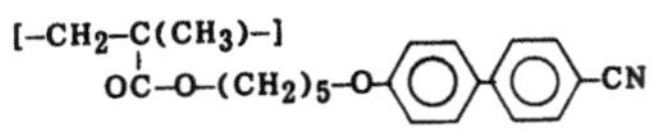

(XXII)

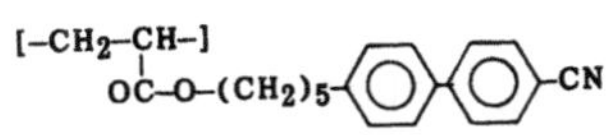

(XXIII)

The analysis of the flow curves of these polymers showed that stable flow is observed for polymer XXIII in the LC phase in a wide range of temperatures up to the glass transition temperature.

For polymer XXII, flow can only be observed in a very narrow temperature range (118-121°C) near T_{cl}. The difference in the rheological behavior of these polymers is most clearly seen when examining the temperature dependences of the viscosity of melts with different rates of deformation (Fig. 4.51).

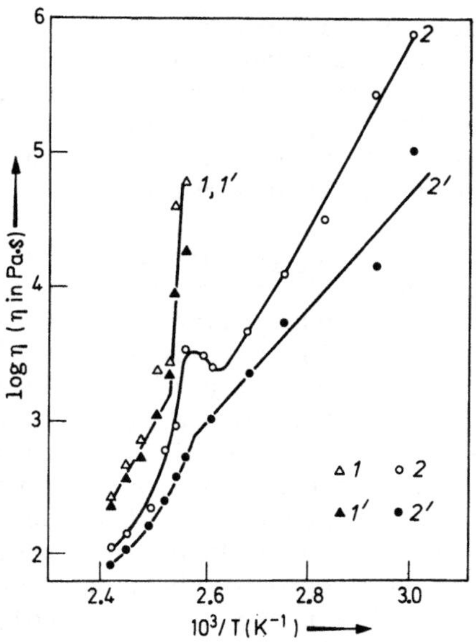

Fig. 4.51. Temperature dependence of the melt viscosity for polymer XXII (1, 1') and polymer XXIII (2, 2') with $\log\dot{\gamma}$ = 3.96 (1, 2) and 1.16 (1', 2') [187].

Note the significant absolute value of the viscosity of polymeric liquid crystals. At T_{cl} (Fig. 4.51), the viscosity of the polymers (~10^3 Pa·sec) is 1-2 orders of magnitude greater than the viscosity of the S_A phase and 4-5 orders of magnitude greater than the viscosity of the nematic phase of low-molecular-weight liquid crystals.

For polymer XXIII, a viscosity maximum is observed in the region of the transition from the LC to the isotropic state with low shear rates $\dot{\gamma}$, and only a break in the dependence of $\log\eta$ on 1/T is observed for polymer XXII in this temperature range. This difference is apparently due to the same causes which determine the difference in the rheological behavior of low-molecular-weight nematic and smectic materials, i.e., during flow, polymer XXII behaves like a smectic liquid crystal, and polymer XXIII behaves like a nematic liquid crystal. The "polymeric character" of liquid crystals is demonstrated by the higher values of the activation energy of viscous flow (E_a) in the mesophase. The value of E_a for

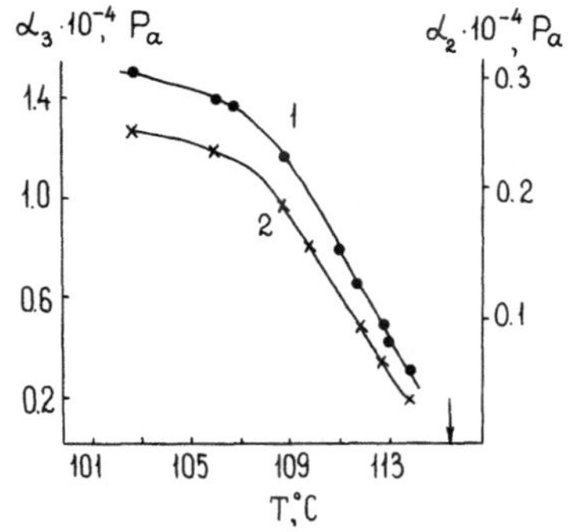

Fig. 4.52. Temperature dependence of Leslie coefficients α_2 (1) and α_3 (2) for polymer XIII. The arrow indicates the N → I transition temperature [274].

polymer XXII is 10^3 kJ/mole, while it is in the range of 80-140 kJ/mole for polymer XXIII.*

Melting of the LC phase observed above T_{cl} causes a sharp change in the character of the flow, manifested by the convergence of the values of E_a for both polymers. In the isotropic phase of polymer XXII, the value of E_a (~140 kJ/mole) is only two times higher than the E_a for polymer XXIII (70-80 kJ/mole). In other words, the transition from the LC phase to the isotropic melt, accompanied by "release" of the mesogenic groups from the mesophase, reduces the difference in the character of the flow of these polymers. The difference in the values of E_a in the isotropic phase is only determined by the difference in the chemical structure of the main chain of these polymers. The values of E_a in this case approach the value of E_a for polybutyl methacrylate and polybutyl acrylate, and these polymers are very similar in structure to polymers XXII and XXIII but contain no mesogenic groups (cf. Section 2.2).

The comparative study of the rotational shear flow behavior of LC polymer XXIII and its low-molecular-weight nematic analog [273-274] provided information on the anisotropy of the rotational viscosity of the LC polymers. The Leslie

---

*This range in E_a values is due to the fact that the value is slightly dependent on the rate of deformation.

coefficients α_2 and α_3 [275] are a quantitative measure of this parameter, and their ratio can be used to judge the form of the structural units which determine the viscous properties of a polymer in the mesophase. A comparison of the temperature dependence of Leslie coefficients α_2 and α_3, determined in [273-274] (Fig. 4.52), with the findings of the rheological studies reported in Fig. 4.51 for the same polymer XXIII, shows the good agreement of the absolute values of the viscosity of the mesophase, and this suggests the possibility of using the Erickson–Leslie theory for melts of LC polymers. As Fig. 4.52 shows, the Leslie coefficients are positive over the entire temperature range studied, and $\alpha_3 > \alpha_2$.

According to the theoretical relations [275], the condition $|\alpha_2| \gg |\alpha_3|$ is usually satisfied for low-molecular-weight nematic liquid crystals, and this corresponds to the oblong shape of the liquid crystal molecules since $\alpha_2/\alpha_3 = a^2b^2$, where b and a are the semiaxes of the ellipsoid of rotation directed perpendicular and along the directrix, respectively. In the case of a LC polymer within the framework of this model in the condition $\alpha_2/\alpha_3 = 0.14$ or $a/b = 0.37$, the structural unit of the polymeric mesophase which determines its viscous properties can be qualitatively approximated by an ellipsoid of rotation whose long axis is perpendicular to the direction of the directrix, i.e., a situation close to the discotic phase occurs.

The study of the rheological properties of LC polymers thus reveals a certain similarity in their behavior to low-molecular-weight liquid crystals. On the other hand, there is also an important difference between them due to the significantly higher absolute values of the viscosity and viscous flow activation energy caused by the contribution of the main chain to the hydrodynamic resistance during flow.

4.7.3.1c. Molecular Mobility in the Solid State. The primary goal of the study of molecular mobility in LC polymers with mesogenic side groups is the clarification of the mechanism of motion of the individual segments of the macromolecules in the main chain: the spacer groups, linkages, mesogenic groups, and the components of their segments.

The more complex structure of LC polymers in comparison to comb-shaped polymers with aliphatic chains significantly complicates the determination of the role of each of these

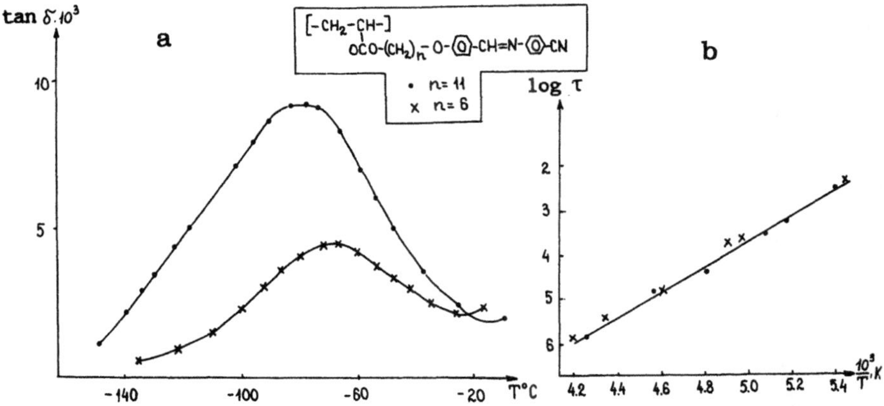

Fig. 4.53. Temperature dependence of tan δ (a) and dielectric relaxation time τ (b) for polymers XXIV (1) and XXV (2) at the frequency of 1 kHz [276].

structural units in molecular mobility. The situation is also complicated by the absence of an adequate number of experimental data which would permit making certain generalizations. In examining the questions concerning molecular mobility in LC polymers, it is useful to recall what De Gennes said about molecular motion in low-molecular-weight liquid crystals: "Very little is known and even less is understood about the dynamics of liquid crystals on the molecular scale" [28]. This is even more applicable to LC polymers, since the data in the literature are occasionally ambiguous and subject to interpretation.

We will follow the local motion of the individual fragments of the macromolecules using, as an example, comb-shaped LC polymers containing Schiff bases as the mesogenic groups

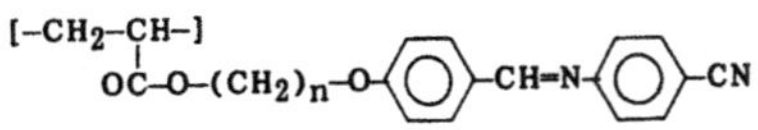

n = 6(N) (XXIV)
n = 11(S) (XXV)

It follows from the data from dielectric studies that a relaxation process related to orientational polarization of the linkages of the ester groups joined to the main chain of the polymer is observed at T = −60 to −80°C for both polymers, regardless of the type of mesophase (Fig. 4.53). This process has approximately the same activation energy for both the nematic and smectic phases (E ≈ 45 J/mole), which is close to the activation energy of comb-shaped polymers containing no mesogenic groups (cf. Chapter 2).

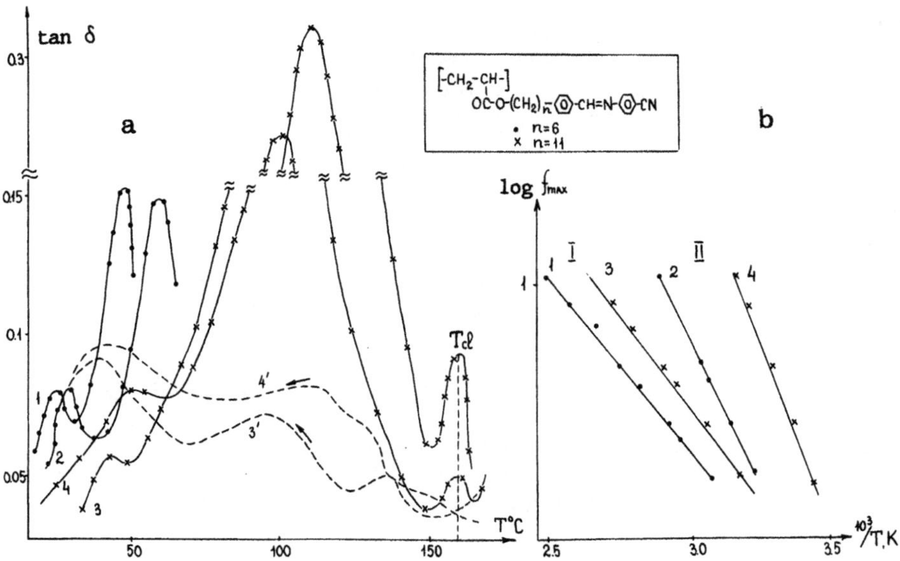

Fig. 4.54. Temperature dependence of tan δ (a) and log f_{max} (b) for the two relaxation regions I and II of polymers XXIV (n = 6) and XXV (n = 11). Frequency: 0.1 (1), 1 (2), 100 (3, 3'), and 300 (4, 4'). Dashed lines 3' and 4' correspond to cooling of the samples [276].

It is interesting that the height of the maximum of tan δ in the polymer with n = 11 is twice that of the polymer with n = 6, although the concentration of polar groups per unit of mass is lower. The absence of a direct correlation between the intensity of the relaxation process and the number of polar groups per unit volume could be caused by the more effective dipole moment of the monomeric unit of polymer XXIV with n = 11 due to a different intramolecular correlation of the ester groups in the smectic phase in comparison to the nematic phase.

In addition to this process, two more relaxation regions and a nonrelaxation region corresponding to melting of the mesophase are observed in the region of the transition from the glassy to the highly elastic state at higher temperatures (Fig. 4.54a). The values of the temperatures at the frequency of 1 Hz were calculated from the dependence of the logarithm of the frequency at the maximum of the peak on the reciprocal temperature (Fig. 4.54b). The values of the activa-

TABLE 4.19. Transition Temperatures, Activation Energy, and Dipole Polarization Relaxation Times Determined by Dielectric Methods for Smectic Cholesterol-Containing Polymers [277]

$[-CH_2-C(CH_3)-]$, with $X-(CH_2)_n-COOChol$ attached to C

X	n	Transition temperature and activation energy (in bulk)			
		T_r, °C	E, kJ/mole	T_g, °C	T_{cl}, °C
CONH	10	50	360	120	180
COO	14	37	190	50	150

Dipole relaxation time in solution (toluene, 40°C) [278-280]

X	n	τ_{dp}, nsec
CONH	11	540
	6	150
COO	10	18
	5	3

tion energies of these processes were also calculated; they are, for regions I and II: region I: 205 kJ/mole (n = 6) and 232 kJ/mole (n = 11); region II: 120 kJ/mole for polymers with n = 6 and 11. Relaxation region I corresponds to thawing of segmental mobility, accompanied by relaxation of the dipole polarization due to the motion of the segments of the main chain and ester group.

The second relaxation region (II), located at higher temperatures, is apparently a result of orientational polarization of the mesogenic groups. The high values of the relaxation times and activation energy of this process are determined by the LC character of side chain ordering and the cooperation of the rotational motion of the side groups above the glass transition temperature. It is difficult to resolve the question of the molecular mechanism of mesogenic group rotation about their long or short axis, as the possibility of segmental motion of the main chain has not been taken into consideration.

As Fig. 4.54 (curve 2) shows, the reverse course of the dependence of tan δ on 1/T upon cooling the polymer indicates a sharp decrease in the intramolecular mobility, apparently due to more perfect LC packing of the mesogenic groups. In addition to these basic relaxation processes, several other relaxation transitions which are still difficult to interpret were observed at some frequencies in [276].

A series of transitions are also observed below the glass transition temperature for polymers containing cholesterol mesogenic groups and forming a smectic mesophase (Table 4.19) [277]. One of them, T_r, corresponds to a relaxation process related to the mobility of the cholesterol units and the ester groups attached to them. A characteristic difference of this process is the value of the activation energy, which is unusually high for local motion. This is apparently due to the greater bulk of the kinetic unit which contains the bulky cholesterol group and possibly the cooperative nature of the polarization process. The value of E in the polymer with the amide bond is higher than in the polymer with the ester group. The dipole polarization relaxation times τ_{dp} for polymers with CO—NH groups are more than one order of magnitude higher than the corresponding values of τ_{dp} for polymers with ester groups.

These data, and the results of structural studies, show that the presence of amide groups, which cause the formation of hydrogen bonds, creates an additional hindrance in the alignment of the cholesterol groups. This results in the fact that the LC state is only realized under certain preparation conditions of the samples for cholesterol-containing polymers with amide groups, while the LC structure is formed under any conditions for cholesterol-containing polymers with ester groups [281-283]. As a consequence, the conditions of LC phase formation in this type of system are determined not only by the thermodynamic but also by the kinetic properties, i.e., the mobility of the side chains bearing mesogenic groups.

The complex processes of the molecular motion of the different polar groups in the side chains significantly complicate the precise assignment of the observed relaxation processes to the motion of the actual segments of the macromolecules, and combination and overlapping of the individual processes frequently occur. Progress in the discovery and understanding of the mechanism of relaxation processes can only be achieved in the rigorous analysis and comparison of structural, thermodynamic, and kinetic data in consideration of all of the molecular characteristics of LC polymers.

The detailed studies of the dielectric behavior of smectic and nematic polymers initiated in this direction [284-288] are intended to reveal the complex mechanism of the structural transformations and relaxation processes which determine the molecular mobility of LC polymers in the solid phase.

4.7.3.2. Copolymers. Copolymerization has been a very fruitful method of obtaining new LC systems. All of the currently known LC copolymers can be divided into two groups: 1) copolymers prepared by copolymerization of two monomers, one of which contains no mesogenic groups; 2) copolymers prepared by copolymerization of two mesogenic monomers (see also Section 4.7.4 on the synthesis of these systems).

Unfortunately, there are almost no systematic studies which follow the change in the physical parameters of the copolymers in a wide range of changes in their compositions. The composition of copolymers has not been specially determined in most published studies and the composition of the initial monomeric mixture has been assumed to be constant, which also complicates the establishment of correlations between the composition and properties of the copolymers. Nevertheless, an analysis of the existing experimental data suggests some conclusions for both the first and second group of copolymers, which is especially important.

4.7.3.2a. Copolymers of Mesogenic and Nonmesogenic Monomers. The study of copolymers of mesogenic monomers with nonmesogenic monomers has established the concentration boundaries of mesophase existence. Using copolymers of cholesterol-containing monomers with alkyl acrylates (A-n) and alkyl methacrylates (MA-n), it was thus shown that the concentration region of mesophase existence is essentially dependent on the length of the alkyl substitutuent C_nH_{2n+1} in the nonmesogenic monomer (Table 4.20) [20]. Although copolymers with butyl acrylate (A-4) or butyl methacrylate (MA-4) exhibit mesomorphism at a concentration of the mesogenic component of less than 20 mole %, copolymers with decyl methacrylate (MA-10) at 75 mole % do not form a mesophase. With a further increase in the length of the alkyl substituent (MA-22), no mesophase is formed even with a 75% concentra-

TABLE 4.20. Copolymers of a Cholesterol-Containing Monomer (CM-II) with n-Alkyl Acrylates (A-n) and n-Alkyl Methacrylates (MA-n) [289]

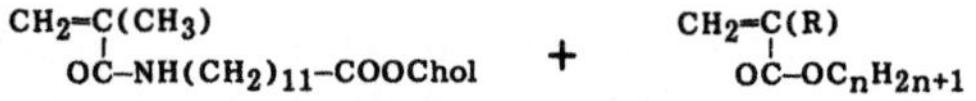

R = H (A-n)
R = CH_3 (MA-n)

Copolymer containing CM-II units, mole %	T_g, °C	T_{cl}, °C
Copolymers of CM-II with A-4		
100	120	180
42	65	160
37	60	140
17	20	100
Copolymers of CM-II with MA-4		
90	115	180
67	105	170
40	85	160
Copolymers of CM-II with MA-10		
75	90	180
58	70	170
25	20	no mesophase
Copolymers of CM-II with A-16		
45	45	100
Copolymers of CM-II with MA-22		
75	70	no mesophase
50	40	

tion of the mesogenic component. The long aliphatic branches shield the mesogenic groups and prevent them from interacting in this case.

The minimum concentration of the mesogenic component necessary in a LC polymer known at present is 9%, which was demonstrated on the example of a copolymer with a polysiloxane main chain [14]:

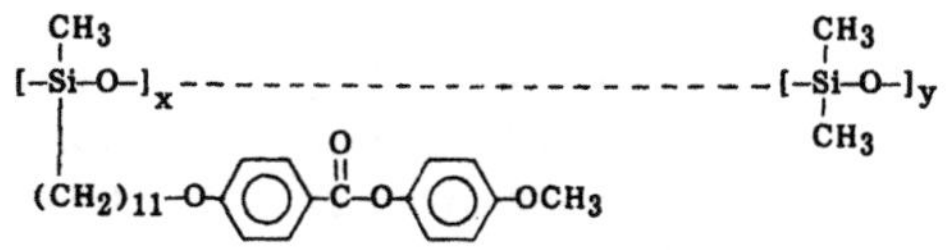

This copolymer has two glass transition temperatures in the region of −114 and −57°C and forms a smectic mesophase with T_{cl} = −15°C.

One of the important advantages of the method of copolymerization of mesogenic and nonmesogenic monomers is the possibility of varying the temperature range of mesophase existence (Table 4.20). The basic mechanism involves a smooth change in the glass transition temperature with a change in the composition of the polymer in the region between the glass transition temperatures of the corresponding homopolymers. In all known cases, the nonmesogenic monomer is selected to decrease the T_g of the copolymer with respect to the T_g of the LC homopolymer. This decrease is obtained both due to an increase in the distance between the bulky nonmesogenic groups ("dilution of the mesogenic groups") and due to a change in the chemical nature of the main chain. For most copolymers, both factors act simultaneously. In addition to the shift in the T_g, a decrease in the clearing point is also usually observed, i.e., the thermal stability of the mesophase decreases.

The introduction of nonmesogenic units into LC homopolymers usually does not cause a change in the type of mesophase. For example, most cholesterol-containing copolymers, like the homopolymers, retain a smectic structure in a wide range of compositions (Table 4.20) [10, 289]. Similar events have also been observed for copolymers containing phenylbenzoate side groups based on acrylates [290] and polysiloxanes [14, 139]. At the same time, of the scores of known copolymers, only a few copolymers exhibiting a change in the type of mesophase on "dilution" of the LC homopolymers with nonmesogenic groups are described in the literature [220]. The introduction of butyl acrylate or isoprene units in smectic cholesterol-containing homopolymers, which increases the flexibility of the main chain, results in a gradual transition from a smectic to a nematic polymer. The opposite picture, the formation of a more ordered smectic mesophase in the copolymer in comparison to the nematic homopolymer, has been found for one of the polysiloxanes (see above, polymers 1 and 2, Table 4.9). At present, it is difficult to explain unambiguously the observed results. However, there is no doubt about the need to conduct a series of systematic studies of copolymers from mesogenic and nonmesogenic components to establish the general mechanisms of the formation and to predict the type of mesophase of the copolymers.

<u>4.7.3.2b. Copolymers of Two Mesogenic Monomers</u>. The method of copolymerization of two mesogenic monomers is now one of the most efficient methods of preparing polymers with

a cholesteric mesophase (see Section 4.7.4). On the other hand, by using copolymerization of smectogenic and nematogenic monomers and studying the properties of copolymers in a wide range of compositions, it is possible to establish the mechanisms of the formation of each type of mesophase. We will demonstrate the efficiency of using this method on the example of copolymers [291] in which the following are succesively varied with identical cyanodiphenyl mesogenic groups: a) the length of the aliphatic spacer groups (n) with the polyacrylate chain unchanged (copolymers of XXVI):

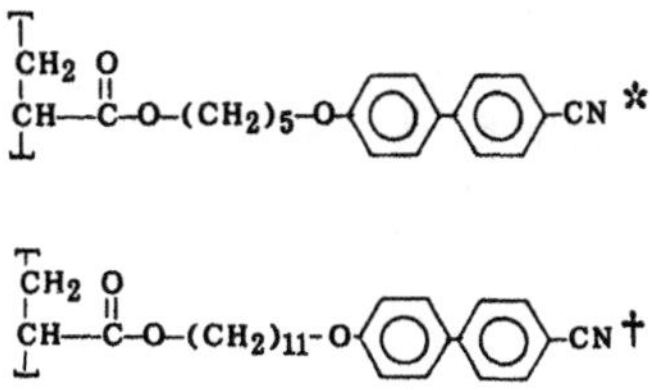

(XXVI)

b) the chemical nature of the main chain with the side groups unchanged (copolymers of XXVII)

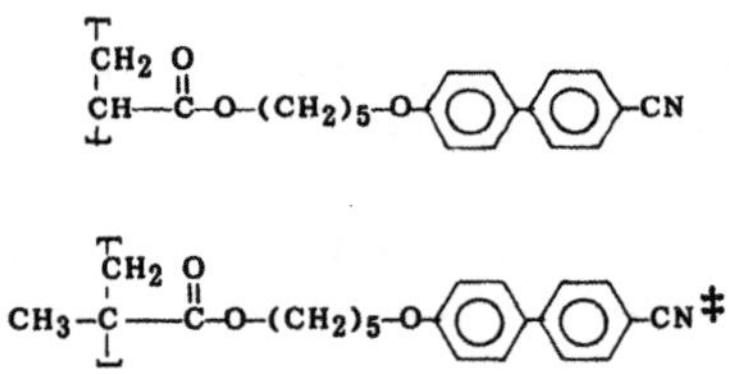

It is important to emphasize that the homopolymers of each of the comonomers in the first case had different types of structures – nematic and smectic – and, in the second case, homopolymers of PA-5 and PM-5 had a similar type of smectic structure. Some of the results of the determination of thermophysical properties of these copolymers are shown in Fig. 4.55. For copolymers of XXVI, the values of T_{cl} and ΔH_{cl} gradually increase with an increase in the concentration of the longer side groups and attain values characteristic of smectic homopolymer PM-11. As Fig. 4.55 and the results of x-ray analysis show, the transition from the nematic to the smectic mesophase takes place with a 30% concentration of

*Homopolymer of PA-5.
†Homopolymer of PA-11.
‡Homopolymer of PM-5.

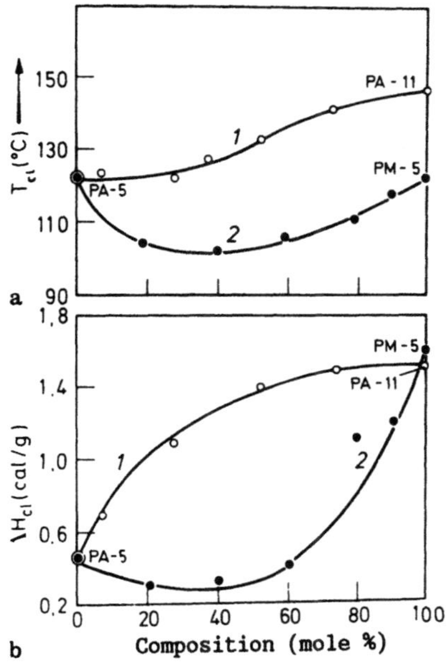

Fig. 4.55. Dependence of T_{cl} (a) and ΔH_{cl} (b) for copolymers XXVI (1) and XXVII (2) on their composition [291].

longer spacer groups. The thickness of the smectic layer formed due to overlapping mesogenic groups increases linearly with an increase in the fraction of smectogenic groups with n = 11. In the case of the comparatively flexible polyacrylate main chain, the type of mesophase formed, as in the case of low-molecular-weight liquid crystals, is determined by the length of the aliphatic segments.

A different picture is observed for copolymer XXVII, where the structure remains smectic in the entire range of compositions, but the flexibility of the main chain gradually changes. As Fig. 4.55 shows, the dependence of T_{cl} and ΔH_{cl} on the composition are curves with a minimum, which could be the consequence of defects in the packing of the side groups due to the introduction of units of a different chemical nature in the main chain.

The glass transition temperatures of these copolymers lie in the region between the T_g of the corresponding homopolymers, as is usually observed for copolymers of mesogenic

TABLE 4.21. Phase Transition Temperatures and Heats in Polysiloxane Copolymers [191]

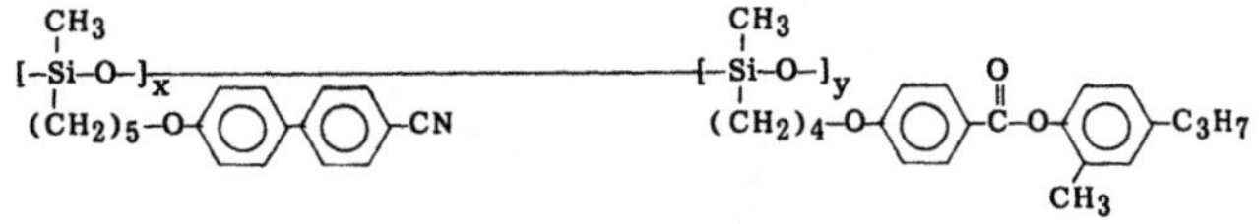

x	y	Type of mesophase	T_g, °C	T_{cl}, °C	ΔH_{cl}, J/g
50	–	S_A	15.0	170	–
35	15	S_A	7.4	151	3.06
25	25	S_A	7.4	122	3.36
16	34	N	4.4	53	0.67

and nonmesogenic monomers. A transition from the S_A smectic mesophase to the nematic mesophase and a decrease in the glass transition temperature were observed in [191] for polysiloxane copolymers of changing comonomer ratios (Table 4.21).

A different situation is observed for copolymers containing CN-substituted and alkoxy-substituted mesogenic groups in the end groups of the side chains of the macromolecules (Table 4.22) [259]. In this case, copolymerization leads to a significant increase in the thermal stability of the smectic mesophase (from 121°C for the homopolymer to 171-177°C for the copolymers). A similar event is also observed for copolymers containing mesogenic groups of the Schiff-base type as the nematogenic comonomer [259]. The increase in the stability of the smectic mesophase could be due to an additional donor—acceptor interaction between the mesogenic groups containing electron-donor ($-OCH_3$) and electron-acceptor (–CN) substituents, similar to what takes place in mixtures of low-molecular-weight compounds. Copolymerization of monomers containing electron-donor and electron-acceptor substituents opens up interesting new possibilities for increasing the thermal stability of the smectic mesophase of polymers.

Copolymerization is thus an effective method for modifying the properties of LC polymers for the purpose of altering their thermal characteristics and creating new types of LC polymers, particularly cholesteric LC polymers.

TABLE 4.22. Phase Transitions of Copolymers [259]

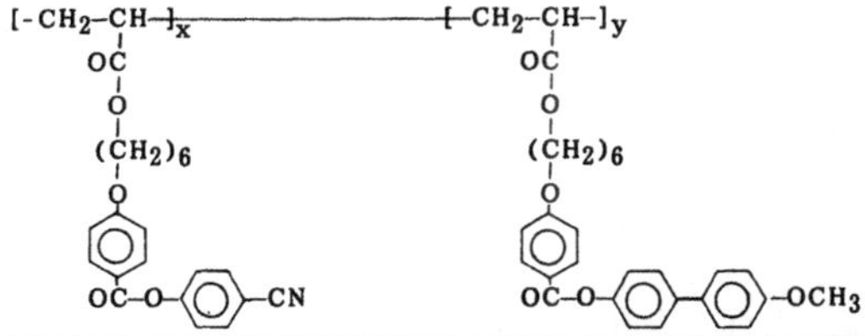

X, mole %	T_g, °C	Type of mesophase and transition temperature
100	60	$N \xrightarrow{109^\circ} I$
50	20	$S \xrightarrow{177^\circ} N \xrightarrow{194^\circ} I$
33	29	$S \xrightarrow{172^\circ} N \xrightarrow{209^\circ} I$
20	45	$S \xrightarrow{171^\circ} N \xrightarrow{227^\circ} I$
0	100	$S \xrightarrow{121^\circ} N \xrightarrow{271^\circ} I$

4.7.4. The Cholesteric Mesophase

Polymers of the cholesteric type are especially interesting LC polymers due to the unique optical properties of cholesteric materials [55].

Attempts to prepare polymeric cholesteric liquid crystals produced no positive results for a long time.

All of the homopolymers prepared by polymerization of cholesterol and cholestanol acrylates and methacrylates described at the beginning of the 1970s (Table 4.23) were rigid, brittle substances with high softening points which did not exhibit LC properties.

It should be noted that in most of the studies from this period, the primary attention was focused on examining the kinetic mechanisms of polymerization of these LC monomers and the determination of the role of the phase state of the monomers in stereospecific processes.

According to the data in [298, 299], some of the polymers reported in Table 4.23 (polymers 1-3 and 5) have a smectic type of structure frozen in the glassy matrix. The side chains of these polymers form layers of the type illustrated in Fig. 4.28 for the smectic A phase of LC polymers. The di-

TABLE 4.23. Cholesterol-Containing Polymers Not Exhibiting Liquid-Crystalline Properties

Monomer No.	Monomer and its structural formula	Phase transition temperature (°C) and type of mesophase of the monomers	Polymerization conditions (bulk, solution, T) and structure of the polymer (small-angle interplanar distance; T_m, T_g, °C)	Reference
1.	Cholesteryl acrylate 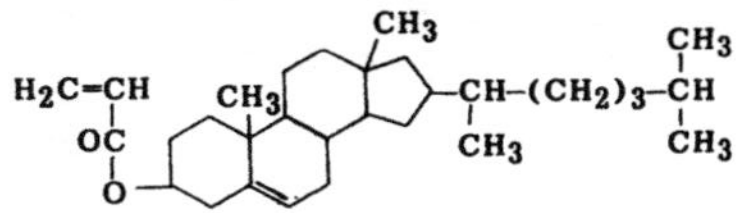	a) $C \xrightarrow{121} I$; $I \xrightarrow{90} Chol \rightarrow C$	Polymerization in a melt yields an insoluble polymer. Polymerization in solution yields a soluble polymer with T_{chem} dec. = 275°C.	[292]
		b) $C \xrightarrow{121} M \xrightarrow{126} I$; I ⇠125⇠ M ⇠110⇠ C	T = 120, 123, 126° Structure of the polymer not studied	[293]
		c) $C \xrightarrow{65} C_2 \xrightarrow{121} Chol$; $\xrightarrow{126} I$		[294, 295]
		d) $C \xrightarrow{81} S \xrightarrow{94} Chol$; $\xrightarrow{120} I$		[154]
		e) $S \xrightarrow{<118.7} N \xrightarrow{125.8} I$		[296]
		f) $C_1 \xrightarrow{58} C_2 \xrightarrow{125.8} M$; M ⇠126⇠ M ⇠100⇠ C_1; $\xrightarrow{127} I$		[297]
		g) $C \xrightarrow{75} H \xrightarrow{125} I$; $\xrightarrow{125} Chol$	d = 33.4 Å	[298, 299]

		h) $C \xrightarrow{125.8} I \xrightarrow{124.8} Chol$	After reprecipitation, the polymer has an amorphous structure	[298, 299]
2.	Cholesteryl methacrylate 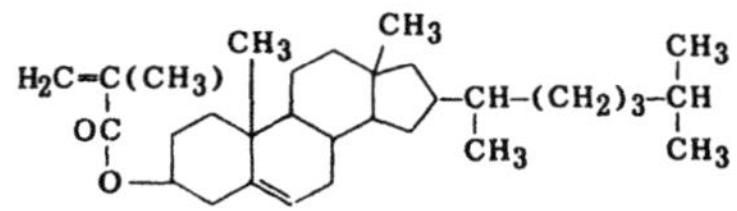	a) $C \underset{90-93}{\overset{107-108}{\rightleftarrows}} Chol \underset{110-111}{\overset{110-112}{\rightleftarrows}} I$	a) –	[300]
		b) $C \xrightarrow{109} I$; $I \xrightarrow{103} Chol$; $Chol \xrightarrow{85} C$	b) –	[301]
		c) $C_1 \xrightarrow{64} C_2 \xrightarrow{115} I$; $C_2 \xrightarrow{111.5} Chol$; $I \xrightarrow{114} Chol$	c) Polymer has no LC properties	[302]
		d) $C \xrightarrow{114.8} I \xrightarrow{111.8} Chol$	Polymer has S structure d = 35.3 Å, T_g = 160°	[298, 299]
3.	Cholestanyl acrylate 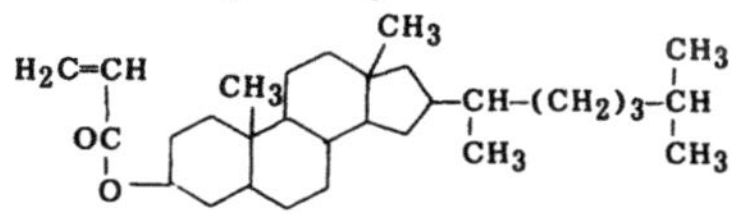	a) $C \xrightarrow{100} I$	–	[292]
		b) $C \xrightarrow{98} Chol \xrightarrow{104} I$	Polymer has no LC properties	[295]
4.	Cholestanyl methacrylate 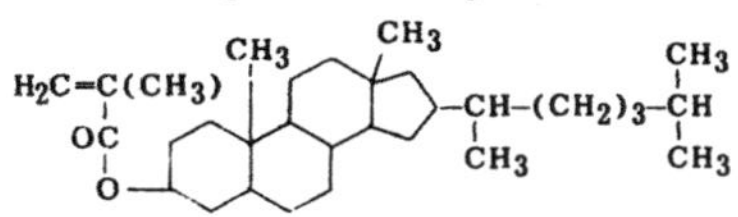	$C \xrightarrow{95} I$; $I \dashrightarrow{95} M$; $M \dashrightarrow{85} C$	Polymer has no LC properties	[295]
		$I \xrightarrow{85} M$	Polymerization conducted in I and LC phase and in solution. T_g = 98°C; all polymers have an amorphous structure	[302]

TABLE 4.23 (continued)

Monomer No.	Monomer and its structural formula	Phase transition temperatures (°C) and type of mesophase of the monomers	Polymerization conditions (bulk, solution, T) and structure of the polymer (small-angle interplanar distance; T_m, T_g, °C)	Reference
5.	Cholesteryl-n-acryloyl hydroxybenzoate 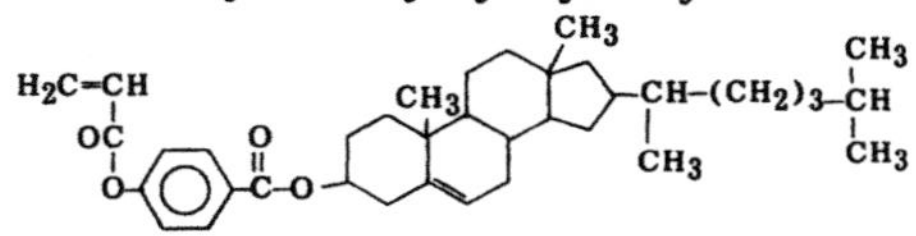	$C \xrightarrow{128} Chol$	Polymer has S structure; d = 45.2 Å and 51.9 Å	[298, 299]

TABLE 4.24. Phase Transition Temperatures of Poly-(N-methacryloyl- and Acryloyl-ω-aminocarboxylic acid) Cholesterol Esters

$[-CH_2-CX-]$
$OC-NH(CH_2)_n-COOChol$ [2, 237, 289]

Polymer	X	n	T_g, °C	T_{cl}, °C
1	CH_3	2	185	250
2	CH_3	5	130	220
3	CH_3	6	130	215
4	CH_3	8	130	200
5	CH_3	10	125	185
6	CH_3	11	120	180
7	H	2	190	>250
8	H	5	165	>250
9	H	11	110	185

rect addition of cholesterol and cholestanol to the main chains of the macromolecules thus actually does not permit obtaining thermotropic LC polymers characterized by a certain temperature range of the fluid LC phase existence.

The formation of the LC phase (and not a LC structure) is only possible if the cholesterol groups are added to the main chain by means of a spacer group of methylene units, i.e., the principle behind the structure of comb-shaped polymers is used. This method of preparing cholesterol-containing LC polymers was first used by Shibaev, Fredizon, and Platé for synthesis of poly-N-acryloyl- and methacryloyl-ω-aminocarboxylic acid cholesterol esters by polymerization of the corresponding monomers (Table 4.24) [2, 237, 289].* As Table 4.24 shows, all of these polymers have relatively high clearing points, but they exhibit LC properties in the

*The synthesis of a polymer similar in structure to polymer 6 (Table 4.24) by chemical addition of cholesterol molecules to poly-N-methacryloylaminolauric acid is also described by Paleos et al. in [134]; however, according to their data, the polymer contained approximately 10% carboxyl groups and did not exhibit LC properties.

TABLE 4.25. Properties of Cholesteric Homopolymers [189, 304]

No.		n	T_g, °C	$T_{S \to Chol}$, °C	T_{cl}, °C	ΔH_{cl}, J/g
1	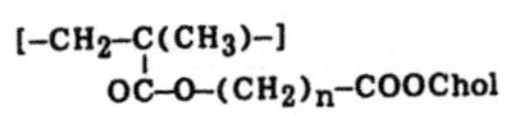	5	85	190	210	2.1
2		10	60	124	158	1.4
3	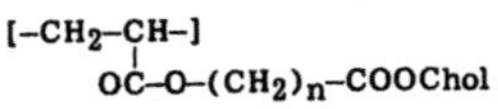	5	55	†	218	4.2
4		10	35	140	148	3.2
5	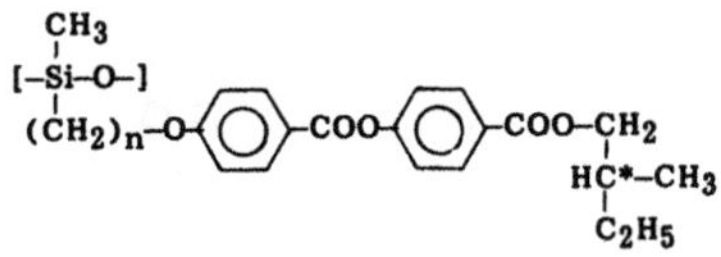	3	2	–	13	–
6		4	−3	17	37	–

*ΔH transition ~0.4 J/g.
†Monotropic cholesteric phase; $T_{S \to Chol}$ is a function of the cooling mode.

region of T_g-T_{cl}. However, structural studies show that these polymers have a smectic type of packing [289]; this is apparently due to a network of hydrogen bonds in these polymers [303] which are capable of fixing the layered structure. However, it is also possible that a different type of structure, nematic or cholesteric, can be realized in the region of high temperatures close to the clearing points where the hydrogen bonds can be destroyed. That this is actually possible is confirmed by the results of a study of structurally similar cholesterol-containing polymers which contain ester COO bonds instead of amide CO—NH groups (polymers 1, 2 and 3, 4, Table 4.25). The polymers in this homologous series form the smectic A phase (see below) in a wide range of temperatures and only form the cholesteric mesophase several degrees below the clearing point. It is important to emphasize that the calorimetric determination of the cholesteric mesophase is made difficult by the very low heat of the smectic—cholesteric transition and the "blurring" of the heat effect due to the polydispersity of these polymers. For this reason, only the existence of the smectic mesophase was communicated in the initial publications on the properties of these cholesterol-containing polymers. In addition, polymers of the acrylic series most easily form the cholesteric mesophase when the samples are cooled (for example, polymer

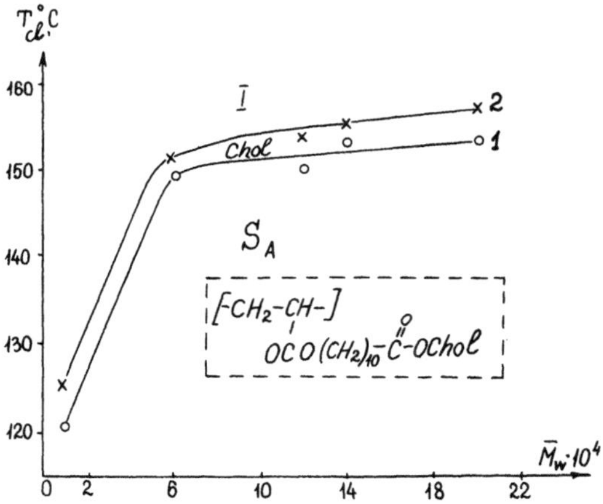

Fig. 4.56. Effect of the molecular weight on the S_A → cholesteric (1) and cholesteric → isotropic melt (2) transition temperatures [304].

3 in Table 4.25). The degree of polymerization also has an important effect on the temperature range of cholesteric mesophase formation (Fig. 4.56). First should be noted the presence of a critical value for the degree of polymerization above which both transition temperatures approach saturation, previously noted for both nematic and smectic polymers and also characteristic of the cholesteric mesophase, as Fig. 4.56 shows. In addition, the temperature range of the cholesteric mesophase is very narrow (~3-4°C), which significantly complicates the study of the properties of these cholesteric polymers.

The formation of the cholesteric mesophase for some polysiloxane homopolymers with chiral mesogenic groups is described by Finkelmann and Rehage [189] (Table 4.25, polymers 5 and 6). The study of the dependence of the transition temperatures on the length of the main chain for oligomeric and polymeric fractions of sample 6 (Table 4.25) showed that, although the formation of the smectic mesophase begins with the dimer, the formation of the cholesteric mesophase begins with the hexamer [189].

Another method of obtaining the cholesteric mesophase in polymers consists of introducing optically active monomeric units (including cholesterol-containing units) into

TABLE 4.26. Cholesteric LC Copolymers Containing Chiral Units

No.	General formula of the copolymer	Concentration of chiral units in the copolymer,* mole %	T_g, °C	T_{cl}, °C	λ max of selective reflection of light, nm (20°C)	Reference
1.1	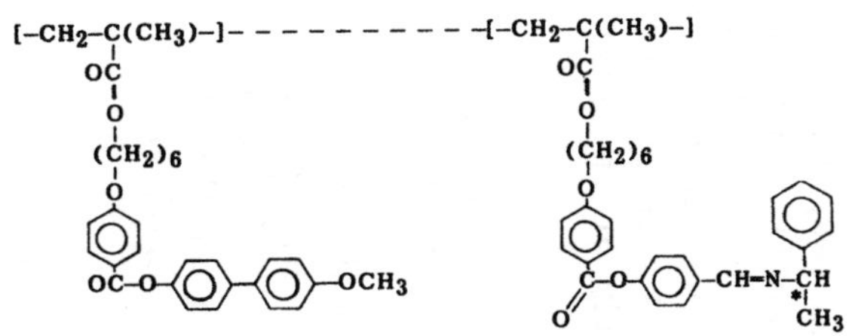	90.6	70	247	1260	[306]
1.2		83.6	73	229	712	
1.3		79.8	77	216	562	
1.4		75.3	80	203	467	
2.1	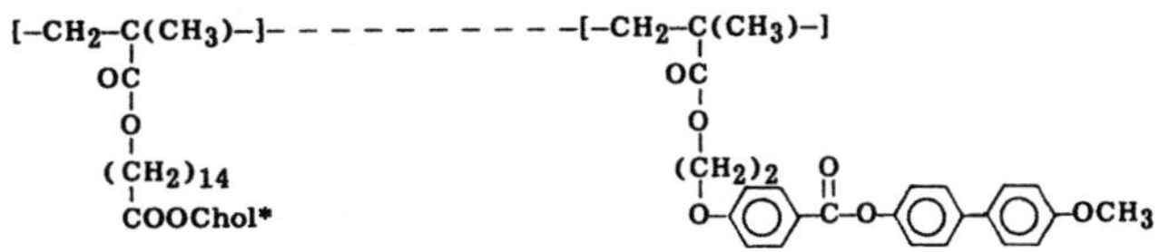	12	132	260	1500	[307]
2.2		16	125	245	1200	
2.3		24	117	238	1100	
3.1	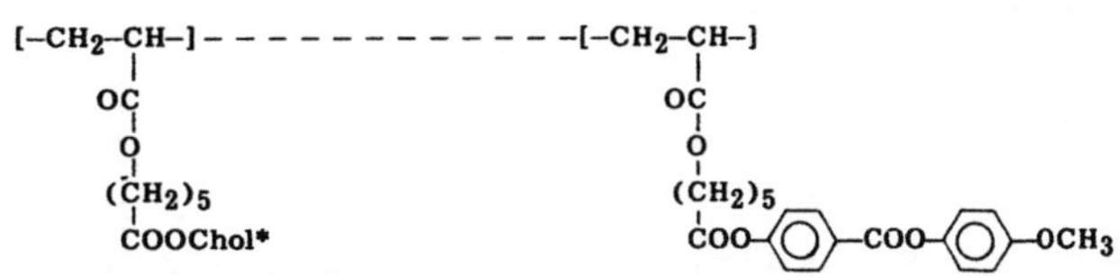	35	48	103	495	[262]

4.1	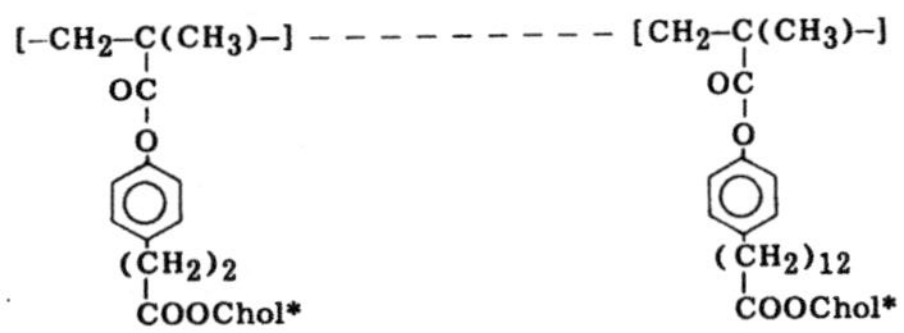		51†	–	209	–	[305]
5.1	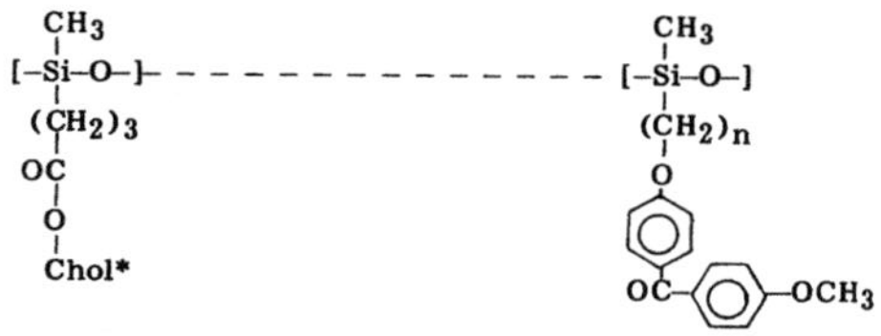	n = 3	3-15	–	–		
5.2		n = 4	5-15†	–	–		[243]
5.3		n = 5	5-15†	–	–		
6.1	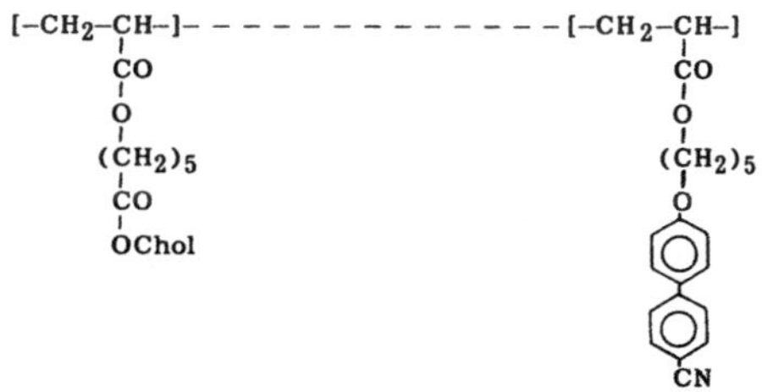		19	50	98		
6.2			28	50	102		
6.3			36	55	105		[308]
6.4			52	55	115		

*The chiral group in the copolymer is indicated by an asterisk.
†Concentration of the chiral monomer in the initial monomeric mixture.

TABLE 4.27. Properties of Cholesterol-Containing Copolymers [262, 308-311]

No.	Copolymer (mole % of cholesterol-containing monomer)	Glass transition temperature, T_g	Clearing point T_{cl}, °C	Wavelength of selective reflection λ_{max}, nm
	Series A CA‡/AM-5			
1	15	20	128	680†
2	21	30	117	530†
3	33	60	95	450†
4	40	80	90	–
5	52	100	–	–
	Series B CA-5/AM-5			
6	20	14	117	560*
7	35	30	103	495*
8	45	40	110	400*
9	65	45	140	–
	Series C CA-10/AM-5			
10	17	20	121	770*
11	21	–**	110	670*
12	28	–**	118	560*
13	39	30	115	460*
	Series D CM-10/AM-5			
14	19	20	122	740*
15	22	20	123	640*
16	33	25	124	515*
	Series E CA-5/AC-5			
17	19	50	98	850*
18	28	50	102	660*
19	36	55	105	555*
20	52	55	150	500*

*λ measured at $T = 0.99T_{cl}$.
†λ measured at 20°C.
‡CA) Cholesteryl acrylate.
**T is below room temperature.

nematic LC polymers [305-311]. Since the cholesteric mesophase is essentially a "twisted" nematic structure, by chemically binding the chiral units in one chain through copolymerization or copolycondensation (in the case of linear polymers) of nematogenic and optically active compounds, it is possible to "twist" the nematic mesophase and obtain copolymers of the cholesteric type. Such monomers as cholesteryl acrylate, whose homopolymer does not form a cholesteric mesophase, can be used as the chiral units. Some characteristics of cholesteric LC polymers are reported in Tables 4.26-4.28.

TABLE 4.28. Phase Transitions and Wavelengths of Selective Light Reflection for Cholesteric Copolymers

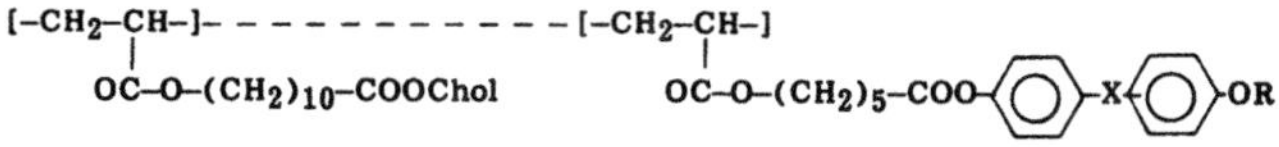

Copolymer	R	X	Mole % CA-10	Phase transition	λ_{max} with $T = T_{S \rightarrow Chol} + 10°C$
1	C_3H_7	$-\mathrm{C(=O)-O}-$	28	$S_A \xrightarrow{100} \mathrm{Chol} \xrightarrow{118} I$	700
2	C_3H_7	$-\mathrm{C(=O)-O}-$	31	$S_A \xrightarrow{98} \mathrm{Chol} \xrightarrow{100} I$	510
3	C_3H_7	$-\mathrm{C(=O)-O}-$	40	$S_A \xrightarrow{80} \mathrm{Chol} \xrightarrow{96} I$	480
4	C_3H_7	$-\mathrm{C(=O)-O}-$	55	$S_A \xrightarrow{85} \mathrm{Chol} \xrightarrow{104} I$	420
5	C_4H_9	$-\mathrm{O-C(=O)}-$	34	$S \xrightarrow{75} \mathrm{Chol} \xrightarrow{122} I$	600

The cholesteric mesophase has been obtained in only one case in copolymerization of approximately equimolar ratios of cholesterol-containing monomers with different lengths of binding groups where each of the homopolymers forms a smectic structure [305].

4.7.4.1. Optical Properties. The selective reflection of incident light is a basic distinctive feature of the cholesteric mesophase which indicates a helical structure in the arrangement of the molecules (Fig. 4.1). This unusual structure of cholesteric materials is the basis for the significant interest in the study of their optical properties. If a homogeneously oriented cholesteric phase is illuminated with light parallel to its optical axis, selective reflection of light of wavelength λ_{max} is observed, as Fig. 4.57 shows. If left-handed circularly polarized light is reflected, these polymers are characterized by a left-handed helix, and if right-handed circularly polarized light is reflected, the polymers have a right-handed helix. It should be noted that most cholesteric polymers are characterized by a left-handed helix (for example, polymers 1-4 in Table 4.25), but there are also some with a right-handed helix (for example, polymer 6 in Table 4.25).

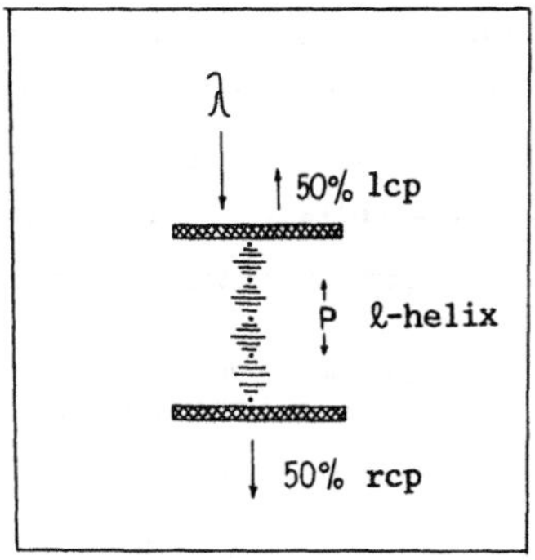

Fig. 4.57. Schematic representation of the selective reflection of left-handed, circularly polarized light (lcp) by the cholesteric mesophase of a polymer with an ℓ-helix. The right-handed circularly polarized light (rcp) passes through the sample.

The value of λ_{max} is correlated with the pitch of the cholesteric helix P of the mesophase by the following relation:

$$\lambda_{max} = \bar{n}D \qquad (4.8)$$

where $\bar{n}$ is the average index of refraction of the cholesteric phase.

The values of λ_{max} for cholesteric polymers can lie in a wide range of wavelengths which includes the IR, visible, and UV regions of the spectrum (Table 4.26). The number of synthesized cholesteric homopolymers described in the literature is very limited, and it is thus not yet possible to establish certain correlations between the molecular structure and optical properties of cholesteric polymers. The basic studies are primarily examinations of the temperature dependence of λ_{max}, i.e., the changes in the pitch of the helix with temperature. In studying the temperature dependence of the pitch of the helix for polysiloxane homopolymers 5 and 6 (Table 4.25) with different spacer lengths, it was found that the values of λ_{max} differ significantly for the same temperature ratio T_{meas}/T_{cl} = 0.97 (where T_{meas} is the temperature of the measurement and T_{cl} is the clearing point); λ_{max} = 490 nm (for polymer 5 with n = 3) and λ_{max} = 645 nm

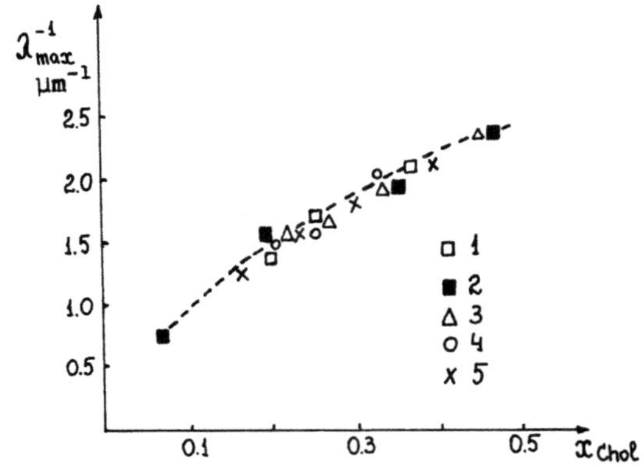

Fig. 4.58. Reciprocal wavelength of reflection λ_{max}^{-1} versus concentration (mole %) of the cholesterol-containing monomer for copolymers of series A (1), B (2), C (3), and D (4) (Table 4.27) and for a low-molecular-weight mixture of cholesterol propionate—butoxybenzylidene-p-butyl aniline (5) [262, 308-311].

(for polymer 6 with n = 4). This means that the shorter spacer in the homopolymers causes stronger twisting of the helix. A similar mechanism was also found for cholesteric copolymers containing cholesterol as the chiral copolymer with variation of the spacer group length in the nematogenic copolymer (polymers 5.1-5.3, Table 4.26).

Since the formation of the cholesteric mesophase in copolymers can be considered to induce a helical structure in a nematic polymer under the effect of chiral units, the following expression is used for characterizing the twisting force of the chiral additive, htp, as in the case of low-molecular-weight nematic—cholesteric mixtures:

$$htp = dP^{-1}/dx_{ch}, \quad x_{ch} \ll 1 \tag{4.9}$$

where x_{ch} is the mole fraction of the chiral additive and P is the pitch of the helix.

In consideration of relation (4.8) and the fact that the average index of refraction $\bar{n}$ = const, it is possible to write

$$htp = \bar{n}d\lambda_{max}^{-1}/dx_{ch} \qquad (4.10)$$

i.e., the value of htp is determined by the slope of the dependence of λ_{max}^{-1} on x_{ch}. By comparing the values of htp for different copolymers, it is possible to estimate the twisting force of the chiral units. Figure 4.58 shows the dependence of λ_{max}^{-1} on the mole fraction of cholesterol-containing units for the series of copolymers in Table 4.27, prepared by copolymerization of cholesterol-containing acrylic and methacrylic monomers

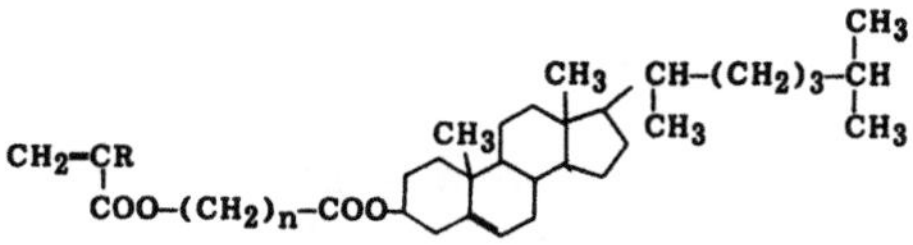

where R = H (CA-n) and R = CH_3 (CM-n) with n = 0 (cholesteryl acrylate), 5, and 10 for different acrylic nematogenic monomers with the same spacer group length.

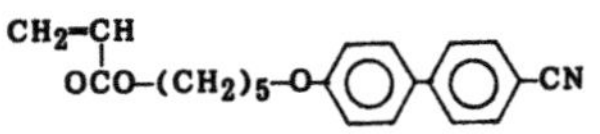

(AC-5)

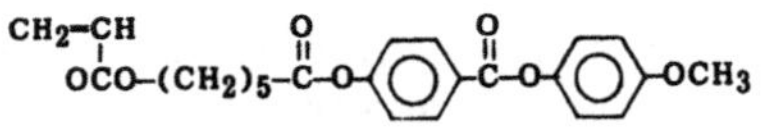

(AM-5)

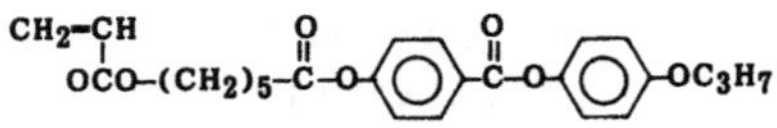

(AP-5)

Figure 4.58 shows that, for most of the copolymers, the experimental points are clustered around one curve, which indicates the similar twisting force of cholesterol-containing monomers CA-5 and CA-10, which is approximately equal to $1\text{-}2\cdot10^{-2}$ nm^{-1}. This value is similar to the values of htp obtained for polysiloxane copolymers and corresponds to the twisting force of low-molecular-weight cholesterol alkanoates in order of magnitude, which can be seen from a comparison of the data for the copolymers and nematic-cholesteric mixtures (points 1-4 and 5). These preliminary data were obtained for a small number of samples and in a limited range of compositions of the copolymer, but they nevertheless suggest that all cholesterol-containing monomers (including cholesteryl acrylate) cause twisting of the cholesteric helix formed in the copolymers according to the same principle

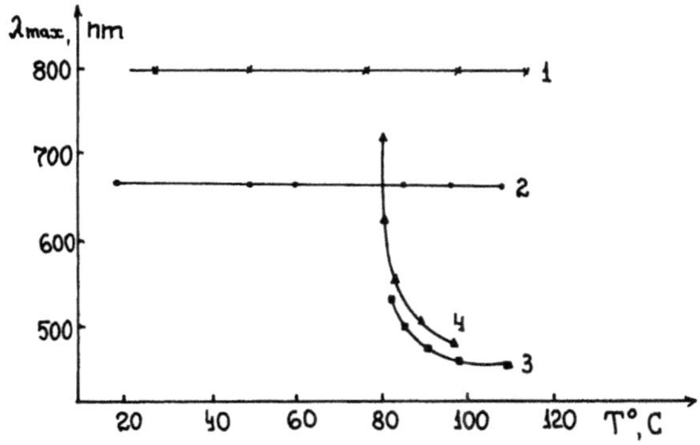

Fig. 4.59. Temperature dependence of λ_{max} for cholesteric polymers: 1) copolymer 10; 2) copolymer 11; 3) copolymer 13 (Table 4.27); 4) copolymer 4 (see Table 4.28).

as in low-molecular-weight mixtures. It is interesting to note that the cholesteric mesophase is obtained even when cholesteryl acrylate (CA) is introduced into the macromolecules of the copolymer (series A), although the homopolymer of CA has an amorphous structure. However, in this case, the cholesteric texture is very defective and is characterized by a broad selective reflection peak, in contrast to the narrow selective reflection peaks in other copolymers. In addition, as Table 4.27 shows, the cholesteric mesophase is not formed (series A) with a concentration of cholesteryl acrylate (CA) above 50 mole %.

The study of the temperature dependence of the pitch of the helix provides important information on the temperature region of selective reflection of light, which is especially important in the practical use of cholesteric polymers for recording thermal fields of different origin. On the other hand, this is necessary for understanding the general mechanisms of the properties of cholesteric polymers and their correlation with molecular structure. Despite the very limited number of studies conducted in this area, the mechanisms found here are of significant interest.

The value of the pitch of the helix is usually weakly dependent on the temperature in low-molecular-weight cholesteric materials in the region far from phase transitions [55]. A sharp temperature dependence of the decreasing pitch of the helix is usually observed near the transition into

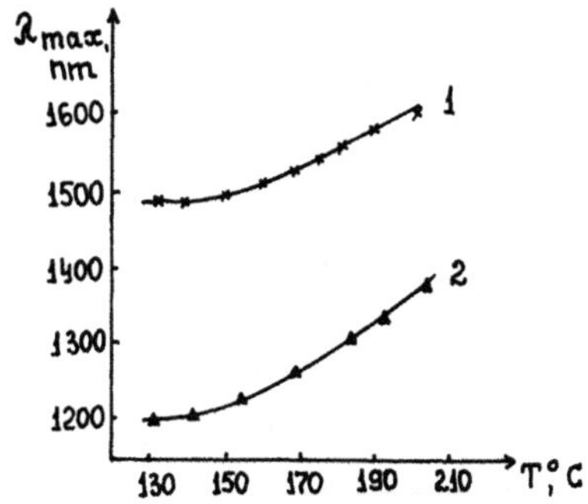

Fig. 4.60. Temperature dependence of λ_{max} for cholesteric polymers (Table 4.26): 1) copolymer 2.1; 2) copolymer 2.2.

the smectic mesophase if it exists. The analysis of the data on the optical properties of cholesteric polymers shows that there are at least three types of temperature dependence of the pitch of the helix.

First, there are monochrome cholesteric materials in which the pitch of the helix does not change through a wide range of temperatures (polymers of series C, Table 4.27). As Fig. 4.59, curves 1 and 2 show, with a small concentration of cholesterol-containing units (under 25 mole %), the copolymers in series C are characterized by a constant value of λ_{max} in a wide range of temperatures from 23 to 110°C. However, for another type of cholesteric polymers, the character of the temperature dependence of λ_{max} changes sharply with an increase in the fraction of chiral units (Fig. 4.59, curve 4). When the polymer is cooled, there is a sharp increase in λ_{max} in the region of 80°C, which probably also indicates the closeness of the cholesteric—smectic phase transition, as in the case of low-molecular-weight cholesteric materials, although this phase transition cannot be established colorimetrically. At the same time, for the series of copolymers containing propoxy and butoxy groups and not methoxy derivatives as the comonomers, whose homopolymers form smectic phases in addition to nematic phases (polymers VIII, IXa, IXb, p. 295), a sharp temperature dependence is observed near the cholesteric—smectic phase transition, as Fig. 4.60 shows for copolymers of CA-10 with butoxy-substituted derivatives (polymer 5, Table 4.28). A comparison of the transition temperatures from the cholesteric to the smectic mesophase (Table

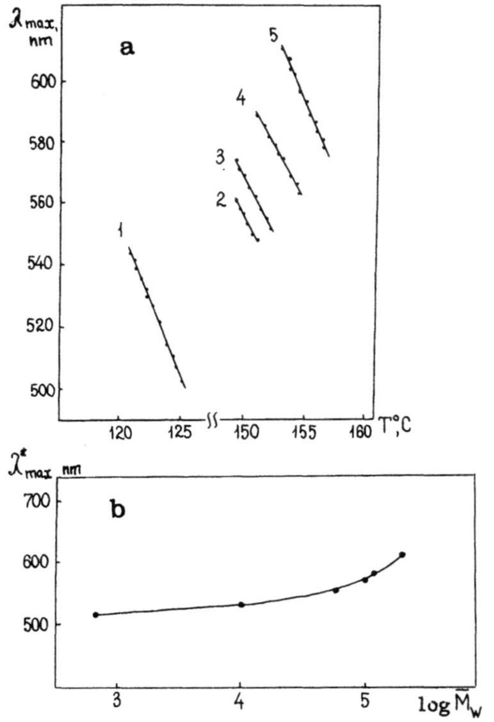

Fig. 4.61. Temperature dependence of λ_{max} for the fractions of cholesterol-containing polymer $[-CH_2-\underset{|}{C}H-]$ $OCO-(CH_2)_{10}-COOChol$ (a) and λ_{max} as a function of the molecular weight of fractions (b) (at $T = T_{S \rightarrow Chol} + 1°C$). Molecular weight of the fractions: 1) monomer; 2) $1.07 \cdot 10^4$; 3) $6.08 \cdot 10^4$; 4) $1.2 \cdot 10^5$; 5) $1.4 \cdot 10^5$; 6) $2.05 \cdot 10^5$ [304].

4.28) with the temperature range of the sharp increase in λ_{max} clearly shows that the increase in λ_{max} takes place in the region of the cholesteric—smectic transition. A temperature dependence of this type is characteristic of most low-molecular-weight cholesteric materials and has been explained within the framework of Keating's theory [313] based on the model of an anharmonic oscillator.

Finally, the inverse course of the temperature dependence of the pitch of the spiral just examined was found for two types of cholesteric polymers (copolymers 2.1-2.3, Table 4.26; Fig. 4.60). The causes of this behavior of cholesteric polymers are just beginning to be elaborated and can partially be explained using theoretical concepts of the anharmonicity of the rotational vibrations of the individual segments and their conformational mobility developed by Lisetski [314].

The effect of the main chain length, i.e., the degree of polymerization, on the optical properties of cholesteric polymers is very interesting. The studies initiated in this direction showed that the region of selective reflection is shifted to the long-wave region of the spectrum with an increase in the degree of polymerization, as Fig. 4.61a shows. The character of the temperature dependence of λ_{max} remains similar for all fractions of the polymer studied. In addition, the value of λ_{max} for the low-molecular-weight monomeric analog has the minimum value, which could be the limit of λ_{max} for the polymeric fractions.

The dependence of λ_{max} on the degree of polymerization (Fig. 4.61b) shows that some untwisting of the helix is observed with an increase in the molecular weight of the polymer, which is either the effect of an increase in the length of the main chain or the effect of the temperature, since an increase in the molecular weight is accompanied by a shift in the high-temperature region. Both factors could act together. In any case, the dependence of the pitch of the helix on the molecular weight of cholesteric polymers should be taken into consideration both in the study of the mechanisms of their optical behavior and in the creation of optical elements in the form of filters and reflectors of IR, visible, and UV light.

The macromolecular nature of cholesteric LC polymers is responsible for an interesting feature related to the possibility of preparing monochrome films. Since the value of the pitch of the spiral in polymeric liquid crystals virtually does not change with a change in the temperature below the glass transition temperature, the cholesteric phase is practically frozen in the glassy matrix. This means that abrupt cooling of polymeric films from the cholesteric mesophase in correspondence with the known temperature dependence of λ_{max} permits fixing their optical characteristics, making it possible to use them at ordinary temperatures as selective monochrome reflectors. On the other hand, these polymeric

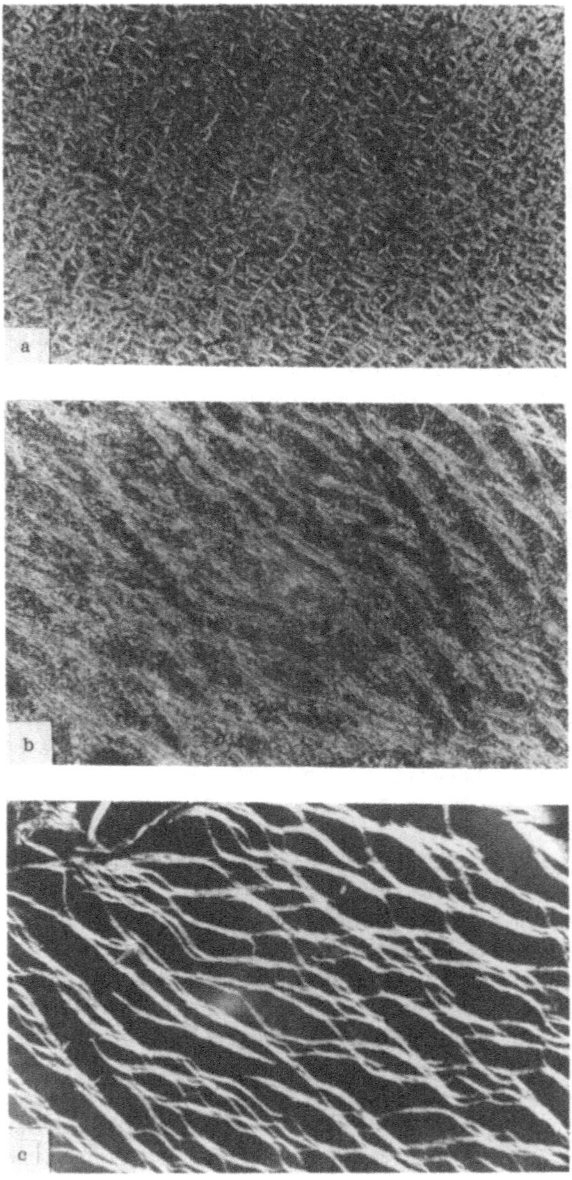

200 μm

Fig. 4.62. Confocal (a), fan-shaped (b), and planar textures with oily streaks (c) of cholesteric polymers for copolymer 5 (Table 4.28).

films have the unusual polarization properties of cholesteric materials, i.e., the capacity to reflect right- and left-handed circularly polarized light differently. The use of these films as circular dichroic optical filters with all of the advantages of polymeric materials is undoubtedly of important scientific and practical interest.

4.7.4.2. The Structure of Cholesteric Polymers. Since most cholesteric polymers have glass transition temperatures above room temperature (except for polysiloxanes), their structure is basically studied in the frozen LC state. Optical microscopic studies show that a confocal fine-grained structure is the most characteristic type of texture of polymeric cholesteric materials (Fig. 4.62a). Prolonged annealing of films with this kind of texture can result in the formation of a fan-shaped texture (Fig. 4.62b) which is also characteristic of low-molecular-weight cholesteric materials. If the polymer samples are deformed in the cholesteric mesophase, for example by shear deformation between glass, the formation of a planar texture with oily streaks can be observed, which is characteristic of the cholesteric mesophase of low-molecular-weight compounds (Fig. 4.62c).

Since some cholesterol-containing homopolymers (Table 4.25) form smectic and cholesteric mesophases, it is useful to begin with an analysis of the smectic packing to examine how the structural organization changes in the transition from the smectic to the cholesteric phase before examining the type of packing in the cholesteric mesophase. As noted above in the section on the structure of smectic polymers, the x-ray diagrams of all cholesterol-containing polymers (Tables 4.24 and 4.25) are characterized by the presence of one diffuse reflection (0.58-0.62 nm) and a series of intense small-angle reflections. In uniaxial orientation of the polymers, the intensity of the diffuse halo increases in the meridional direction, and the small-angle reflections are transformed into equatorial arcs. This distribution of the intensity of the reflections is primarily determined by the perpendicular arrangement of the side groups with respect to the axis of the main chains. The position of the small-angle reflections is a function of the length of the spacer group, which indicates the formation of layered structures with a layer thickness determined by the length of the side group. Schematic representations of the packing of cholesterol-containing polymers are shown in Fig. 4.63 [217]. Polymers 2-6, 8, and 9 (Table 4.24) form antiparallel single-

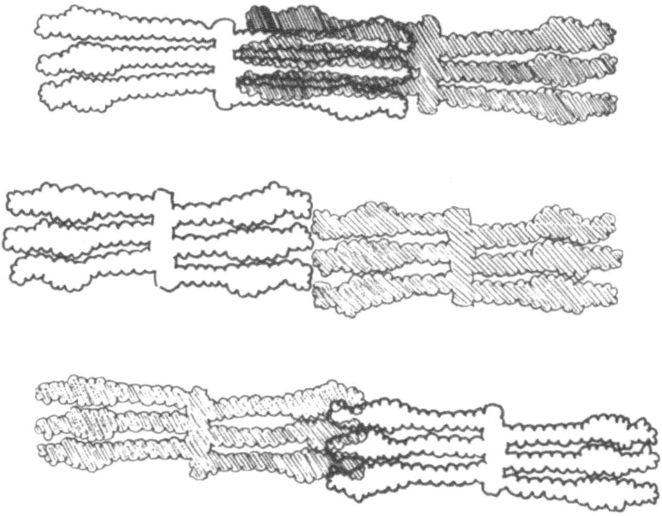

Fig. 4.63. Schematic representation of the packing of macromolecules of the polymers: $[-CH_2-CX-]$ with side group $OCO-(CH_2)_n-C(=O)-O-Chol$

a) single-layer (PChM-14, PChA-10, and PChM-10); b) two-layer (PChA-5 and PChM-5); c) intermediate (PChA-10 and PChM-10). The shaded molecules lie in the plane parallel to the plane of the figure [217].

layer packing of the side groups so that the cholesterol groups of one macromolecule are surrounded by the methylene chains of the neighboring macromolecules (Fig. 4.63a). In the case of other cholesterol-containing polymers (polymers 1 and 7, Table 4.24, and polymers 1 and 3, Table 4.25), parallel two-layer packing occurs (Fig. 4.63b). There are two types of packing in polymers with a long spacer group (polymers 2 and 4 in Table 4.25): antiparallel single-layer (Fig. 4.63a) and intermediate, where the cholesterol groups of the neighboring macromolecules partially overlap with the alkyl substituents on the seventeenth carbon atom of the cholesterol backbone (Fig. 4.63c). This is the structural organization of cholesterol-containing polymers at room temperature, i.e., in the S_A smectic mesophase. On the example of acrylic and methacrylic polymers with a spacer length of 10 methylene units (polymers 2 and 4 in Table 4.25), we will now examine what happens when they are heated, i.e., in the transition to the cholesteric mesophase.

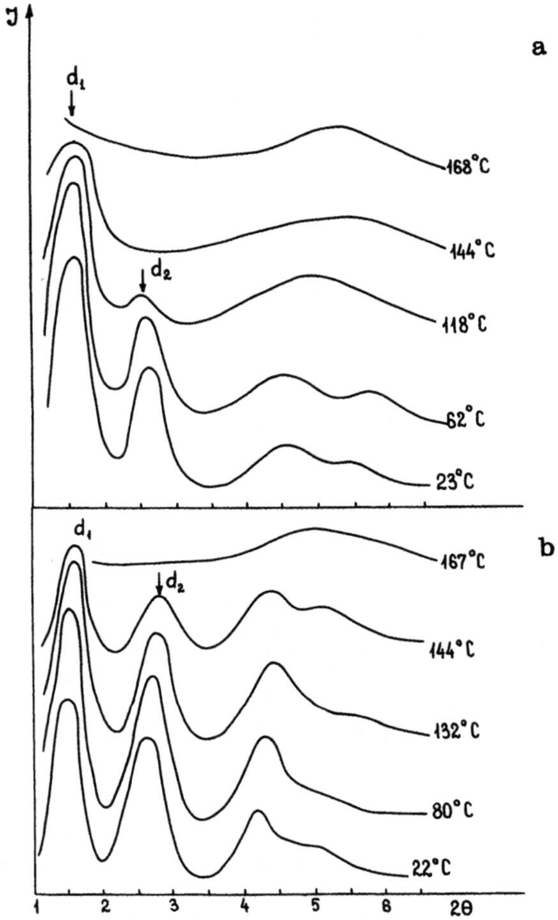

Fig. 4.64. X-ray diffraction patterns for polymers $[-CH_2-C(CH_3)-]$ $OCO-(CH_2)_{10}-COOChol$ (a) and $[-CH_2-CH-]$ $OCO-(CH_2)_{10}-COOChol$ (b) at different temperatures [315].

As the analysis of the small- and large-angle x-ray data shows, an increase in the temperature results in a decrease in the perfection of both the layered and the intralayer packing of these polymers. The perfection of the short-range order in the packing of the mesogenic side groups deteriorates above the region of the smectic—cholesteric transition, as shown by a decrease in the intensity of the wide-angle maximum and a decrease (by ~20%) in the correlation length in the direction perpendicular to the normal of the plane of the smectic layer. At the same time, the data in Fig. 4.64 indicate that layered structures characteristic of the smectic mesophase are preserved in both polymers in the range of the existence of the cholesteric mesophase. As Fig. 4.64c indicates, two types of layered packing, one type with partial overlapping of the end "tails" of the cholesterol nuclei (reflection d_1) and another type with complete overlapping of the side chains (reflection d_2; cf. Figs. 4.63 and 4.64), exist in the smectic mesophase of both polymers. An increase in the temperature results in a decrease in the intensity of all small-angle reflections for both the acrylic and methacrylic polymers, indicating a decrease in the perfection of the layered ordering in the transition of the polymer to the cholesteric mesophase. However, the layered structures and thickness of the layers are even preserved in the range of the cholesteric mesophase existence. The layered packing with partial overlapping of the tails of the cholesterol groups totally disappears in the polymer with a methacrylate chain, but the packing is preserved with complete overlapping of the side chains; in the acrylate polymer, traces of the corresponding layered structures are clearly preserved. Above the clearing point, the layered structures disappear, and a diffuse halo in the region of $2\theta = 5°$ is preserved on the scattering curve, which can be correlated with the so-called correlation hole effect [221]. The presence of layered structures which do not hinder the formation of a cholesteric supermolecular helix is thus a feature of the structural organization of cholesteric polymers. Nevertheless, the construction of a real model of the cholesteric polymer in consideration of the conformational structure of the polymer chain requires additional data and is a matter for future investigation.

4.8. BEHAVIOR OF LIQUID-CRYSTALLINE POLYMERS IN ELECTRIC AND MAGNETIC FIELDS

One of the most characteristic and unique properties of low-molecular-weight liquid crystals is their capacity to orient in external fields: mechanical, electric, and magnetic. This property determines the broad possibilities of using liquid crystals in a number of technological areas. The electro- and magnetooptics of liquid crystals are now an independent and important practical field in the physics of the condensed state of a substance [28-32].

Electrooptical and magnetooptical studies of liquid-crystalline polymers have just begun, and their history dates back no more than five years.

The first studies were conducted on LC polymers with mesogenic side groups at the Institute of Organic Chemistry in Mainz [259, 316-317] and at Moscow State University [263, 284, 318-324]; somewhat later, studies on the effect of an electric field on LC polymers were also conducted on polymers containing mesogenic groups in the main chains of the macromolecules [325-327].

It is known that the appearance of electrooptical effects in low-molecular-weight liquid crystals is only possible for compounds characterized by a high constant dipole moment and high anisotropy of polarizability of the molecules. The anisotropy of the dielectric constant $\Delta\varepsilon = \varepsilon_{\parallel} - \varepsilon_{\perp}$, equal to the difference in the dielectric constants of a liquid crystal measured along an external field ($\varepsilon_{\parallel}$) and perpendicular to the field ($\varepsilon_{\perp}$), is an overall measure of these two parameters. For long molecules whose constant dipole moment is directed along the axis of the molecules, the value of $\Delta\varepsilon$ is almost always greater than zero. The absence of compounds with a high value of $\Delta\varepsilon$ among the LC polymers synthesized up to the end of the 1970s was apparently also one of the basic problems which prevented the study of electrooptical events in LC polymers.

In view of this ability of low-molecular-weight liquid crystals to orient in an electric field, a series of acrylic, mathacrylic, and siloxane polymers and copolymers of the smectic and nematic types containing mesogenic groups with nitrile substituents of the following type were recently synthesized:

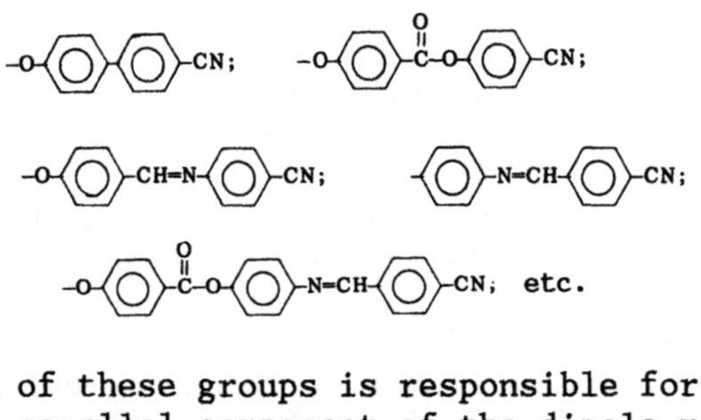

The presence of these groups is responsible for the high value of the parallel component of the dipole moment in the direction of the main chain and the positive value of the anisotropy of the dielectric constant ($\Delta\varepsilon$) of these polymers.

All of the currently known electrooptical effects in polymeric liquid crystals can be divided into two groups. First, there are so-called orientational effects which are not caused by the direct passage of current but only by the effect of the electric field (field effect) on the LC polymers. The second group of electrooptical effects concerns the events caused by the anisotropy of the conductivity ($\Delta\sigma$) of liquid crystals and are called electrohydrodynamic effects.

4.8.1. Orientational Effects

4.8.1.1. The Concept of Electro- and Magnetooptical Effects in Low-Molecular-Weight Liquid Crystals. Due to the anisotropy of the dielectric constant ($\Delta\varepsilon$) and the diamagnetic susceptibility ($\Delta\chi$), the free energy of the set of molecules in a nematic liquid crystal has a minimum at a rigorously determined orientation of the axes of the mesogenic molecules with respect to the field. Unit vector $\vec{L}$, which characterizes the predominant orientation of the long axes of the mesogenic molecules at each point of a liquid crystal, is called the directrix. Elastic forces which attempt to return the structure of a liquid crystal to the equilibrium state, given by the initial boundary conditions (usually by the walls of the electrooptical cell), oppose the orientation of the directrix. The deformation of the layer in the liquid crystal is the result of the competition of the two types of effects, dielectric and elastic, and this results in a change in its optical properties which is easily established from the change in the degree of polarization due to the high value of the birefringence Δn of the liquid-crystalline compound.

The group of effects known and described for nematic liquid crystals in electric E and magnetic H fields [29] are schematically illustrated in Fig. 4.65: these are S, B, and T effects, corresponding to splay, bend, and twist deformations. The initial orientation of the liquid-crystal molecules is given as shown in Fig. 4.65. Reorientation is the result of the interaction of the field with the dipole moment of the medium and is not related to the conductivity of the substance. When the value of $\Delta\varepsilon$ is positive, the directrix tends to be established along the field (Fig. 4.65a), and when the value is negative, it is perpendicular to the field

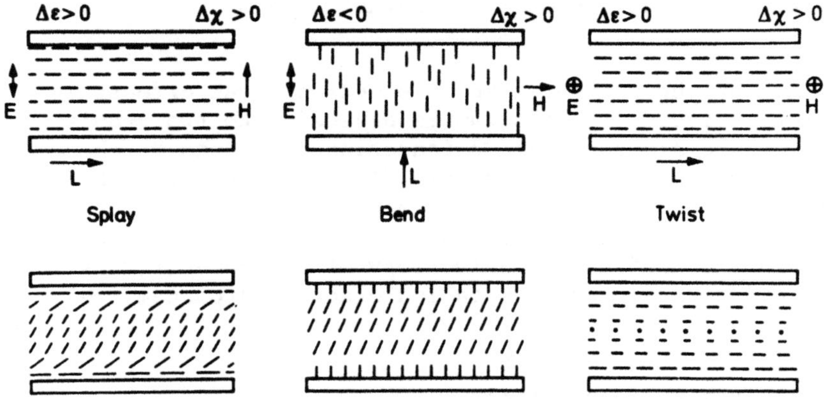

Fig. 4.65. The three principal types of orientational effects induced by electric E and magnetic H fields corresponding to splay, bend, and twist deformation of a low-molecular-weight nematic liquid crystal. The initial geometries of the molecules are shown at the top [29].

(Fig. 4.65b). Since $\Delta\chi$ is greater than zero, the direction of orientation in a magnetic field always coincides with the direction of the field. Reorientation causes elastic deformations of the liquid crystal which are a hindering factor. For this reason, the process of orientation has a threshold character, and its onset corresponds to the condition where the torque of the electric forces becomes equal to the torque of the elastic forces. The corresponding values of the threshold voltage of the electric field U_0 or the intensity of the magnetic field H_0 are determined by the following equations:

$$U_0 = \pi \sqrt{\frac{4\pi K_{ii}}{\Delta\varepsilon}} \tag{4.11}$$

$$H_0 = \frac{\pi}{d} \sqrt{\frac{4\pi K_{ii}}{\Delta\chi}} \tag{4.12}$$

where K_{ii} are the elasticity constants corresponding to the different types of deformation of the liquid crystal: splay and bend deformations with S and B effects (Fig. 4.65a, b) and twist deformation with the T effect (Fig. 4.65c); d is the thickness of the liquid-crystal layer.

The kinetic characteristics of the orientation are not only dependent on the elastic constants of the liquid crystal but also its viscosity. If the orientation time or (as it is called) the switching-on time of the electrooptical effect τ_{on} is 10^{-3} sec, the relaxation after switching off takes place significantly more slowly (switching-off time $\tau_{off} \approx 10^{-2}$-$10^{-1}$ sec) corresponding to the equations

$$\tau_{on} = 4\pi\gamma d^2/\Delta\varepsilon U^2 - 4\pi^3 K_{ii} \quad (4.13)$$

$$\tau_{off} = \gamma d^2/\pi^2 K_{ii} \quad (4.14)$$

where γ is the rotational viscosity coefficient of the nematic phase.

4.8.1.2. Comb-Shaped Liquid Crystalline Polymers

4.8.1.2a. The Fredericks Transition (S Effect). The studies of orientational effects began with [316-324] where the possibility of orientation of LC polymers in alternating and direct electric fields was first demonstrated on the example of polymers and copolymers with the structural units indicated below:

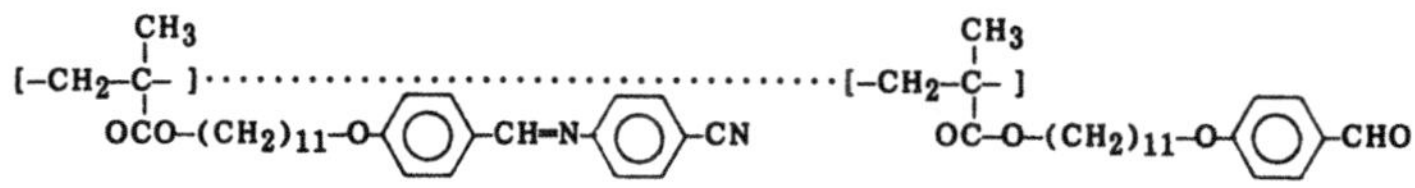

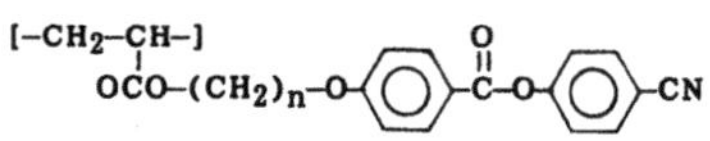
n = 2, 6

The studies initially conducted with this type of polymer and then with polymers of different structure [317-324] showed that their optical properties actually change significantly under the effect of an electric field. The orientational process is established by the change in the optical characteristics of LC films (transparency, birefringence) placed between two glass plates with a transparent conducting coating. Although a nematic polymer exhibits positive anisotropy of the dielectric constant ($\Delta\varepsilon > 0$), orientation of the mesogenic groups in the direction of the electric field (homeotropic orientation) takes place in the liquid-crystalline phase. The orientation is illustrated in Fig. 4.66a, where the optical transmission I (arbitrarily set at 100%) of the initial film of a nematic polymer with an optically anisotropic texture (in crossed polarizers) actually decreases to zero on application of a low-frequency field.

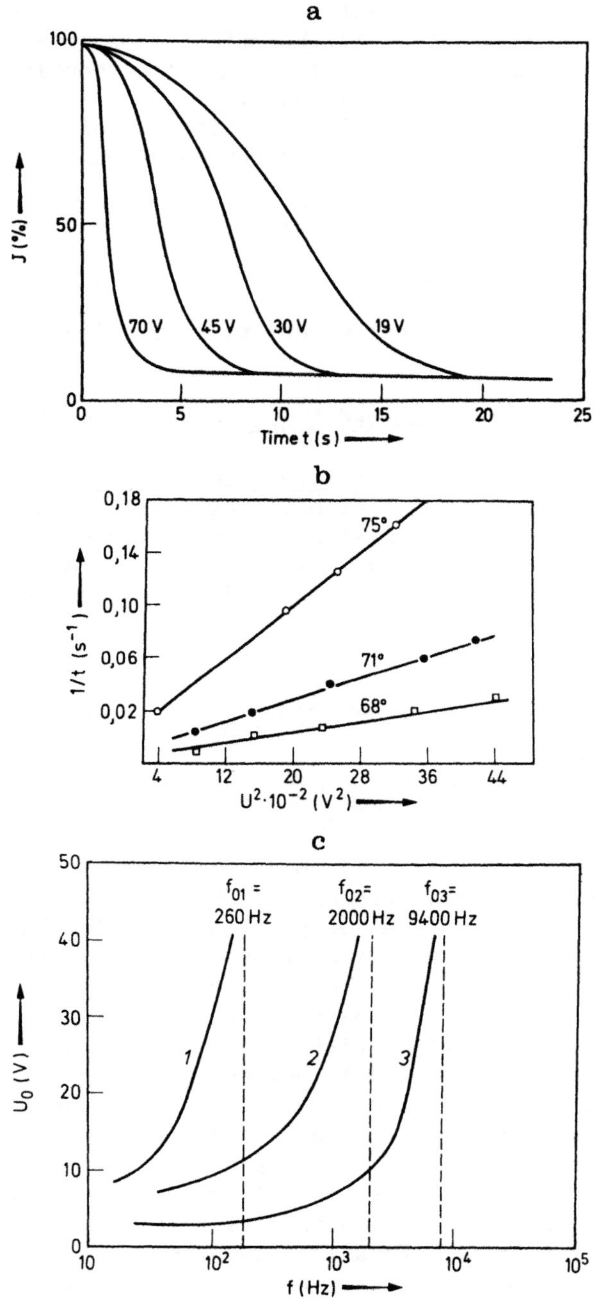
a
100
50
0
J (%)
70 V
45 V
30 V
19 V
0 5 10 15 20 25
Time t (s)
b
0,18
0,14
0,10
0,06
0,02
1/t (s^{-1})
75°
71°
68°
4 12 20 28 36 44
$U^2 \cdot 10^{-2}$ (V^2)
c
50
40
30
20
10
0
U_0 (V)
f_{01} = 260 Hz
f_{02}= 2000 Hz
f_{03}= 9400 Hz
1
2
3
10 10^2 10^3 10^4 10^5
f (Hz)

As Fig. 4.66a shows, the rate of the decrease in the light intensity under the effect of an electric field (f = 50 Hz) is a function of the value of the applied voltage (U) to a significant degree and increases sharply with an increase in the values of U. It was found that the value of the reciprocal of the time of the change in the light transmission (with a fixed value of I = 50%) is proportional to the square of the voltage (Fig. 4.66b), and the dependence of the threshold voltage U_0 on the frequency of the applied field is asymptotic (Fig. 4.66c). The analysis of these results formed the basis for drawing an analogy between the observed electro-optical effect and the S effect (Fredericks effect) mentioned above, known for low-molecular-weight nematic liquid crystals with $\Delta\varepsilon > 0$ (Fig. 4.65). However, in contrast to the latter, the oriented structure obtained in an electric field for LC polymers can be fixed by cooling the polymer below T_g. The structure of such a polymeric film corresponds to the structure of a uniaxial positive single crystal with an optical axis which coincides with the direction of the mesogenic groups and the vector of the electric field voltage (Fig. 4.66).

The similarity in the behavior of polymeric and low-molecular-weight liquid crystals is also evident from the dependence of the orientation of the mesogenic groups on the frequency of the applied field. The dependence $U_0 = \varphi(f)$ shown in Fig. 4.66 indicates the existence of some frequency f_0 at which the sign of $\Delta\varepsilon$ changes, which is characteristic of low-molecular-weight liquid crystals with $\Delta\varepsilon > 0$. Figure 4.67a, b clearly demonstrates the change in the orientation of the mesogenic groups with a change in the frequency of the electric field.

Fig. 4.66. Electrooptical behavior of a nematic polymer

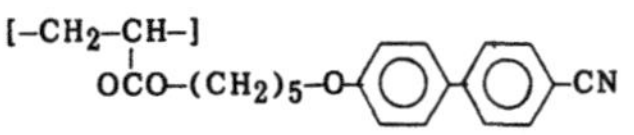

; T_g = 50°C, T_{cl} = 77°C, P_w = 20 [319]. a) Optical transmission as a function of time t at different voltages (f = 50 Hz, T = 75°C, crossed polarizers); b) reciprocal of the rise time as a function of the square of the voltage at different temperatures; c) threshold voltage U_0 as a function of the frequency f at 55 (1), 63 (2), and 68°C (3).

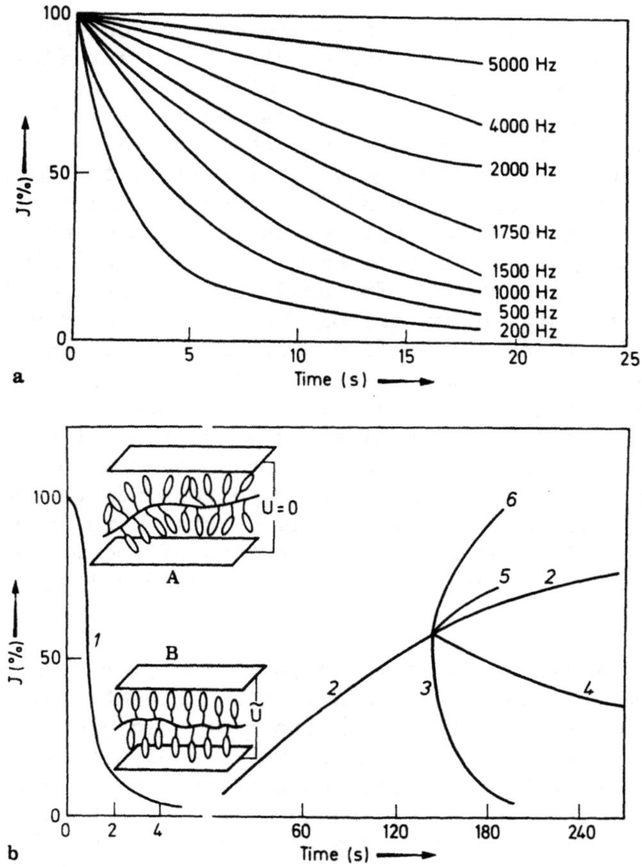

Fig. 4.67. Effect of the electric field frequency on the electrooptical behavior of a nematic polymer

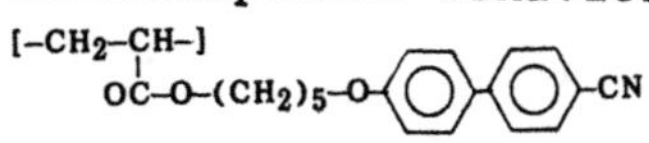

(P_W = 20) and diagram of the orientation of the mesogenic groups before (a) and after (b) application of the electric field: a) optical transmission as a function of time at different frequencies (U = 30 V, T = 75°C); b) optical transmission as a function of time on application of an electric field with U = 85 V (f = 50 Hz); relaxation on switching the electric field off (2) and on application of an electric field (U = 80 V) of different frequency, f = 1 (3), 5 (4), 7 (5), and 20 kHz (6) during the relaxation process [319].

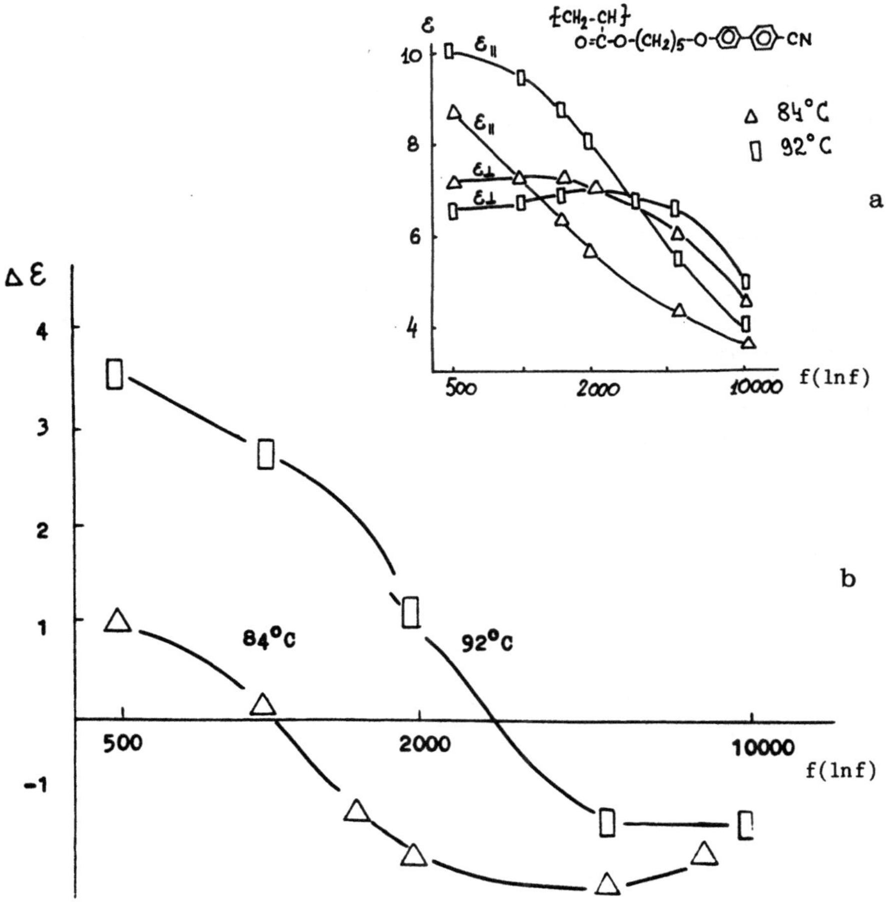

Fig. 4.68. a) Frequency dispersion of the parallel $\varepsilon_{\parallel}$ and perpendicular $\varepsilon_{\perp}$ components of the dielectric constant at different temperatures; b) frequency dispersion of the anisotropy of the dielectric constant (these data were obtained by S. V. Belyaev at our request).

Curve 1 in Fig. 4.67b corresponds to the formation of a homeotropic structure under the effect of a low-frequency field (f = 50 Hz, U = 85 V). Switching off the field perturbs the homeotropic orientation and the system is disoriented (Fig. 4.67b, curve 2). When a field of different frequency is switched on again (τ = 135 sec), different orientation of the mesogenic groups can be observed. Although the effect of a field with $f < f_0 = 6$ kHz results in orientation of the side groups along the direction of the field, when $f > f_0$ the side groups "attempt" to unfold across the direction of the force lines of the electric field. This event is due to a change in the sign of $\Delta\varepsilon$ at f_0 and is an example of the frequency addressing of the Fredericks effect. This effect of changing the sign of $\Delta\varepsilon$ and its components $\varepsilon_{\perp}$ and $\varepsilon_{\parallel}$ is directly visible in Fig. 4.68 for a polyacrylic derivative of hydroxycyanobiphenyl. This specific polymeric feature is seen in this case by a decrease in the frequency of the sign change of $\Delta\varepsilon$, apparently due to the high viscosity of the LC melt of the polymer. As a consequence, it is possible both to alter the orientation of the groups in the polymer in the mesophase by varying the frequency and voltage of an alternating electric field and to fix the macroscopic structure of polymeric films by cooling them below T_g.

In studying the orientation of comb-shaped polymers, the question arises as to whether the side groups are oriented independently of the polymer chains or if the interaction of the liquid crystal (formed by the packing of the side groups) with the external field also involves a change in the position of the main chains of the macromolecules, forcing them to align so that the side groups are oriented along the field according to the sign of $\Delta\varepsilon$ or $\Delta\chi$ of the liquid crystal.

Answering this question requires a rigorous analysis of the elastic constants of the mesophases of comb-shaped polymers and the kinetics of the orientation processes. However, it should be noted that obtaining experimental information on the threshold characteristics of electro- and magnetooptical effects and consequently the elastic constants of LC polymers involves a number of experimental difficulties. The problem primarily lies in the creation of an initial homogeneous planar texture resulting from the orienting effect of the surface. For polysiloxane nematic derivatives, it was possible in [328] to obtain planar oriented layers, measure the dynamics of the Fredericks effect, and estimate the bend constant K_{11}. As Fig. 4.69, which shows the temperature dependence of K_{11} measured for polymeric and low-molecular-weight nematic materials, indicates, the good agreement of the values of K_{11} unexpectedly indicate the absence of any "contribution" of the polymer chain to the elastic characteristics of a LC polymer.

It is also possible that the absence of any contribution of the polymer chain is related to the elevated flexibility

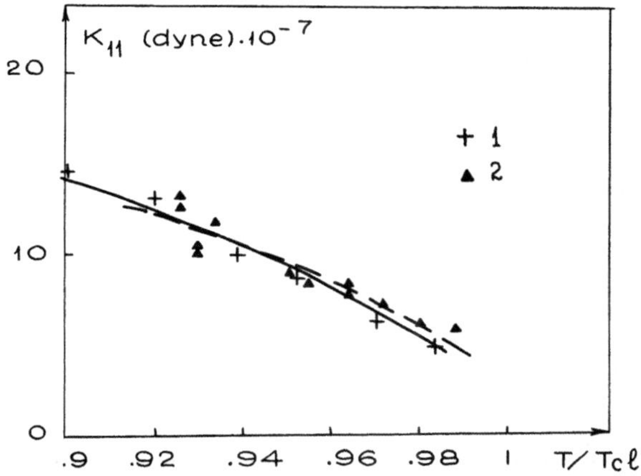

Fig. 4.69. Temperature dependence of the elastic constant K_{11} for the polymer

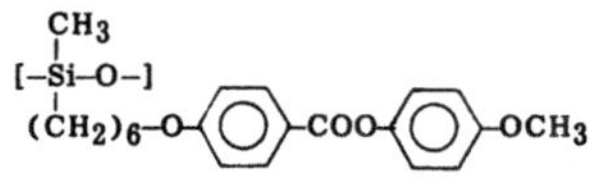

(1) and low-molecular-weight nematic liquid crystal

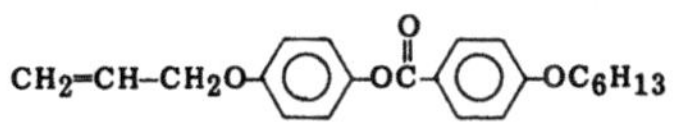

(3) [328].

and mobility of the polymethylsiloxane chain in this case, which creates conditions for the manifestations of elasticity characteristic of the LC phase alone, together with the sufficient spatial distance of the mesogenic group (methylene bridge length n = 6). However, it can be imagined that a reduction in the distance between the mesogenic fragment and the main chain due to a decrease in the length of the aliphatic bridge should significantly increase the contribution of the polymer chain to the elasticity of the liquid crystal. The change in the effective value of the reorientation voltage, which has low values (~3-5 V) similar to the corresponding values of the threshold voltages for low-molecular-weight materials for polymers 2 and 4 with n = 6 (Table 4.29), is correlated with the difference in the elastic constants in [329]. Based on the known relation of the threshold voltage and the elastic constant [Eq. (4.11)] and assuming that the values of the order parameter S and $\Delta\varepsilon$ for polymers with the same mesogenic group are similar, the following ratios were obtained in [260] for the elastic constants of nematic polymers 1-4 with a different length of the methylene fragment

TABLE 4.29. Threshold Voltage U_0, Orientation Time τ_{on}, and Relaxation Time τ_{off} for Liquid-Crystalline Polymers of the Series [260]

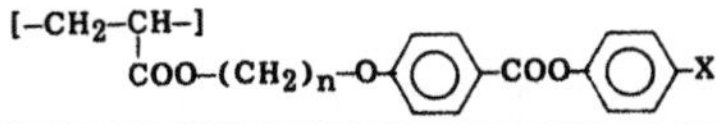

No.	X	n	ΔT*	U_0, V	τ_{on}, sec	τ_{off}, sec
1.	CN	2	22	60	14	87
2.	CN	6	76	5	3	3
3.	$=CN-C_6H_4CN$	2	84	23	3	3
4.	$=CH-C_6H_4CN$	6	120	4	1	4

*$\Delta T = T_{cl} - T_{meas}$.

(Table 4.29):

$$K_{11}^{1}:K_{11}^{2} \sim 100:1; \qquad K_{11}^{3}:K_{11}^{4} \sim 36:1.$$

The hypothetical nature of this conclusion, due to the absence of the real threshold characteristics, is confirmed by the more rigorous quantitative data in [186]. The increase in the threshold voltage of the electrooptical effect for polysiloxane copolymers

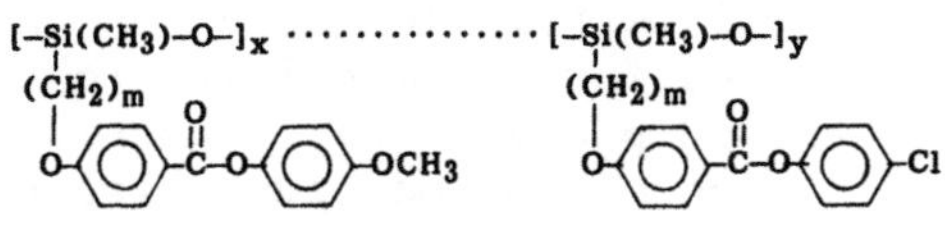

with a decrease in the length of the aliphatic segment is attributed to a change in the elastic characteristics of LC polymers.

The role of chemical binding of the mesogenic groups and the chain structure of the macromolecules is revealed to an even greater degree in the analysis of the kinetics of orientational processes.

The study of the kinetic mechanisms of the orientational effect conducted for polymeric cyanodiphenyl and azomethine derivatives with a comb-shaped structure [330] showed that the curves of the change in the transparency of an initially unoriented sample in time reflect the relaxation process which takes place in the polymer in an external field and are described by equations of the type

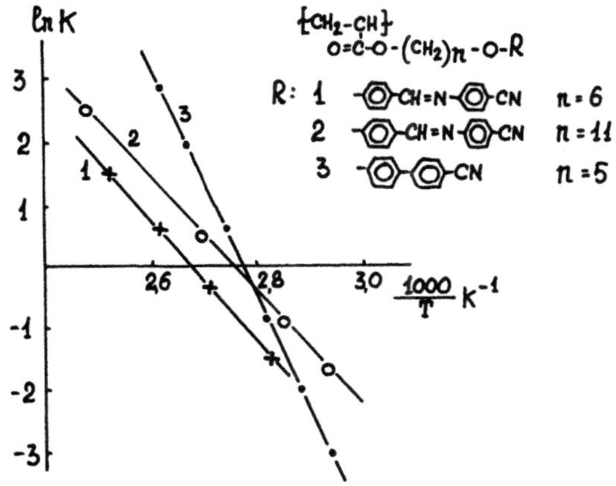

Fig. 4.70. Temperature dependence of the rate constant of the orientation process for different comb-shaped LC polymers.

$$1 - \theta = e^{-k\tau^n} \qquad (4.15)$$

where θ is the degree of completion of the process, τ is the time, and k and n are constants (n = 1 in a wide range of temperatures and is not dependent on the degree of polymerization). Constant k plays the role of the rate constant of the orientation process and is strongly dependent on the temperature.

The exponential character of this dependence (Fig. 4.70) indicates the activation nature of the orientation process. It can be postulated that the rate of orientation is correlated with the temperature primarily through the viscosity coefficient. A strong temperature dependence of the rotational viscosity coefficient γ for a comb-shaped nematic polymer with a polysiloxane main chain is also observed in [328]. Such a strong temperature dependence of the rate constant of orientation and the rotational viscosity γ found in [328, 330-331] is reflected in the high values of the activation energy of orientation (Table 4.30) and the activation energy of the rotational viscosity, which is approximately 100 J/mole for a polysiloxane polymer.

It is evident that the values of E_a for polyacrylic polymers (Table 4.30) are weakly dependent on the degree of polymerization for polymers of the same chemical structure,

TABLE 4.30. Activation Energy (E_A) and Times of Orientation ($\tau_{1/2}$) for Polymers with Different Degrees of Polymerization $\bar{P}_w$

No.	Polymer	$\bar{P}_w$	$\tau_{1/2}$,* sec	E_A, kJ/mole
1	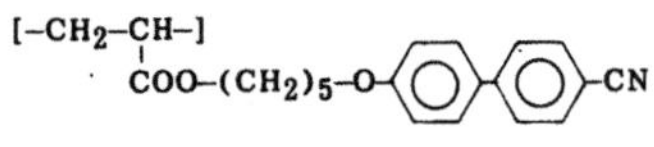	70	0.54	180
		140	0.8	170
		550	3.0	170
		2000	14.4	–
2	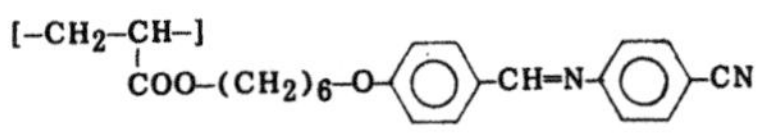	5-10	4.0	–
		250	18.0	90
		350	25.0	100
		1200	510.0	100

*The values of time $\tau_{1/2}$ were obtained with U = 200 V and $\Delta T = T_{cl} - T_{exp}$ = 10°C for polymer 1; U = 200 V and ΔT = 45°C for polymer 2.

and this indicates the relative invariability of the mechanism of the orientation process with a change in the degree of polymerization by 1-2 orders of magnitude. Significantly higher than the known values of the activation energy of rotational viscosity of low-molecular-weight nematic materials, the values of E_a reported in Table 4.30 are close to the E_a of different segmental processes characteristic of polymers of the corresponding structure (activation energy of viscous flow [187], activation energy of dipole polarization: α process [169]).

All of the above forms the basis for suggesting that macromolecular segments are involved in orientational motion. In this case, too, the segmental mobility of the macromolecules is responsible for the kinetic control of the orientation of the polymer in an external field. However, the fact that an increase in the degree of polymerization results in an increase in the orientation time means that there is also a change in the position of the macromolecules as a whole. For this reason, it is not possible to speak of the forward motion of the macromolecules, but instead their rotation around the center of gravity. Since the polymer chain has a dipole moment, an external field can only create a rotational moment of force.

It is possible to conclude that the mobility of the macromolecules in a LC melt is essentially determined by their conformational state, which in turn affects the orientation under the effect of the field. Thus, addition of 20-50 mole % methacrylic units to an acrylic polymer is sufficient for increasing the reorientation barrier by 1.5-2 times; progressing to polymethacrylate main chains significantly limits the possibilities of orientation, and the complete orientation of these homopolymers is not attained even in fields with an intensity of up to 10^6 V/cm. It is very obvious that the kinetic characteristics of the field effect reflect the mechanism of orientational motion, which also includes the motion of the macromolecules. This determines the significantly higher times of switching on and switching off, which are a function of the degree of polymerization. A decrease in the time characteristics of the field effects in polymers can be achieved by progressing to polymers of lower molecular weights [186, 332, 181], introducing highly miscible low-molecular-weight systems in polymeric liquid crystals [333] and, finally, selecting the appropriate field characteristics (voltage and frequency of the electric field [334]).

On the whole, it is possible to say that, in the case of comb-shaped polymers, the tendency of the side chains, which form the LC phase with the corresponding dielectric and diamagnetic anisotropy, toward orientation is the driving force of orientational processes. However, the interaction of the side groups with the external field causes movement of the main chains of the macromolecules, which is also responsible for the specificity of the behavior of polymeric liquid crystals not only in electric but also in magnetic fields.

4.8.1.2b. Orientation in a Magnetic Field. The use of an electric field is not the only effective means of acting on the structure of a LC polymer. A magnetic field has a similar effect, although very few studies have been conducted in this area [269-270, 272, 328, 329]. The study of orientation in a magnetic field is interesting both with respect to the method of orientation of LC polymers and for obtaining information on the polymer structure and the calculation of a number of physical parameters (such as the degree of order, the magnetic susceptibility, the rotational viscosity, and the elastic constants of a LC polymer). A magnetic field with H = 1.5-8 T is usually used for orientation. As with the effect of an electric field, the mesogenic groups of the LC polymer undergo orientation. These

TABLE 4.31. Characteristics of the Fredericks Effect in a Magnetic Field [Eq. (4.11)] [328]

Liquid-crystalline compound	H_0, G	τ, sec	$k_{11}\cdot 10^7$, dyne	γ_1, P
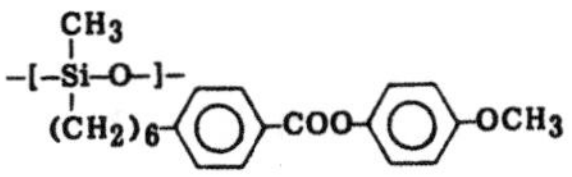	877	1800	7.8	$2.9\cdot 10^2$
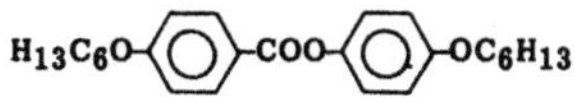	1250	0.93	15.8	0.248
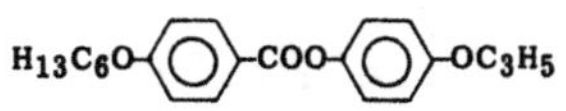	964	3.48	9.4	0.660

groups are arranged along the direction of the effect of the magnetic field (since the anisotropy of the diamagnetic susceptibility $\Delta\chi$ is always greater than 0; cf. Fig. 4.65), which was convincingly demonstrated in [269] in a study of the structure of oriented polymer 1 reported in Table 4.18.

The order parameter S of this polymer in a wide range of temperatures was calculated using the value of the splitting of the partially resolved triplet, which arises due to the direct additional interaction of the protons in the benzene ring (mesogenic groups [269]) with the side components (see Table 4.18 and Fig. 4.49). The temperature dependence of S was analyzed above in Section 4.7.3.1a.

Magnetooptical studies of a polysiloxane polymer and low-molecular-weight analogs of its mesogenic side groups were conducted in [328], and, using Eq. (4.12), the values of the relaxation time τ of the Fredericks field effect, the rotational viscosity γ_1, and the elastic constants K_{11} (splay) were calculated using the threshold values of the magnetic field H_0 (Table 4.31). The similarity of the values of $\Delta\chi$ and those of K_{11} indicate the significant similarity of the structure of the polymeric and low-molecular-weight nematic materials. At the same time, the significant difference in the values of the relaxation time τ of the Fredericks effect is a reflection of the significantly higher value of the rotational viscosity coefficient.

The study of deuterium-substituted derivatives of comb-shaped polymers is important for understanding the mechanisms of orientational processes and the conformational and structural rearrangements which take place under the effect of a magnetic field [272].

Partial deuteration of comb-shaped macromolecules in the methylene segments and in the mesogenic fragments can be used to estimate the order parameters related to the flexible and rigid fragments of the side chains. Binding of the mesogenic groups to the polymer chains with a flexible bridge of sufficient length ($n > 4$-6) insignificantly affects the order parameter of mesogenic groups in the LC phase, although the values of S decrease in the order low-molecular-weight liquid crystal → oligomer → polymer, as the data in [272] suggest. However, the order parameter of the methylene "bridge" (spacer) decreases by almost two times (in comparison to the order parameter of the mesogenic group), indicating the strong "perturbing" effect of the polymer chain and a significant increase in the fraction of gauche conformers in the methylene bridge over the number of gauche isomers in the end groups of molecules of low-molecular-weight liquid crystals. The study of LC polymers with "labeled" main chains would definitely be of use in the estimation of the order parameter and conformational state of the backbone of the macromolecule during the structural transformations which take place in electromagnetic fields. The preparation and study of such polymers should be considered a problem for the near future.

The possible orientation of liquid-crystalline polymers in electric and magnetic fields widens the prospects for studying both the structure of these polymers and the mechanisms of orientation and structural rearrangement of low-molecular-weight liquid crystals, where these events take place extremely rapidly and in some cases are very difficult to record.

4.8.1.2c. The "Guest–Host" Effect. The so-called "guest–host" effect, well known for low-molecular-weight liquid crystals, is one of the orientation events in LC polymers in an electric field. In the case of LC polymers, the polymer matrix plays the role of host and a dye whose molecule is stretched plays the role of guest, while an absorption oscillator is parallel (or perpendicular) to the long axis of the molecule [216, 267, 268, 333-335]. Experiments on the "guest–host" effects in nematic polymers using molecules of dichroic dyes either covalently bound to the polymer [335] (type I) or mechanically introduced [267] (type II) have been described:

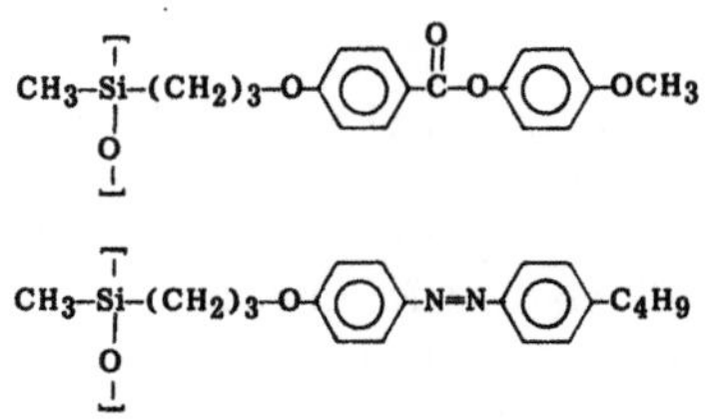

Type I

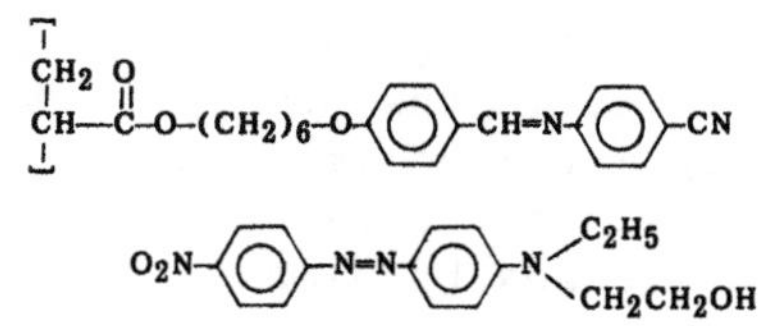

Type II

The mesogenic groups of a polymer oriented in a strong external field (mechanical or electric) force the orientation of the dye molecules, causing the appearance of, or a change in, color as a function of the type of dye and mesogenic groups (sign of $\Delta\varepsilon$) and the parameters of the field of force (frequency, voltage). The polymeric character of such a liquid crystal allows the required structure to be "frozen" in the glassy matrix by cooling the mesophase.

The use of the "guest–host" effect makes it possible to obtain information on the structural organization of a LC polymer by estimating the value of the order parameter S (see Section 4.7.3.1a). The long (several years) preservation of structural order (constancy of S) in the region of the glassy state is an important advantage of LC polymers over low-molecular-weight liquid crystals.

The possibility of effectively controlling the orientation of the side groups in polymer and dye molecules in the LC phase by changing the parameters of the electric field along with fixation of the oriented structure in the glassy state thus allows these LC systems to be used to obtain polymeric materials (color indicators of polarizers) with certain required optical properties.

<u>4.8.1.2d. Optical Recording of Information (Thermal Addressing</u>. The creation of devices for recording and optical display of data is one of the applications of the elec-

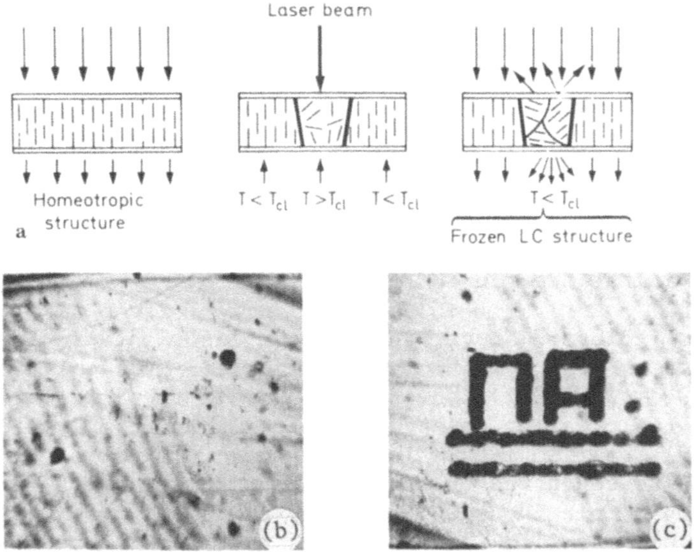

Fig. 4.71. Scheme of thermal recording using a film of a homeotropically oriented polymer (a) and thermal recording of letters on the polymer film

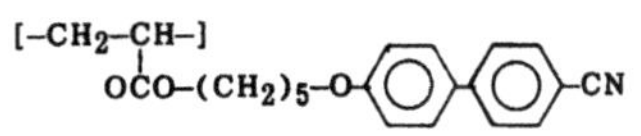
at 105°C (film thickness 40 μm): b) initial homeotropic texture; c) texture with laser thermally recorded letters [338, 339].

trooptical properties of LC polymers. The first use of LC polymers for this purpose is described in studies by young Soviet scientists at Moscow University [338-339]. The basic scheme of recording data in an oriented layer of a polymeric liquid crystal is shown in Fig. 4.71a.

Local hot spots are created with a laser beam on a transparent film of a homeotropically oriented liquid crystal. The liquid crystal changes into an isotropic melt at these sites, its orientation is perturbed (Fig. 4.71), and a nontransparent monodomain homeotropic texture and a light-scattering polydomain texture are formed there on cooling. Some information is thus recorded on the transparent film which can be "erased" under the effect of an electric field. This type of recording is usually called thermal recording or thermal addressing. For this to take place, it is necessary that the oriented state be stable without an electric field for a relatively long time, i.e., the rate of relaxation of orientation must be sufficiently small. Low-molecular-weight nematic liquid crystals, which are rapidly disoriented after the electric field is switched off, do not satisfy this requirement. For this reason, the "smectic—nematic" transition is most frequently used in thermal recording devices operating in low-molecular-weight liquid crystals, since the homeotropic texture is sufficiently stable in the smectic LC state.

As indicated above, the rate of disorientation of nematic polymers with a sufficiently high degree of polymerization (DP = 200-2000) is very low even in the immediate vicinity of the clearing point, and this allows LC polymers to be used as a film matrix for recording information. Recording can be conducted by exposure of a homeotropically oriented sample to a focused laser beam at temperatures near T_{cl}. If the polymer film is illuminated with a defocused laser beam, dark contrast spots are exhibited on the screen at the sites corresponding to the effect of the focused beam. An example of recorded letters on a film of a LC polymer is shown in Fig. 4.71c. The symbols obtained in this way are "erased" by switching on an alternating electric field for 2-3 sec.

The information recorded with the laser lasts for a long time if the polymer sample is cooled below T_g. This property of LC polymers is valuable for creating devices for long storage of recorded data. From this point of view, LC polymers differ advantageously from the low-molecular-weight liquid crystals used for these purposes. The storage time of recorded information in devices based on the latter is limited to several days in most cases. These processes demonstrate the possible use of the LC state for controlling the structural and optical properties of polymeric materials.

4.8.1.2e. The Structural Transition Induced by an Electric Field. In addition to the electrooptical process related to a field effect examined above, a new electrooptical effect which has no analogs among low-molecular-weight liquid crystals has been discovered in LC polymers [340]. It consists essentially of pronounced perturbation of the homeotropic orientation caused by an electric field in films of comb-shaped liquid-crystalline polymers of the nematic type. Such polymers include the LC

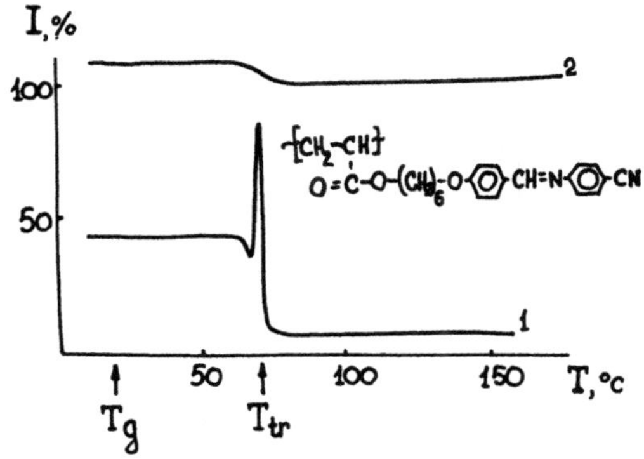

Fig. 4.72. Temperature dependence of optical transmission measured in crossed polarizers (1) and with no analyzer (2) (with cooling in an electric field, U = 200 V, d = 12 μm) [337].

polymers shown in Fig. 4.72, which have an even number of methylene groups in the spacer. Perturbation of the homeotropic orientation and the appearance of birefringence take place when the polymer is cooled in a very narrow temperature range (T_t). The analysis of the dielectric behavior of polymer films, conducted using a derivative of a Schiff base with m = 6 V [70], showed that the optically observed transition is of a very pronounced structural-phase nature: the temperature position of the maximum of the dielectric loss tangent is not a function of the frequency of the electric field (Fig. 4.73). The transition is reversible on heating and cooling the polymers in an electric field; the transition temperature decreases with an increase in the voltage (Fig. 4.74a) [17]. This indicates that the field which induces the initial orientation significantly stabilizes it. This effect has no analogs in low-molecular-weight systems, and its detailed mechanism is not yet clear. However, as Fig. 4.74b shows, a relatively broad range of temperatures in which a sudden decrease in the orientation time of LC polymers is observed, indicating a change in the mechanism of orientation [in Eq. (4.15), the index n differs from 1 and is equal to ~2-2.5), precedes the temperature of the transition. It was suggested that this transition is related to a cooperative change in the conformational state of the main chains of the macromolecules and possibly of the segments added to it which are chemically bound with their mesogenic groups oriented in the electric field.

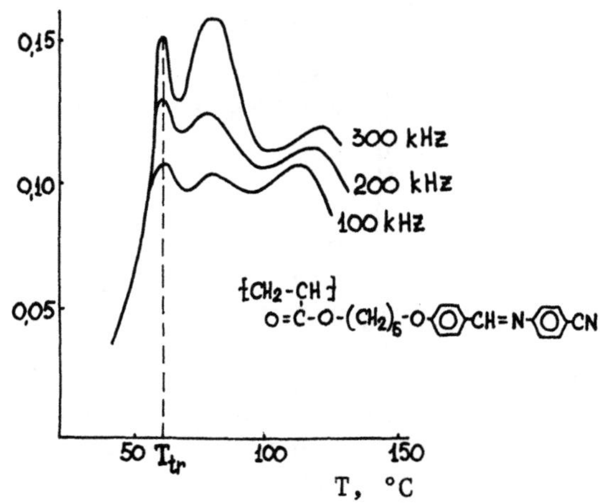

Fig. 4.73. Temperature dependence of tan δ in a previously oriented sample of polymer [337].

4.8.2. Electrohydrodynamic Effects

All of the events observed in LC polymers in electric fields examined above are not directly caused by passage of the current, but rather purely by a field effect.

Another group of effects consists of events of electrohydrodynamic instability caused by movement of a liquid in an external electric field. In liquid crystals, this instability only arises with sufficiently high impurity conduction of the substance. There are two mechanisms of the onset of electrohydrodynamic instability in nematic liquid crystals.

One of them, the "isotropic" mechanism, is not specific for liquid crystals and can be observed with any values of $\Delta\varepsilon$ and anisotropy of conduction $\Delta\sigma$. In this case, the electrohydrodynamic process consists of the movement of the liquid in an electric force gradient which arises due to the inhomogeneity of the charge and field distribution in the bulk of the sample [28, 29].

Anisotropy of conduction is the cause of the second type of instability according to the Carr—Helfrich model. The spatial separation of the charges in the electric field re-

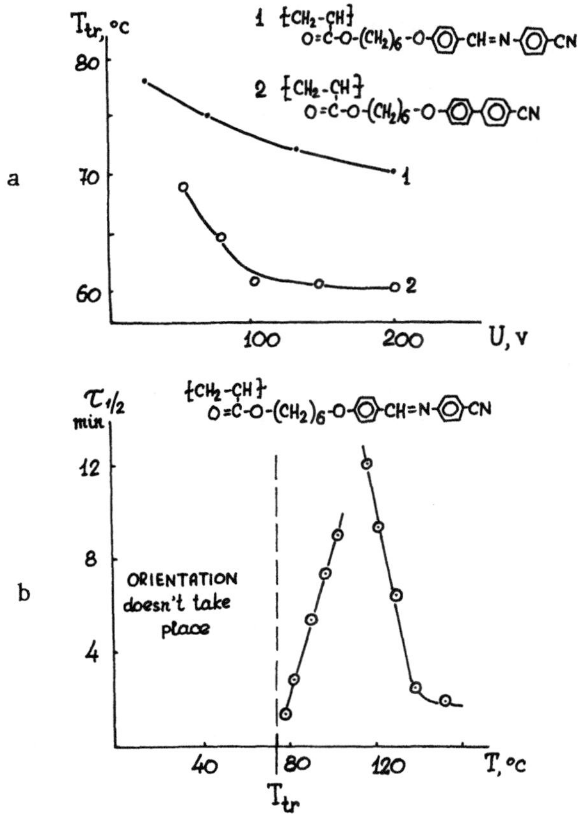

Fig. 4.74. Dependence of the structural transition temperature on the electric field voltage (a) and temperature dependence of the orientation time in an electric field (b) [337].

sults in the appearance of an electric force moment which creates cylindrical currents of the liquid. As a result, the orientation of the liquid-crystal molecules (with $\Delta\varepsilon < 0$ and $\Delta\sigma > 0$) is anomalous, and this effect is observed optically either as a characteristic periodic texture of the so-called Kapustin—Williams domains or as chevron structures. The domain structures are observed in fields insignificantly higher than the threshold fields. The motion becomes turbulent and is accompanied by intense light scattering (dynamic light scattering effect) with a subsequent increase in the voltage due to an increase in the flow rate of the currents in the liquid.

The formation of domain structures resembling Kapustin-Williams domains was first observed for LC polymers in linear macromolecules by Krigbaum et al. [325-327]. The appearance of domains for comb-shaped copolymers of the following type is described in [260]:

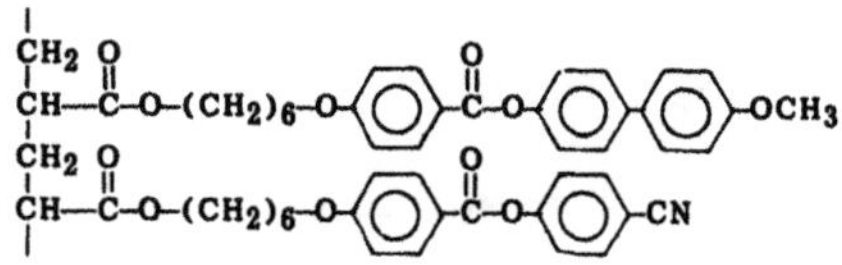

Finding electrohydrodynamic instabilities can primarily be considered an unambiguous test of the presence of a thermotropic nematic phase, and the appearance of Kapustin-Williams domains and fluctuating domains in the conduction mode indicates that these polymers are characterized by negative anisotropy of the dielectric constant ($\Delta\varepsilon < 0$) and positive anisotropy of conduction ($\Delta\sigma > 0$). The times of the development of domain structures in liquid-crystalline polymers, in contrast to low-molecular-weight liquid crystals, are many tens of minutes, due to the high viscosity of the polymer melts.

The appearance of domain structures in an electric field is not found for comb-shaped nitrile-containing liquid-crystalline homopolymers of the type

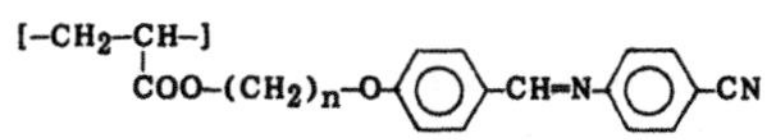

(n = 6, 11), as should be expected for liquid crystals with a high positive $\Delta\varepsilon$.

When an alternating electric field is applied to a homeotropically oriented layer of a LC melt of these polymers in a certain temperature range (close to T_{cl}), the homogeneous orientation is destroyed, accompanied by a sharp decrease in the transparency of the polymer film. This perturbation of the orientation is caused by electrohydrodynamic (EHD) instability related to the turbulent motion ("boiling") of the currents of the LC melt of the polymer. When the polymer sample is cooled, the motion gradually attenuates, and the transparency of the samples increases due to the formation of a homogeneous homeotropic orientation. This process is reversible, and the reverse transition to conditions

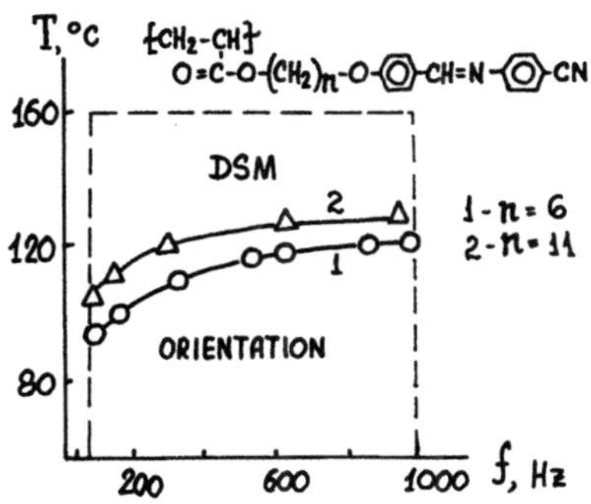

Fig. 4.75. Dependence of the temperature of the appearance of EHD instability (T_{EHD}) in homeotropically oriented polymer films as a function of the electric field frequency (U = 200 V, d = 12 μm) [337].

of EHD instability is observed when homeotropically oriented films are heated above a certain temperature (T_{EHD}) (Fig. 4.75). The appearance of EHD instability is characteristic of higher temperatures where the mobility of ionic impurities is higher. An increase in the frequency of the electric field causes narrowing of the temperature region of the EHD process, as well as a shift in the values of T_{EHD} toward higher temperatures, as Fig. 4.75c shows.

The nature of the dependence of T_{EHD} on the field frequency (Fig. 4.75) allows two-frequency switching of the optical properties of the polymer films to be conducted (Fig. 4.76). At temperature T, where $T_{EHD}^{f_1} < T < T_{EHD}^{f_2}$, it is possible to go from a homeotropic orientation to the EHD mode by varying the frequency of the electric field (f_1 or f_2). The initial homeotropically oriented film, characterized by high transparency, begins to strongly scatter light when a low-frequency field is switched on, changing into the mode of EHD instability. The transparency is decreased 5- to 6-fold. The times of switching on of the dynamic light scattering effect in polymers are slightly longer than the orientation times and are a function of the frequency of the external field. As in linear LC polymers, these processes are observed when the viscosity of the liquid-crystalline melt is not too high. Although electrohydrodynamic processes are observed in a relatively wide range of temperatures for

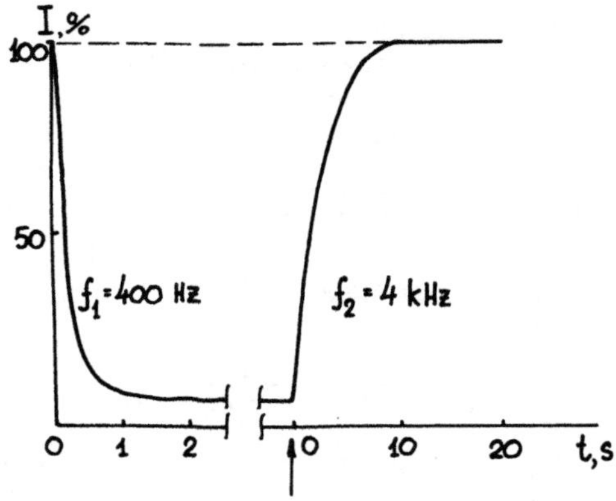

Fig. 4.76. Regulation of the optical transparency of polymer films 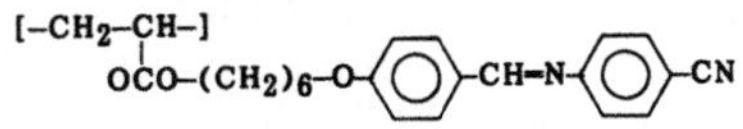 by discrete changes in the electric field frequency, f_1 = 400 Hz and f_2 = 4 kHz. The arrow indicates switching of frequencies f_1 and f_2 [337].

a polymer with a degree of polymerization of $\bar{P}_w$ ~ 100-200, for a higher-molecular-weight sample with $\bar{P}_w$ ~ 1500, dynamic light scattering in an electric field of the same frequency (f = 50 Hz) is only seen a few degrees before the transition into the isotropic phase, and is much less pronounced. The formation of turbulent eddies was observed by Simon and Coles in smectic polymethylsiloxane derivatives of hydroxycyanobiphenyl [342].

What is the mechanism of the electrohydrodynamic instability in polymeric liquid crystals? The processes of the formation of domain structures in linear polymers can apparently be described within the framework of the classic Carr–Helfrich–Williams model. In nitrile-containing polymeric Schiff bases, these events are also the result of passage of an ionic electric current. However, due to the high positive $\Delta\varepsilon$, the EHD instability in these comb-shaped polymers should not be controlled by the mechanism indicated above; rather, a process of the "isotropic type" [29] apparently takes place in this case, since the character of the EHD instability is not dependent on the type of mesophase (smectic or nematic) of the liquid crystal.

As the above indicates, the observed mechanisms of the electro- and magnetooptical behavior of polymers are analogous to a series of mechanisms in low-molecular-weight liquid crystals. However, it should be emphasized that this similarity is more qualitative in nature. The quantitative characteristics such as the reorientation times and stresses, the viscosity of the mesophase, and the activation energy of orientational processes differ significantly due to the chain structure of the macromolecules.

In examining the possible practical use of the capacity of polymers for structural rearrangements under the effect of external fields, it can be concluded that these systems cannot compete with low-molecular-weight liquid crystals where a high-speed response is required. However, they offer a unique opportunity for creating new optical materials whose properties are formed by external fields in the region of the LC state and are then fixed in the glassy state for an extended time.

In addition to the possibilities of creating materials for recording and optical display of data, preparation of dichroic polarizers, selective filters, and reflectors based on LC polymers, the cholesteric—nematic transition recently observed in polymeric cholesteric materials [343-344] is also of definite interest, as it allows polymer films to be obtained which selectively reflect light in a fixed range of wavelengths by the successive deformation and uncoiling of the helix with an electric field (Fig. 4.77).

Despite the fact that the well-studied mechanisms of the behavior of low-molecular-weight liquid crystals serve as a good basis for research on the behavior of LC polymers in electric and magnetic fields, research on more complex systems such as comb-shaped polymers suggests that new features and mechanisms specific to the polymeric state of these interesting systems alone will appear.

4.9. BEHAVIOR OF LIQUID-CRYSTALLINE POLYMERS WITH MESOGENIC SIDE GROUPS IN DILUTE SOLUTIONS

Due to the specific polymeric nature of liquid-crystalline polymers, fundamental information on their molecular characteristics – molecular weight, conformational state, polymer chain flexibility, mobility, intramolecular optical

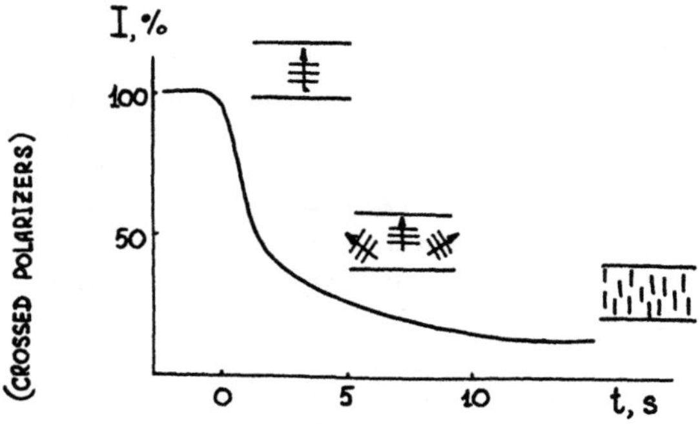

Fig. 4.77. Uncoiling of the cholesteric helix of a cholesteric polymer [344].

anisotropy, and other parameters — can be obtained from studying the behavior of dilute solutions of these compounds. This field of research is of independent interest and has been examined in a number of reviews, including the monograph by Tsvetkov [345] and articles by Burshtein and Shibaev [281] and Anufrieva et al. [346-347]. For this reason, we will only examine questions which are directly related to the initial stages of formation of the liquid-crystalline phase in solutions of polymers.

The establishment of orientational order of the mesomorphic type, caused by the interaction of the side chains, is a distinctive feature of solutions ot liquid-crystalline polymers with mesogenic side groups.

The theoretically and experimentally elaborated analysis of the dependence of the optical anisotropy on the length of the side chain conducted by Tsvetkov et al. [345, 348-349] (cf. Chapter 3) allowed conclusions to be drawn concerning the elevated equilibrium rigidity of the side groups of comb-shaped polymers, even those containing no mesogenic groups [poly(n-alkyl acrylate)s, poly(n-alkyl methacrylate)s], which significantly exceeds the rigidity of the main chains of the macromolecules. The most distinct increase in the rigidity of the side groups and the manifestation of elevated "crystal-like" order occurs for polymers containing mesogenic groups directly attached to the main chain. Studies of the dynamooptical and electrooptical properties of solutions of such LC polymers (Table 4.32) show the unusually high values of the segmental anisotropy and the dipole moments for the macromolecules. These values are usually more than

TABLE 4.32. Segmental Anisotropy $(\alpha_1-\alpha_2)$, Kerr Constant (K), Relaxation Time (τ), and Number of Monomeric Units (ν) in the Kuhn Segment for Some Comb-Shaped LC Polymers and Monomers in Different Solvents [348-350]

No.	Polymer	$(\alpha_1-\alpha_2)\cdot 10^{25}$, cm^3 (solvent)	$K\cdot 10^{10}$, $cm^5\cdot g^{-1}\cdot (V/300)^{-2}$	$\tau\cdot 10^5$, sec	ν
1	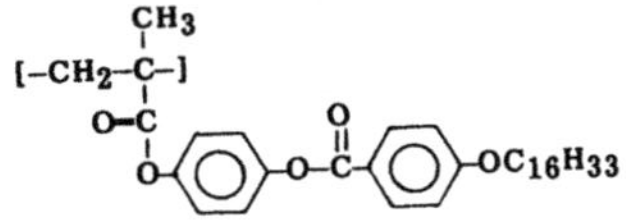	−3100 (CCl_4)	−10	[illegible]	24
		−1600 (benzene)	− 2.2	–	–
		−4200 (benzene + heptane, 52:48)	−40	–	–
2.	Monomer of polymer I	–	+ 0.21	–	–
3.	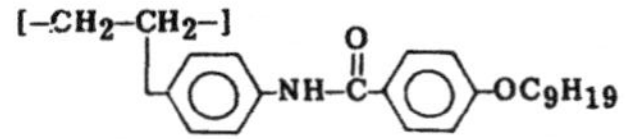	−2500	−140	0.2	20
4.	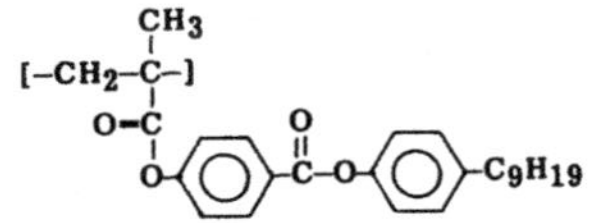	−2700 (CCl_4)	−(8.0-3.0)	–	25
5.	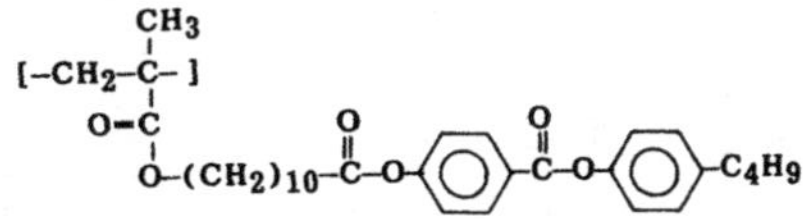	−90 (dioxane)	−(0.8±0.2)	–	26

one order of magnitude higher than the corresponding values for the usual flexible-chain polymers. The birefringence of solutions of comb-shaped LC polymers, which is higher in value and negative in sign (Table 4.32), indicates that not only an axial orientational but also a polar orientational order is characteristic of these macromolecules.

As Table 4.32 shows, the Kerr constants for polymers 1, 3, and 4 and monomer 2 not only differ in value but also in sign. The high degree of polar orientational order with a comparatively high equilibrium flexibility of the main chain (the value of the Kuhn segment is 20-25 monomeric units) is associated with very unusual kinetic properties. The relaxation times τ have a value of 10^{-4}-10^{-5} sec, i.e., 5-6 orders of magnitude higher than the corresponding values of τ characteristic of flexible-chain polymers. This means that important segments of the macromolecules which contain hundreds to thousands of monomeric units are oriented in an electric field.

Intramolecular order of the mesomorphic type where the mesogenic side groups form a mobile liquid-crystalline structure thus takes place in LC polymers of this type.

The orientation of the polar groups in these polymers in an electric field takes place according to the mechanism of large-scale motion characteristic of rigid-chain polymers. However, the introduction of spacer groups into the macromolecules (polymer 5, Table 4.32) sharply alters the character of the intramolecular aggregation and the mechanism of orientation in an electric field.

As a comparison of the values of the optical anisotropy and the Kerr constant for polymers 1, 3, 4, and 5 shows (Table 4.32), the separation of the mesogenic groups from the main chain by a polymethylene spacer results in a sharp decrease in the correlation in the orientations of the side groups and main chain and, as a consequence, leads to deterioration of the intramolecular ordering.

The significantly lower value of the Kerr constant and other factors indicate the small-scale intramolecular mechanism of orientation in an electric field in this case. The size of the kinetic units for a polymer with a spacer group is smaller, and their mobility is significantly higher than for polymers without spacer groups. At the same time, the decrease in the degree of coherence (correlation) between the main chain and the mesogenic fragments is responsible for the autonomy of their behavior, and facilitates the appearance of LC order between the side groups of the different macromolecules (see Section 4.4). In essence, the data reported in Table 4.32 for polymer 5 are quantitative confirmation of the independence of the behavior of the main and side chains, which enables the realization of thermotropic mesomorphism in the bulk.

One more important circumstance should be emphasized again, as it is directly related to the initial stages of the formation of the mesophase in solutions of polymers. This concerns the interaction between the molecules of the LC polymer and the solvent. The conformation of the macromolecules is very sensitive to the thermodynamic quality of the solvent, which significantly affects the character of the intramolecular crosslinking. For example, chain twisting in a thermodynamically poor solvent results in a sharp increase in the degree of intramolecular orientational order in the side chains, manifested by an increase in the optical anisotropy and the values of the Kerr constants of solutions of LC polymers (cf. values of $(\alpha_1 - \alpha_2)$ and K for polymer 1 in Table 4.32 [179-180]).

By altering the thermodynamic quality of the solvent (by using different solvents, mixtures of solvents, or by altering the temperature), it is thus possible to significantly affect the level of intra- and/or intermolecular aggregation of the side groups of LC polymers, which also affects the formation of the supermolecular organization of the mesophase in bulk polymer samples.

This was most clearly demonstrated in the study of the conformational behavior and intramolecular mobility (IMM) of cholesterol-containing polymers in dilute solutions with wide variation in the thermodynamic quality of the solvent [281-283, 356-359] and temperature. The method of polarized luminescence is one of the successfully recommended methods of estimating the IMM (see also Chapter 3.2) [346, 347]. This method allows relatively direct information to be obtained on both the rotational mobility of the macromolecules as a whole and the mobility of the main and side chains due to the introduction of luminescing anthrylacylhydroxymethane groups, so-called luminescing labels (LL). The structures of the macromolecules with LL which provide information on

TABLE 4.33. Values of the Relaxation Times τ_w, τ_{mn}, and τ_s of Macromolecules of Cholesterol-Containing Polymers PCMA-n and PCMO-n in Different Solvents at 25°C [282, 283, 356, 357]

Solvent	PCMA-n			PCMO-n			
	n = 11		n = 6	n = 14	n = 10		n = 5
	τ_{mn}, nsec	τ_s, nsec	τ_w, nsec	τ_w, nsec	τ_{mn}, nsec	τ_s, nsec	τ_w, nsec
Heptane	490		490	400	600	200	150
Heptane +0.2% TFAA	47	–	–	–	–	–	–
Cyclohexane	130		130	50	62	18	200
Toluene	100	3.3	66	28	25	3.2	27
Chloroform	22	1.6	22	16	16	3.0	17

the mobility of the main (LL-I) and side chains (LL-II) are shown below:

(LL-I)

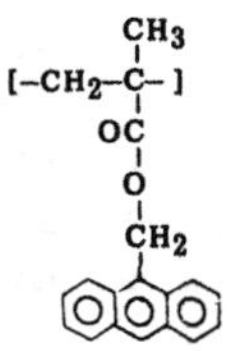

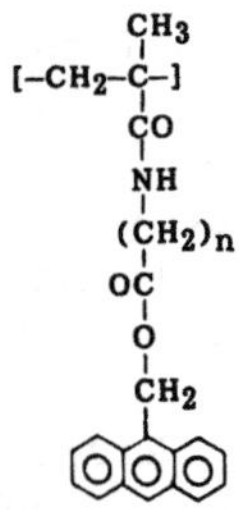

(LL-II)

The times characterizing the mobility of the main (τ_{mn}) and side (τ_s) chains of the polymers and the time of rotational diffusion of the macromolecule as a whole (τ_m) are correlated by the relation [358]:

$$\frac{1}{\tau_w} = \frac{1}{\tau_m} + \frac{1}{\tau_{mn(s)}}$$

The study of a homologous series of cholesterol-containing LC polymers of the methacrylic series

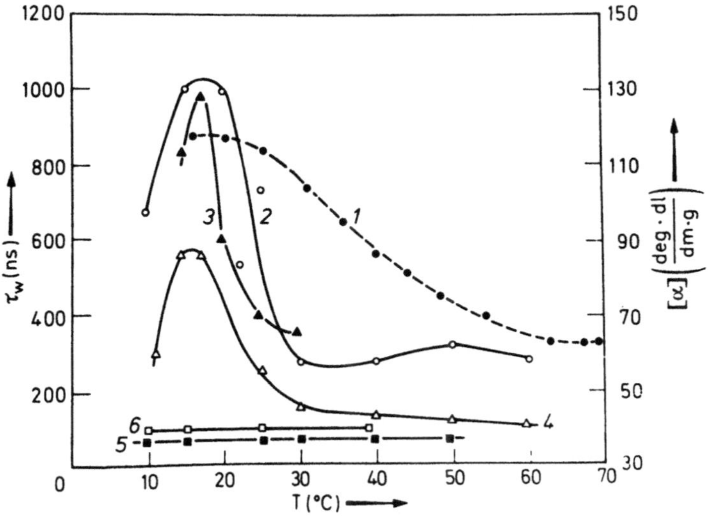

Fig. 4.78. Temperature dependence of [α] (1) and τ_w (2-6) for solutions of $[-CH_2-C(CH_3)-]$ with $CO-NH(CH_2)_{11}-COOChol$ side group (1, 2) and copolymers of $[-CH_2-C(CH_3)-]$ with $CO-NH-(CH_2)_{11}-COOChol$ side group with butyl methacrylate (MA-4) containing 10 (3), 25 (4), and 60 mole % (5) of MA-4 and poly(cetylmethacryloylamino-lauric acid) (6) in heptane [356].

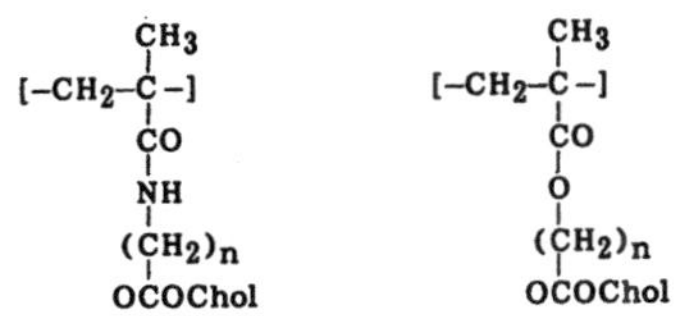

n = 5-14

(PCMA-n) (PCMO-n)

with luminescing labels showed that mesophase nuclei are formed in dilute solutions of these polymers under certain conditions due to the intramolecular ordering of the cholesterol units [282, 283, 346, 347, 356-358].

The values of the relaxation time characterizing the intramolecular mobility of the different sections of these cholesterol-containing polymers with LL in different solvents

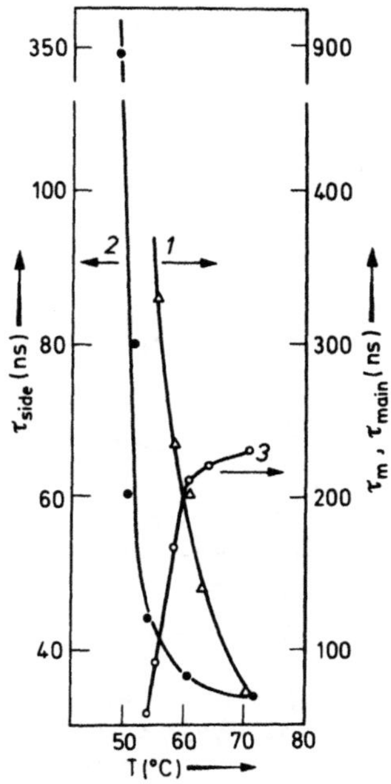

Fig. 4.79. Effect of the temperature on the mobility of the main chain (1), side chain (2), and the rotational mobility of the macromolecule as a whole (3) for the polymer $[-CH_2-C(CH_3)-]$ with side group $OCO-(CH_2)_{10}-COOChol$ in heptane [358].

are reported in Table 4.33.* As Table 4.33 indicates, the values of the relaxation times measured in different solvents for the same polymer differ strongly, reflecting the specific features of their conformational state and intramolecular organization.

The analysis of these data together with the results of studying the optical activity [282-283], hydrodynamic behavior, and light scattering data [357, 359] of solutions of these polymers showed that domains with ordering of the mesogenic groups are formed in macromolecules of PCMA-n and PCMO-n

*The relaxation times, measured in solvents of different viscosity, are normalized to the same value as that of the viscosity of the solvent here and below: 0.38 cP ($0.38 \cdot 10^{-3}$ $N \cdot sec/m^2$).

in saturated aliphatic hydrocarbons (i.e., "poor" solvents) (see Table 4.33, high values of the relaxation times in heptane). The formation of even more compact intramolecular structures occurs in a relatively narrow temperature range as a cooperative conformational coil–globule transition. Figure 4.78 (curves 1 and 2) and Fig. 4.79 (curves 1 and 2) graphically show how this transition takes place in the macromolecules of polymers PCMA-11 and PCMO-10.

The formation of domains with ordered mesogenic (optically active) groups is accompanied by an increase in the optical activity of a solution of the polymer (Fig. 4.78, curve 1) with a simultaneous pronounced increase in the relaxation times τ_w, reflecting a decrease in the IMM (Fig. 4.78, curve 1 up to a temperature of 20°C). The intramolecular hindrance of the main and side chains simultaneously increases, as shown by an increase in the values of τ_{mn} and τ_s for polymer PCMO-10 (Fig. 4.79, curves 1 and 2). At the same time, the presence of a maximum in the curve of the temperature dependence of τ_w (Fig. 4.78, curve 2) reflects the transition of the macromolecules into the compact globular state. The decrease in τ_w with a further decrease in the temperature cannot be explained in terms of the IMM alone and requires consideration of the mobility of the macromolecule as a whole (τ_m). The pronounced decrease in the time characterizing the rotational mobility of the macromolecule affects the contraction of the chains of cholesterol-containing polymers (Fig. 4.79, curve 3). The sharp decrease in the intrinsic viscosity $[\eta]$ and the dimensions of the macromolecules $(\bar{R})^{1/2}$ (Fig. 4.80, curves 1 and 2) to values characteristic of protein globules is also due to the above.

A decrease in the mobility of the side chains is a necessary condition for the formation of domains with ordered arrangement of the mesogenic groups, the nuclei of the mesophase (Table 4.33). If the mobility of the side chains is high and $\tau_s \approx 2$-3 nsec, ordering does not take place. If the mobility of the side groups decreases and τ_s increases by one order of magnitude or more, domains with an ordered arrangement of the mesogenic groups and a compact globular structure are formed.

The dependence of the temperature range of ordering on the length of the side chain is also related to the effect of the mobility of the side chains on the formation of mesophase nuclei. More significant cooling of solutions of cholesterol-containing polymers with a longer spacer group is required

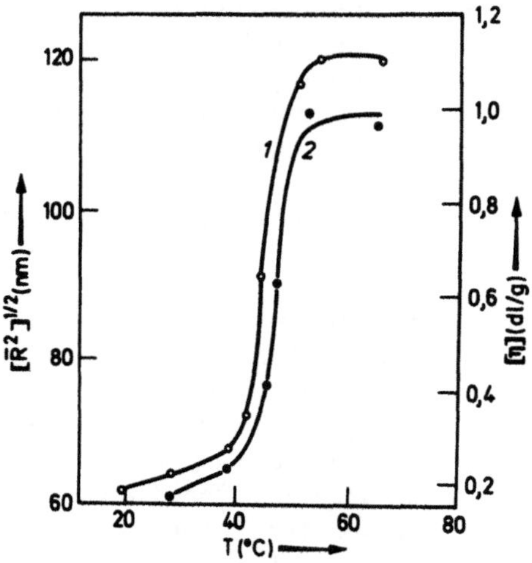

Fig. 4.80. Temperature dependence of [η] (1) and $(\bar{R}^2)^{1/2}$ (2) of $[-CH_2-C(CH_3)-]$ with side group $OCO-(CH_2)_{10}-COOChol$ in heptane [357].

for the formation of mesophase nuclei. In the series of PCMO-n polymers, the ordering range is 308-313°K for PCMO-14, 323-333°K for PCMO-10, and the internal structure is formed at even higher temperatures for PCMO-5 [358].

The detailed analysis of the amplitude of high-frequency motions of the luminescing groups of anthracene added to the main chain of a polymer (type LL-I) and the estimation of the contribution of these motions to the relaxation spectrum allowed two stages in the formation of the globular structure to be distinguished [360]. A so-called isotropic-liquid globule with loose packing of the units and isotropic arrangement of the mesogenic groups is formed in the first stage, and the temperature range of the formation of the isotropic-liquid globule is not dependent on the molecular weight of the polymer. The second stage begins at lower temperatures: the motion of the side groups becomes more hindered (the values of τ_s increase sharply). Ordering of the mesogenic groups takes place; a liquid-crystalline globule with close packing of the units and kinetic parameters corresponding to the condensed state is formed. The temperature range of the formation of the LC globule is dependent on the molecular weight of the polymer and is shifted toward lower temperatures with a decrease in the molecular weight. It can be dem-

TABLE 4.34. Values of Correlation Parameter g, Relaxation Time τ_{dp}, and Activation Energy U of Dipole Polarization Processes for Some Comb-Shaped Polymers of the Methacrylic Series [279-281]

$[-CH_2-C(CH_3)-]$ with R on C

R	g	τ_{dp}, nsec (toluene, 25°C)	U, kJ/mole
$-COO-(CH_2)_{17}-CH_3$	0.6	20	31.0
$-CONH-(CH_2)_{17}-CH_3$	2.0	250	42.0
$-CONH-(CH_2)_{11}-COO-$Chol	2.6	1090	52.5

onstrated that the formation of a LC globule is related to the critical concentration of mesogenic groups necessary for the formation of mesophase nuclei [360].

The study of the causes of the formation of intramolecular structures shows that the interaction of the cholesterol groups, which is intensified with a decrease in the temperature, plays a significant role. Dilution of the sequence of the cholesterol-containing monomeric units by butyl methacrylate units (copolymers of CMA-11 with butyl methacrylate) gradually results in the appearance of a conformational transition (Fig. 4.78, curves 3-5).

The decisive role of the cholesterol groups in the formation of a compact globular structure is confirmed by the study of the temperature dependence of τ_w for solutions of a polymer (PCMA-11) in which the cholesterol mesogenic groups are substituted by aliphatic cetyl groups. As Fig. 4.78 (curve 6) shows, the globular structure is not formed in solutions of this polymer.

Hydrogen bonds formed between amide groups in the macromolecules of PCMA-n also play a significant role in the stabilization of intramolecular structures [303, 359]. For example, addition of such a strong hydrogen bond competitor as trifluoroacetic acid (TFAA) to solutions of PCMAA-11 in heptane results in a sharp increase in the IMM (with addition of 0.2% TFAA, the value of τ_{mn} decreases by more than one order of magnitude) and no intramolecular structure is formed.

The study of dipole polarization relaxation processes and the dipole moments of cholesterol-containing polymers and copolymers [278-281, 361] confirms the presence of intramolecular organization of the mesogenic groups in solution. For example, this is indicated by both the high values of the relaxation time (τ) and activation energy (U) of dipole polarization processes and the high values of correlation parameter g, which is a relative measure of hindrance of internal rotation in the macromolecules (Table 4.34).

The results of studying the character of intramolecular structural-ordering in solutions of LC polymers examined above are also directly related to the formation of structural order in bulk samples of these polymers.

The segments of macromolecules with an ordered arrangement of the cholesterol groups formed in poor solvents can also be nuclei for the formation of supermolecular order in films prepared from these solvents. The studies of the structure of films of a cholesterol-containing polymer PCMA-11, prepared from different solvents, confirmed these conclusions. Films of PCMA-11 prepared from solutions in chloroform and toluene (low relaxation times, see Table 4.33) are optically isotropic, and films prepared from solutions in heptane (high values of the relaxation times, see Table 4.33) exhibit optical anisotropy, indicating a difference in their conformational state. A high degree of ordering of the side chains, established from x-ray studies, in comparison to optically isotropic films is characteristic of optically anisotropic films. The observed difference in the structure in the solid phase of LC polymers is the consequence of their different conformational state in solution; this opens up interesting prospects for controlling the structure of LC polymers in the initial stages of formation of mesophase nuclei in solutions.

The above experimental data on the formation of compact globular structures and coil—globule transitions are in good agreement with the theoretical study of Grosberg [362], who examined macromolecules in which the rod-shaped mesogenic groups were included in the flexible chain in the form of pendant side groups. The theoretical basis of the possible formation of an intramolecular structure of the globular type and the coil—globule transition, similar to our experimental study, was the result of calculations [282, 358].

The data on the examination of the initial stages of structural ordering in solutions of LC polymers briefly reported in this section are important from two points of view. First, the study of these processes provides quantitative information on the molecular parameters and IMM of LC polymers which are the basis for understanding and predicting the physicochemical behavior of polymeric liquid crystals in the condensed state. Second, understanding the mechanism of the formation of intramolecular structures in dilute solutions opens up broad possibilities for the investigation of the formation of lyotropic LC systems based on polymers with mesogenic side groups, an area of research which has barely been started [363].

REFERENCES

1. V. P. Shibaev, Doctoral Dissertation, Moscow State University, Moscow (1974).
2. Ya. S. Freidzon, V. P. Shibaev, and N. A. Platé, Proceedings of the 3rd All-Union Conference on Liquid Crystals [in Russian], Ivanova (1974), pp. 214-215.
3. V. P. Shibaev, Ya. Freidzon, and N. A. Platé, Proceedings of the 11th Mendeleev Congress on General and Applied Chemistry [in Russian], Vol. 2, Nauka, Moscow (1975), p. 164.
4. V. P. Shibaev and N. A. Platé, Vysokomol. Soedin., A19, 923-973 (1977).
5. Yu. B. Amerik and V. A. Krentsel', Chemistry of Liquid Crystals and Mesomorphic Polymer Systems [in Russian], Nauka, Moscow (1981).
6. A. Blumstein (ed.), Mesomorphic Order in Polymers and Polymerization in Liquid-Crystalline Media, American Chemical Society Symposium Series, No. 74, ACS, Washington, D.C. (1978).
7. A. Blumstein (ed.), Liquid-Crystalline Order in Polymers, Academic Press, London (1978).
8. S. P. Papkov and V. G. Kulichikhin, Liquid-Crystalline State of Polymers [in Russian], Khimiya, Moscow (1977).
9. J. McIntyre and A. Mulburn, Br. Polym. J., 13, 5-12 (1981).
10. V. P. Shibaev, in: Advances in Liquid Crystal Research and Applications, Vol. 2, L. Bata (ed.), Pergamon Press, Oxford; Akademiai Kiado, Budapest (1980), pp. 869-914.
11. V. P. Shibaev and N. A. Platé, Zh. Vses. Khim. Ova. Mendeleeva, 28, No. 2, 165-176 (1983).
12. N. A. Platé and V. P. Shibaev, Makromol. Chem. Suppl., 6, 3-26 (1984).

13. M. Gordon and N. Platé (eds.), Liquid Crystal Polymers I-III, Adv. Polym. Sci., 59, 60/61 (1984).
14. H. Ringsdorf and A. Schneller, Br. Polym. J., 13, 43-46 (1981).
15. H. Finkelmann, in: Polymer Liquid Crystals, A. Cifferi, W. Krigbaum, and R. Meyers (eds.), Academic Press, New York (1982).
16. R. V. Tal'roze, V. P. Shibaev, and N. A. Platé, Vysokomol. Soedin., A25, 2467-2487 (1983).
17. N. A. Platé, R. V. Tal'roze, and V. P. Shibaev, Pure Appl. Chem., 56, 403-416 (1984).
18. A. Blumstein (ed.), Polymeric Liquid Crystals, Plenum Press, New York (1985).
19. N. A. Platé, R. V. Tal'roze, and V. P. Shibaev, Makromol. Chem., Suppl., 8, 47-61 (1984).
20. G. Hardy, K. Nyitrai, and F. Cser, Abstracts of the 5th Liquid Crystal Conference of Socialist Countries [in Russian], Vol. 2, Part 1, Odessa (1983), p. 98.
21. F. Cser, J. Phys. (France), 40, No. C3, 459-466 (1979).
22. T. Asada and S. Onogi, Polym. Eng. Rev., 3, 323-353 (1983).
23. Proceedings of the 1st All-Union Symposium on Liquid-Crystalline Polymers [in Russian], Suzdal (December, 1982).
24. F. Cser, K. Nytrai, and G. Hardy, in: Advances in Liquid Crystal Research and Applications, Vol. 2, L. Bata (ed.), Pergamon Press, Oxford; Akademiai Kiado, Budapest (1980), pp. 845-867.
25. Contributed Papers, International Single Topic Symposium on Noncrystalline Order in Polymers, Naples, Italy (May, 1985).
26. I. G. Chistyakov, Liquid Crystals [in Russian], Nauka, Moscow (1966).
27. G. W. Gray, in: Liquid Crystals and Plastic Crystals, Vol. 1, G. W. Gray and P. A. Winsor (eds.), Ellis Horwood, Chichester (1974).
28. P. G. De Gennes, The Physics of Liquid Crystals, Clarendon Press, Oxford (1974).
29. L. M. Blinov, Electro- and Magnetooptics of Liquid Crystals [in Russian], Nauka, Moscow (1980).
30. S. Chandrasekhar, Liquid Crystals, Cambridge University Press, Cambridge (1977).
31. S. A. Pikin, Structural Transformations in Liquid Crystals [in Russian], Nauka, Moscow (1981).
32. G. W. Gray, in: Polymer Liquid Crystals, A. Cifferi, W. Krigbaum, and R. Meyer (eds.), Academic Press, New York (1982), pp. 1-33.

33. A. Elliot and E. Ambrose, Discuss. Faraday Soc., 17, 246-254 (1950).
34. C. Robinson, Trans. Faraday Soc., 52, 571-580 (1958); Tetrahedron, 13, 219-234 (1961).
35. K. Wissbrun, Br. Polym. J., 12, 163-169 (1980).
36. M. G. Dobb and J. E. McIntyre, Adv. Polym. Sci., 60/61, 61-98 (1984).
37. Ch. K. Ober, J. I. Jin, and R. W. Lenz, Adv. Polym. Sci., 60/61, 103-146 (1984).
38. A. Ya. Malkin and S. P. Papkov (eds.), Orientational Phenomena in Solutions and Melts of Polymers [in Russian], Khimiya, Moscow (1980).
39. A. Cifferi and I. Ward (eds.), Ultra-High Modulus Polymers, Applied Science Publishers, London (1979).
40. S. P. Papkov, Adv. Polym. Sci., 59, 75-102 (1984).
41. R. D. Gilbert and P. A. Patton, Prog. Polym. Sci., 9, 115-131 (1983).
42. V. G. Kulichikhin, A. Ya. Malkin, and S. P. Papkov, Vysokomol. Soedin., A26, 451-471 (1984).
43. B. Donald, Polym. News, 9, 169-170 (1984).
44. F. Reinitzer and Wilnez, Monatsh. Chem., 85, 421-426 (1888).
45. D. Demus, H. Demus, and H. Zaschke, Flüssige Kristalle in Tabellen, VEB Dtsch. Verlag Grundstoffind., Leipzig, DDR (1976).
46. D. Demus and H. Zaschke, Flüssige Kristalle in Tabellen II, VEB Dtsch. Verlag Grundstoffind., Leipzig, DDR (1980).
47. G. Friedel, Ann. Phys., 19, 273-281 (1922).
48. D. Demus and L. Richter, Textures of Liquid Crystals, VEB Dtsch. Verlag Grundstoffind., Leipzig, DDR (1980).
49. S. Chandrasekhar, B. Sadashiva, and K. Suresh, Pramana, 9, 471-480 (1977).
50. C. Destrade, M. Mondon, and J. Malthete, J. Phys. (France) Coll., 40, No. C3, 17-29 (1979).
51. G. W. Gray, in: Advances in Liquid Crystals, G. Brown (ed.), Academic Press, London (1976), pp. 1-25.
52. W. Haas, Mol. Cryst. Liq. Cryst., 94, 1-31 (1983).
53. G. Sprokel (ed.), Physics and Chemistry of Liquid Crystal Devices, Plenum Press, New York-London (1980).
54. G. M. Zharkova (ed.), Properties and Applications of Liquid-Crystal Temperature Indicators [in Russian], Institute of Theoretical and Applied Mechanics, Novosibirsk (1980).
55. V. A. Belyakov and S. S. Sonin, Optics of Cholesteric Liquid Crystals [in Russian], Nauka, Moscow (1982).

56. A. Griffin and A. Johnson (eds.), Liquid Crystals and Ordered Fluids, Vol. 4, Plenum Press, New York (1984).
57. I. Uematsu and Y. Uematsu, Adv. Polym. Sci., 59, 37-74 (1984).
58. E. Samulski, in: Liquid-Crystalline Order in Polymers, A. Blumstein (ed.), Academic Press, London (1978), pp. 167-190.
59. N. V. Usoltseva, Zh. Vses. Khim. Ova. Mendeleeva, 28, No. 2, 156-165 (1983).
60. V. A. Kargin and G. L. Slonimskii, Zh. Fiz. Khim., 15, 1022-1026 (1941).
61. Yu. K. Ovchinnikov, G. S. Markova, and V. A. Kargin, Vysokomol. Soedin., A11, 329-338 (1969).
62. E. B. Bokhian, Yu. K. Ovchinnikov, G. S. Markova, and V. A. Kargin, Vysokomol. Soedin., A13, 1805-1812 (1971).
63. E. B. Bokhian, Yu. K. Ovchinnikov, G. S. Markova, and N. V. Bakeev, Vysokomol. Soedin., A16, 1974-1982 (1974).
64. Yu. K. Ovchinnikov, N. N. Kuzmin, Yu. A. Makhnovskii, and N. F. Bakeev, Vysokomol. Soedin., A26, 629-633 (1984).
65. C. Oseen, Trans. Faraday Soc., 29, 883-894 (1933).
66. H. Zocher, Trans. Faraday Soc., 29, 945-952 (1933).
67. A. S. Sonin, Introduction to the Physics of Liquid Crystals [in Russian], Nauka, Moscow (1983).
68. G. Natta and P. Corradini, Nuovo Cimento Suppl., 15, 9-15 (1960).
69. V. I. Poddubnyi (Poddubny), V. K. Lavrentyev, V. A. Shevelev, N. K. Mikhailova, A. V. Sidorovich, and V. G. Baranov, Polym. Bull., 11, 41-45 (1984).
70. A. I. Kitaigorodskii, Dokl. Akad. Nauk SSSR, 124, 861-864 (1959).
71. A. I. Kitaigorodskii, Molecular Crystals [in Russian], Nauka, Moscow (1971), p. 110.
72. P. Smit, Kolloid Z. Z. Polym., 250, 27-38 (1972).
73. J. Zerbi, M. Guzzoni, and E. Ciampelli, Spectrochim. Acta, A23, 301-308 (1967).
74. Ya. A. Zubov, Doctoral Dissertation, Physical Chemistry Research Institute, Moscow (1975).
75. M. Yasuniva and T. Takemura, Polymer, 15, 661-678 (1974).
76. D. Bassett, Polymer, 17, 460-471 (1976).
77. A. Zachariades and J. Logan, J. Polym. Sci., Polym. Phys. Ed., 21, 821-830 (1983).
78. C. Beatty, J. Pochan, M. Froix, and D. Hinman, Macromolecules, 8, 547-558 (1975).
79. I. N. Makarova, I. M. Petrova, Ya. K. Godovskii, B. D. Lavrukhin, and A. A. Zhdanov, Dokl. Akad. Nauk SSSR, 269, 1226-1230 (1983).

80. V. M. Litvinov, B. D. Lavrukhin, V. S. Papkov, and A. A. Zhdanov, Proceedings of the 5th Liquid-Crystal Conference of Socialist Countries (October 10-15, 1983) [in Russian], Part 1, Odessa (1983), pp. 114-115.
81. Yu. K. Godovskii, N. I. Makarova, I. M. Petrova, and A. A. Zhdanov, Proceedings of the 5th Liquid-Crystal Conference of Socialist Countries (October 10-15, 1983) [in Russian], Vol. 2, Part 1, Odessa (1983), pp. 118-119.
82. C. Desper and N. Schneider, Macromolecules, 9, 424-432 (1976).
83. C. Desper, M. Alexander, P. Sagalyn, and N. Schneider, Polym. Preprints, 18, 73-76 (1977).
84. N. B. Sokolskaya, V. V. Kochervinskii, V. V. Kireev, V. V. Korshak, Yu. V. Selenev, and A. K. Bittirova, Dokl. Akad. Nauk SSSR, 258, 911-914 (1981).
85. P. Ho and M. Williams, Polym. Eng. Sci., 21, 233-246 (1981).
86. P. J. Flory, Proc. R. Soc. London, A234, 73-85 (1956).
87. P. J. Flory and G. Ronca, Mol. Cryst. Liq. Cryst., 54, 289-296 (1979).
88. P. J. Flory, in: Adv. Polym. Sci., 59, 1-36 (1984).
89. G. K. Elyashevich, V. G. Baranova, and S. Ya. Frenkel', J. Macromol. Sci., 13B, 255-289 (1979).
90. V. I. Poddubnyi (Poddubny), G. K. Elyashevich, V. G. Baranov, and S. Ya. Frenkel', Polym. Eng. Sci., 20, 206-211 (1980).
91. G. K. Elyashevich, Adv. Polym. Sci., 43, 205-245 (1982).
92. T. G. Litvina, G. K. Elyashevich, and V. G. Baranov, Vysokomol. Soedin., B24, 387-391 (1982).
93. G. K. Elyashevich, Doctoral Dissertation, Institute of Macromolecular Compounds, Leningrad (1984).
94. B. Wunderlich and J. Grebowicz, Polym. Prep., 24, 290-291 (1983).
95. B. Wunderlich and J. Grebowicz, in: Adv. Polym. Sci., 60/61, 1-59 (1984).
96. W. Jackson, Br. Polym. J., 12, 154-161 (1980).
97. B. Griffin and M. Cox, Br. Polym. J., 12, 147-154 (1980).
98. A. Blumstein, K. Sivaramakrishnan, R. Blumstein, and S. Clough, Polymer, 23, 47-53 (1982).
99. B. Jo, R. Lenz, and J. Jin, Makromol. Chem., Rapid Commun., 3, 23-26 (1982).
100. B. Jo, J. Jin, and R. Lenz, Eur. Polym. J., 18, 233-239 (1982).
101. A. Griffin and S. Havens, J. Polym. Sci., Polym. Lett. Ed., 18, 259-263 (1980).

102. L. Strzelecki and D. Van Luyen, Eur. Polym. J., 16, 299-304 (1980).
103. A. Blumstein, S. Vilasager, S. Ponrathnam, S. Clough, and R. Blumstein, J. Polym. Sci., Polym. Phys. Ed., 20, 877-892 (1982).
104. A. Roviello and A. Sirigu, Eur. Polym. J., 15, 423 (1979).
105. A. Yu. Bilibin, A. V. Tenkovtsev, O. N. Piraner, and S. S. Skorokhodov, Vysokomol. Soedin., A26, 2570-2578 (1984).
106. S. Yamakawa, Y. Shuto, and F. Yamamoto, Electron. Lett., 20, No. 5, 199-201 (1984).
107. S. Deteresa, S. Allen, R. Farris, and R. Porter, J. Mater. Sci., 19, 57-72 (1984).
108. A. Blumstein, H. Schmidt, O. Thomas, G. Kharas, R. Blumstein, and H. Ringsdorf, Mol. Cryst. Liq. Cryst. Lett., 92, 271-280 (1984).
109. R. Blumstein, E. Stickles, M. Gauthier, and A. Blumstein, Macromolecules, 17, 177-183 (1984).
110. E. Samulski, M. Gauthier, R. Blumstein, and A. Blumstein, Macromolecules, 17, 479-483 (1984).
111. A. Blumstein, O. Thomas, J. Asrar, P. Makris, S. Clough, and R. Blumstein, J. Polym. Sci., Polym. Lett. Ed., 22, 13-19 (1984).
112. W. Krigbaum, A. Cifferi, and J. Preston, US Patent No. 4,412,059 (Priority 08.20.80).
113, J. Asrar, H. Toriumi, J. Watanabe, W. Krigbaum, A. Cifferi, and J. Preston, J. Polym. Sci., Polym. Phys. Ed., 21, 1119-1131 (1983).
114. D. Acierno, F. La Mantia, G. Polizzotti, A. Cifferi, W. Krigbaum, and R. Kotek, J. Polym. Sci., Polym. Phys. Ed., 21, 2027-2036 (1983).
115. W. Krigbaum, R. Kotek, T. Ishikawa, and H. Hakemi, Eur. Polym. J., 20, 225-235 (1984).
116. A. Roviello, S. Santagata, and A. Sirigu, Makromol. Chem., Rapid Commun., 5, 141-143 (1984).
117. P. Janelli, A. Roviello, and A. Sirigu, Eur. Polym. J., 18, 753-760 (1982).
118. A. Roviello, S. Santagata, and A. Sirigu, Makromol. Chem., Rapid Commun., 5, 209-215 (1984).
119. C. Noel, F. Lauprete, C. Friedrich, B. Fayolle, and L. Bosio, Polymer, 25, 805-814 (1984).
120. C. Noel, C. Friedrich, L. Bosio, and C. Strazielle, Polymer, 25, 1281-1290 (1984).
121. A. I. Grigoriev, G. N. Matveeva, N. A. Andreeva, A. Yu. Bilibin, S. S. Skorokhodov, and V. E. Eskin, Vysokomol. Soedin., B26, 143-146 (1984).

122. B. Z. Volchek, N. S. Kholmuradov, A. Yu. Bilibin, and S. S. Skorokhodov, Vysokomol. Soedin., A26, 328-334 (1984).
123. Yu. V. Molchanov, A. F. Privalov, A. Yu. Bilibin, and S. S. Skorokhodov, Proceedings of the 5th Liquid-Crystal Conference of Socialist Countries (October 10-15, 1983), Vol. 2, Part 1, Odessa (1983), p. 111.
124. J. Majnusz, J. Catala, and R. Lenz, Eur. Polym. J., 19, 1043-1046 (1983).
125. H. Dicke and R. Lenz, J. Polym. Sci., Polym. Chem. Ed., 21, 2581-2588 (1983).
126. A. Donald, C. Viney, and A. Windle, Polymer, 24, 155-159 (1983).
127. A. Donald and A. Windle, Colloid Polym. Sci., 261, 793-799 (1983).
128. A. Donald, Philos. Mag., A47, L13-L17 (1983).
129. G. Mitchell and A. Windle, Polymer, 24, 1513-1520 (1983).
130. T. M. Birshtein and B. I. Kolegov, Vysokomol. Soedin., A25, 2519-2525 (1983).
131. S. V. Vasilenko, V. P. Shibaev, and A. R. Khokhlov, Makromol. Chem., Rapid Commun., 3, 917-920 (1982).
132. S. V. Vasilenko, A. R. Khokhlov, and V. P. Shibaev, Macromolecules, 17, 2270-2275, 2275-2279 (1984).
133. H. Kamogawa, J. Polym. Sci., B10, 7-9 (1972).
134. G. Paleos, G. Margomenou-Leonidopoulous, S. Eilippakis, and A. Malliaris, J. Polym. Sci., Polym. Chem. Ed., 20, 2267-2271 (1982).
135. H. Finkelmann and G. Rehage, Makromol. Chem., Rapid Commun., 1, 31-35 (1980).
136. W. Kreuder and H. Ringsdorf, Makromol. Chem., 4, 807-815 (1983).
137. G. Aguilera, B. Hisgen, H. Ringsdorf, and A. Schneller, Freiburger Arbeitstagung Flüssigkristalle, Freiburg (1981).
138. H. Finkelmann, H. Kock, and G. Rehage, Makromol. Chem., Rapid Commun., 2, 317-322 (1981).
139. H. Anzinger, J. Schmitt, and M. Mutter, Makromol. Chem., Rapid Commun., 2. 637-643 (1981).
140. Yu. B. Amerik, I. I. Konstantinov, and B. A. Krentsel', Dokl. Akad. Nauk SSSR, 165, 1097-2001 (1965).
141. Yu. B. Amerik and B. A. Krentsel' (Krentzel), J. Polym. Sci., C, No. 16, 1383-1393 (1967).
142. G. Hardy, F. Cser, N. Fedorova, and M. Batky, Acta Chim. Acad. Sci. Hung., 94, 275-281 (1977).
143. G. Hardy, K. Nyitrai, F. Cser, G. Cselik, and J. Nagy, Eur. Polym. J., 5, 133-139 (1969).

144. C. Paleos, T. Laronge, and M. Labes, Chem. Commun., 18, 1115-1118 (1968).
145. E. Steiger and H. Dietrich, Mol. Cryst. Liq. Cryst., 16, 279-286 (1972).
146. E. Perplies, H. Ringsdorf, and J. Wendorff, Makromol. Chem., 175, 553-561 (1974).
147. L. Strzeleski and L. Liebert, Bull. Soc. Chim. Fr., Ser. C, No. 2, 597-608 (1973).
148. L. Strzeleski, L. Liebert, and D. Vacogne, Bull. Soc. Chim. Fr., Ser. C, No. 12, 2849-2852 (1974).
149. L. Strzeleski and L. Liebert, Bull. Soc. Chim. Fr., Ser. C, No. 2, 605-609 (1973).
150. K. Sigiyama and T. Nakaya, Makromol. Chem., 178, 271-279 (1977).
151. A. Blumstein, R. Blumstein, G. Murphy, C. Wilson, and J. Billard, in: Liquid Crystals and Ordered Fluids, Vol. 2, Plenum Press, New York (1974), pp. 277-291.
152. V. P. Shibaev, V. M. Moiseenko, and Ya. S. Freidzon, Proceedings of the International Symposium on Macromolecular Chemistry, Tashkent [in Russian], Vol. 5, Nauka, Moscow (1978), pp. 7-8.
153. Yu. B. Amerik and B. A. Krentsel', in: Advances in Polymer Chemistry and Physics [in Russian], Khimiya, Moscow (1973), pp. 97-126.
154. B. A. Krentsel' and Yu. B. Amerik, Vysokomol. Soedin., A13, 1358-1371 (1971).
155. P. G. De Gennes, Phys. Lett., Ser. A, 28, 725-727 (1969).
156. S. Clough, A. Blumstein, and E. Hsu, Macromolecules, 9, 123-127 (1976).
157. L. Liebert and L. Strzelecki, Compt. Rend. Acad. Sci., Ser. C, 276, 647-651 (1973).
158. L. Liebert and L. Strzelecki, Bull. Soc. Chim. Fr., Ser. C, No. 2, 603-604 (1973).
159. H. Lorkowski and F. Reuther, Plaste Kautschuk, 23, 81-87 (1976).
160. L. Strzelecki and L. Liebert, Colloque sur les Cristaux Liquides, Pont-Mousson, France (June 27, 1971).
161. A. Blumstein, R. Blumstein, S. Clough, and E. Hsu, Macromolecules, 8, 73-78 (1975).
162. Y. Tanaka, M. Hitotsuganagi, Y. Shimura, A. Okada, and T. Sakurada, Makromol. Chem., 177, 3035-3040 (1976).
163. S. Clough, A. Blumstein, and A. De Vries, in: Mesomorphic Order in Polymers and Polymerization in Liquid-Crystalline Media, A. Blumstein (ed.), ACS Symposium Series, No. 74, Washington, D.C. (1978), pp. 1-11.

164. Y. Bauligand, P. Ciadis, L. Liebert, and L. Strzelecki, Mol. Cryst. Liq. Cryst., 25, 233-248 (1974).
165. A. Blumstein, S. Clough, L. Patel, R. Blumstein, and E. Hsu, Macromolecules, 9, 243-248 (1976).
166. V. P. Shibaev, R. V. Tal'roze, F. I. Karakhanova, A. V. Kharitonov, and N. A. Platé, Dokl. Akad. Nauk SSSR, 225, 632-635 (1975).
167. R. V. Tal'roze, F. I. Kharakhanova, T. I. Borisova, L. L. Burshtein, N. A. Nikonorova, V. P. Shibaev, and N. A. Platé, Vysokomol. Soedin., A20, 1835-1844 (1978).
168. V. P. Shibaev, R. V. Tal'roze, F. I. Karakhanova, and N. A. Platé, J. Polym. Sci., A17, 1671-1684 (1979).
169. N. A. Platé and V. P. Shibaev, J. Polym. Sci., Macromol. Rev., 8, 117-253 (1974).
170. P. Magagnini, A. Marchetti, F. Matera, G. Pizzirani, and G. Turcxi, Eur. Polym. J., 10, 585-594 (1974).
171. V. Frosin, G. Levita, D. Lupinacci, and P. Magagnini, Mol. Cryst. Liq. Cryst., 66, 341-348 (1981).
172. P. Magagnini, Macromol. Chem. Suppl., 4, 223-238 (1981).
173. V. P. Shibaev, Ya. S. Freidzon, and N. A. Platé, Dokl. Akad. Nauk SSSR, 337, 1412-1415 (1976).
174. V. P. Shibaev, Ya. S. Freidzon, and N. A. Platé, USSR Inventor's Certificate No. 525,709 (03.10.1975).
175. V. P. Shibaev, N. A. Platé, and Ya. S. Freidzon, in: Mesomorphic Order in Polymers and Polymerization in Liquid-Crystalline Media, A. Blumstein (ed.), ACS Symposium Series, No. 74, ACS, Washington, DC (1978), pp. 33-55.
176. H. Finkelmann, H. Ringsdorf, and J. Wendorff, in: Mesomorphic Order in Polymers and Polymerization in Liquid-Crystalline Media, A. Blumstein (ed.), ACS Symposium Series, No. 74, ACS, Washington, DC (1978), pp. 22-32.
177. H. Finkelmann, H. Ringsdorf, and J. Wendorff, Makromol. Chem., 179, 273-279 (1979).
178. I. N. Shtennikova, T. V. Peker, G. F. Kolbina, V. R. Petrov, V. S. Grebneva, I. I. Konstantinov, and Ya. B. Amerik, Vysokomol. Soedin., A24, 2047-2058 (1982).
179. I. N. Shtennikova, V. N. Tsvetkov, G. F. Kolbina, E. V. Korneeva, and S. V. Bushin, Proceedings of the 5th Liquid-Crystal Conference of Socialist Countries, Odessa (October 10-15, 1983) [in Russian], Vol. 2, Part 1, Odessa (1983), pp. 124-125.
180. V. N. Tsvetkov and I. N. Shtennikova, Am. Chem. Soc., Polym. Prepr., 24, 280-281 (1983).

181. S. G. Kostromin, R. V. Tal'roze, V. P. Shibaev, and N. A. Platé, Makromol. Chem., Rapid Commun., 3, 803-806 (1982).
182. V. P. Shibaev, S. G. Kostromin, and N. A. Platé, Eur. Polym. J., 18, 651-659 (1982).
183. J. Frenzel and G. Rehage, Makromol. Chem., Rapid Commun., 1, 129-135 (1980).
184. G. Rehage and J. Frenzel, Br. Polym. J., 14, 173-178 (1982).
185. J. Frenzel and G. Rehage, Makromol. Chem., 184, 1685-1703 (1983).
186. H. Finkelmann and G. Rehage, in: Adv. Polym. Sci., 60-61, 90-172 (1984).
187. V. P. Shibaev, V. G. Kulichikhin, S. G. Kostromin, N. V. Vasilieva, L.P. Braverman, and N. A. Platé, Dokl. Akad. Nauk SSSR, 263, 152-155 (1982).
188. H. Kock, H. Finkelmann, W. Gleim, and G. Rehage, Am. Chem. Soc., Polym. Prepr., 24, 300-301 (1983).
189. H. Stevens, G. Rehage, and H. Finkelmann, Macromolecules, 17, 851-856 (1984).
190. B. Hahn, J. H. Wendorff, M. Portugall, and H. Ringsdorf, Colloid. Polym. Sci., 259, 875-887 (1981).
191. P. A. Gemmell, G. W. Gray, and D. Lacey, Polym. Preprints, 24, 253-254 (1983); Mol. Cryst. Liq. Cryst., 108 (1985) (in press).
192. Z. A. Roganova, A. L. Smolyansky, S. G. Kostromin, and V. P. Shibaev, Eur. Polym. J., 21 (1985) (in press).
193. P. Corradini and M. Vacatello, Mol. Cryst. Liq. Cryst., 97, 119-130 (1983).
194. M. R. Kuzma, Mol. Cryst. Liq. Cryst., 101, 351-365 (1983).
195. A. T. Bosch, P. Maissa, and P. Sixow, Am. Chem. Soc. Polym. Prepr., 24, 246-247 (1984).
196. P. Maissa, A. T. Bosch, and P. Sixow, J. Polym. Sci., Polym. Lett. Ed., 21, 757-756 (1983).
197. S. V. Vasilenko, A. R. Khokhlov, and V. P. Shibaev, Vysokomol. Soedin., A26, 606-614 (1984).
198. R. P. Matheson and P. Flory, Macromolecules, 14, 954-962 (1981).
199. A. Blumstein, R. Blumstein, M. M. Gauthier, O. Thomas, and J. Asrar, Mol. Cryst. Liq. Cryst., 92, 87-93 (1983). (1983).
200. A. Abe, Macromolecules, 17, 2280-2287 (1984).
201. A. Roviello and A. Sirigu, Makromol. Chem., 183, 895-904 (1982).
202. W. R. Krigbaum, J. Watanabe, and T. Ishikawa, Macromolecules, 16, 1271-1279 (1983).

203. V. P. Shibaev, S. V. Vasilenko, and A. R. Khokhlov, Proceedings of the 10th International Liquid-Crystal Conference, York, United Kingdom (July 15-21, 1984), p. D41.
204. S. V. Vasilenko, A. R. Khokhlov, and V. P. Shibaev, Dokl. Akad. Nauk SSSR, 282 (1985) (in press).
205. S. V. Vasilenko, A. R. Khokhlov, and V. P. Shibaev, Makromol. Chem., 186 (1985) (in press).
206. H. Sackmann and D. Demus, Mol. Cryst. Liq. Cryst., 21, 239-246 (1973).
207. C. Noël and J. Billard, Mol. Cryst. Liq. Cryst., 41, 269-272 (1980).
208. A. Griffin and S. Havens, J. Polym. Sci., Polym. Lett. Ed., 18, 259-263 (1980).
209. G. Hardy, F. Cser, and K. Nyitrai, Abstracts of the 4th International Liquid-Crystal Conference of Socialist Countries [in Russian], Vol. 1, Tbilisi (1981), p. 371.
210. C. Noël, C. Friedrich, F. Rauprete, J. Billard, L. Bosio, and C. Strazielle, Polymer, 25, 263-273 (1984).
211. H. Sackmann and D. Demus, Mol. Cryst., 2, 81-87 (1966).
212. W. L. McMillan, Phys. Rev., A4, 1238-1244 (1971); Phys. Rev., A6, 936-943 (1972).
213. S. G. Kostromin, V. P. Shibaev, and N. A. Platé, Makromol. Chem., Rapid Commun., 6 (1985).
214. B. Krucke, S. G. Kostromin, H. Zashke, and V. P. Shibaev, Acta Polym., 36 (1985) (in press).
215. S. G. Kostromin, Candidate's Dissertation, Moscow State University, Moscow (1982).
216. V. V. Sinitsyn, R. V. Tal'roze, V. P. Shibaev, and N. A. Platé, Abstracts of the 4th International Liquid Crystal Conference of Socialist Countries [in Russian], Vol. 2, Tbilisi (1981), p. 213.
217. Ya. S. Freidzon, N. I. Boiko, V. P. Shibaev, and N. A. Platé, Mol. Cryst. Liq. Cryst., 80, 211-218 (1982).
218. R. V. Tal'roze, V. V. Sinitsyn, V. P. Shibaev, and N. A. Platé, Mol. Cryst. Liq. Cryst., 80, 211-218 (1982).
219. S. G. Kostromin, V. V. Sinitsyn, R. V. Tal'roze, V. P. Shibaev, and N. A. Platé, Makromol. Chem., Rapid Commun., 3, 809-814 (1982).
220. V. P. Shibaev and N. A. Platé, in: Adv. Polym. Sci., 60/61, 173-252 (1984).
221. Yu. S. Lipatov, V. V. Tsutruk, and V. V. Shilov, J. Macromol. Sci. Rev. Macromol. Chem. Phys., C24, No. 2, 173-238 (1984).
222. V. V. Tsukruk, V. V. Shilov, Yu. S. Lipatov, I. I. Konstantinov, and Yu. B. Amerik, Acta Polym., 33, 63-69 (1982).

223. V. V. Tsukruk, V. V. Shilov, and Y. S. Lipatov, Acta Polym., 36 (1985).
224. J. H. Wendorff, H. Finkelmann, and H. Ringsdorf, in: Mesomorphic Order in Polymers and Polymerization in Liquid-Crystalline Media, A. Blumstein (ed.), ACS Symposium Series, No. 74, ACS, Washington, D.C. (1978), pp. 12-21.
225. H. Finkelmann, H. Ringsdorff, W. Siol, and J. H. Wendorff, in: Mesomorphic Order in Polymers and Polymerization in Liquid-Crystalline Media, A. Blumstein (ed.), ACS Symposium Series, No. 74, ACS, Washington, D.C. (1977), pp. 22-32.
226. H. Kelker and U. Wirzing, Mol. Cryst. Liq. Cryst., 49, 175-177 (1979).
227. H. Finkelmann, M. Happ, M. Portugal, and H. Ringsdorf, Makromol. Chem., 179, 2541-2544 (1978).
228. A. Blumstein, Macromolecules, 10, 872-877 (1977).
229. S. B. Clough, A. Blumstein, and A. De Vries, Polymer Prepr., 18, 1-2 (1977).
230. V. P. Shibaev, A. V. Kharitonov, Ya. S. Freidzon, and N. A. Platé, Vysokomol. Soedin., A21, 1849 (1979).
231. Ya. S. Freidzon, V. P. Shibaev, A. V. Kharitonov, and N. A. Platé, in: Advances in Liquid-Crystal Research and Applications, Vol. 2, L. Bata (ed.), Pergamon Press, Oxford, Akademiai Kiado, Budapest (1980), p. 899.
232. S. G. Kostromin, V. V. Sinitsyn, R. V. Tal'roze, and V. P. Shibaev, Vysokomol. Soedin., A26, 335-344 (1984).
233. P. A. Gemmell, G. W. Gray, D. Lacey, A. K. Alimoglu, and A. Ledwith, Polymer, 26 (1985).
234. Yu. S. Lipatov, V. V. Tsukruk, V. V. Shilov, V. S. Grebneva, I. I. Konstantinov, and Yu. B. Amerik, Vysokomol. Soedin., B23, 818-822 (1981).
235. I. I. Konstantinov, V. S. Grebneva, Yu. B. Amerik, and A. A. Sitnov, Abstracts of the 4th International Liquid-Crystal Conference of Socialist Countries [in Russian], Vol. 2, Tbilisi (1981), p. 214.
236. V. P. Shibaev, N. A. Platé, and Ya. S. Freidzon, J. Polym. Sci., Polym. Chem. Ed., 17, 1655-1670 (1979).
237. Ya. S. Freidzon, V. P. Shibaev, N. N. Kustova, and N. A. Platé, Vysokomol. Soedin., A22, 1083-1090 (1980).
238. V. P. Shibaev, V. M. Moiseenko, and N. A. Platé, Makromol. Chem., 181, 1381-1392 (1980).
239. V. A. Gudkov, I. G. Chistyakov, B. K. Vainshtein, and V. P. Shibaev, Kristallografiya, 27, 537-546 (1982).
240. V. P. Shibaev, V. M. Moiseenko, N. Yu. Lukin, and N. A. Platé, Dokl. Akad. Nauk SSSR, 237, 401-404 (1977).

241. V. P. Shibaev, V. M. Moiseenko, A. L. Smolyanskii, and N. A. Platé, Vysokomol. Soedin., A23, 1969-1980 (1981).
242. I. I. Konstantinov, J. Phys. (France), 40, No. C3, 475-478 (1979).
243. H. Finkelmann and H. Rehage, Makromol. Chem., Rapid Commun., 1, 733-736 (1980).
244. M. V. Kozlovskii and V. P. Shibaev, Abstracts of the 1st All-Union Symposium on Liquid-Crystalline Polymers [in Russian], Suzdal (1982), pp. 47-48.
245. V. P. Shibaev, M. V. Kozlovsky, L. A. Beresnev, L. M. Blinov, and N. A. Platé, Polym. Bull., 12, 299-301 (1984).
246. Yu. S. Lipatov, V. V. Tsukruk, V. V. Shilov, I. I. Konstantinov, and Yu. B. Amerik, Vysokomol. Soedin., A23, 1533-1543 (1981).
247. Yu. S. Lipatov, V. V. Tsukruk, V. V. Shilov, V. S. Grevneva, I. I. Konstantinov, and Yu. B. Amerik, Abstracts of the 4th International Liquid-Crystal Conference of Socialist Countries [in Russian], Vol. 2, Tbilisi (1981), p. 191.
248. Yu. S. Lipatov, V. V. Tsukruk, V. V. Shilov, V. S. Grebneva, I. I. Konstantinov, and Yu. B. Amerik, Vysokomol. Soedin., B23, 818-822 (1982).
249. R. Zentel and H. Ringsdorf, Makromol. Chem. Rapid Commun., 5, 393-398 (1984).
250. Yu. S. Lipatov, V. V. Tsukruk, and V. V. Shilov, Ukr. Fiz. Zh., 28, 543-549 (1983).
251. V. V. Shilov, V. V. Tsukruk, and Yu. S. Lipatov, J. Polym. Sci. Polym. Phys. Ed., 22, 41-47 (1984).
252. C. G. Vonk, J. Appl. Crystallogr., 11, 541-546 (1978).
253. V. P. Shibaev, "Block copolymers," in: Encyclopedia of Polymers [in Russian], Vol. 1, Moscow (1972), pp. 270-280.
254. V. V. Tsukruk, V. V. Shilov, and Yu. S. Lipatov, Polym. Commun., 24, 260-262 (1983).
255. P. Zugenmaier, Makromol. Chem. Suppl., 6, 31-39 (1984).
256. Z. A. Roganova, A. L. Smolinskii, M. V. Kozlovskii, R. V. Tal'roze, and V. P. Shibaev, Vysokomol. Soedin., A27, 477-485 (1985).
257. L. M. Blinov and L. A. Beresnev, Usp. Fiz. Nauk, 143, 391-428 (1984).
258. E. Hsu and A. Blumstein, J. Polym. Sci. Lett., 15, 129-132 (1977).
259. H. Ringsdorf and R. Zentel, Abstracts of Communications of the 27th International Symposium on Macromolecules, Vol. 2, Strasbourg, France (1981), pp. 969-973.
260. H. Ringsdorf and R. Zentel, Makromol. Chem., 183, 1245-1248 (1982).

261. S. G. Kostromin, V. P. Shibaev, and N. A. Platé, Abstracts of the 4th International Liquid-Crystal Conference of Socialist Countries [in Russian], Vol. 2, Tbilisi (1981), p. 185.
262. A. M. Mousa, Ya. S. Freidzon, V. P. Shibaev, and N. A. Platé, Polym. Bull., 6, 485-488 (1982).
263. V. P. Shibaev, S. G. Kostromin, R. V. Tal'roze, and N. A. Platé, Dokl. Akad. Nauk SSSR, 259, 1147-1152 (1981).
264. H. Finkelmann and D. Day, Makromol. Chem., 180, 2269-2273 (1979).
265. V. N. Tsvetkov, Acta Physicochim. URSS, 16, 132-138 (1942).
266. K. Wassmer, E. Ohmes, G. Kothe, M. Portugall, and H. Ringsdorf, Makromol. Chem. Rapid Commun., 3, 281-284 (1982).
267. V. V. Sinitsyn, Candidate's Dissertation, Moscow State University, Moscow (1982).
268. J. Van Dusen, D. J. Williams, and G. R. Meredith, Proceedings of the 28th International Macromolecular Symposium, Amherst, Mass. (1982), pp. 814-815.
269. M. V. Piskunov, S. G. Kostromin, L. B. Stroganov, V. P. Shibaev, and N. A. Platé, Makromol. Chem. Rapid Commun., 3, 443-446 (1982).
270. H. Geib, B. Hisgen, U. Pschorn, H. Ringsdorf, and H. Spiess, J. Am. Chem. Soc., 104, 917-918 (1982).
271. H. Finkelmann, H. Bentack, and G. Rehage, J. Chim. Phys., 80, 163-170 (1983).
272. Ch. Boeffel, B. Hisgen, U. Pschorn, H. Ringsdorf, and H. W. Spiess, Israel J. Chem., 23, 388-394 (1983).
273. L. M. Blinov, S. V. Yablonsky, V. P. Shibaev, and S. G. Kostromin, Proceedings of the 9th International Liquid-Crystal Conference, Bangalore, India (1982), p. 275.
274. S. V. Yablonsky, L. M. Blinov, S. G. Kostromin, and V. P. Shibaev, Kristallografiya, 29, 984-989 (1984).
275. W. J. Helfrich, J. Chem. Phys., 53, 2267-2275 (1970).
276. T. I. Borisova, L. L. Burshtein, N. A. Nikonorova, R. V. Tal'roze, and V. P. Shibaev, Vysokomol. Soedin., A27 (1985) (in press).
277. T. I. Borisova, L. L. Burshtein, N. A. Nikonorova, Ya. S. Freidzon, V. P. Shibaev, and N. A. Platé, Vysokomol. Soedin., A26, 1506-1512 (1984).
278. T. I. Borisova, L. L. Burshtein, T. P. Stepanova, and V. P. Shibaev, Vysokomol. Soedin., 21, 828-835 (1979).
279. T. I. Borisova, L. L. Burshtein, T. P. Stepanova, Ya. S. Freidzon, V. P. Shibaev, and N. A. Platé, Vysokomol. Soedin., A24, 1103-1109 (1982).

280. T. I. Borisova, L. L. Burshtein, T. P. Stepanova, Ya. S. Freidzon, A. V. Kharitonov, and V. P. Shibaev, Vysokomol. Soedin., B24, 451-456 (1982).
281. L. L. Burshtein and V. P. Shibaev, Vysokomol. Soedin., 24A, 3-19 (1982).
282. V. P. Shibaev, Yu. S. Freidzon, I. M. Agranovich, V. D. Pautov, E. V. Anufrieva, and N. A. Platé, Dokl. Akad. Nauk SSSR, 232, 401-404 (1977).
283. E. V. Anufrieva, V. D. Pautov, Ya. S. Freidzon, and V. P. Shibaev, Dokl. Akad. Nauk SSSR, A19, 755-760 (1977).
284. H. Kresse and R. V. Tal'roze, Makromol. Chem., Rapid Commun., 2, 369-375 (1981).
285. H. Kresse, S. G. Kostromin, and V. P. Shibaev, Makromol. Chem., Rapid Commun., 3, 509-512 (1982).
286. H. Kresse and V. P. Shibaev, Z. Phys. Chem. (Leipzig), 264, 161-164 (1983).
287. H. Kresse and V. P. Shibaev, Makromol. Chem., Rapid Commun., 5, 63-66 (1984).
288. H. Ringsdorf, G. Strobl, and R. Zentel, Am. Chem. Soc. Polym. Prepr., 24, 308 (1983).
289. V. P. Shibaev, Ya. S. Freidzon, and N. A. Platé, Vysokomol. Soedin., A20, 82-95 (1982).
290. V. P. Shibaev, V. M. Moiseenko, N. Yu. Lukin, N. A. Kuznetsov, Z. A. Roganova, A. L. Smolyanskii, and N. A. Platé, Vysokomol. Soedin., A20, 2122-2137 (1978).
291. S. G. Kostromin and E. Yu. Kozlova, Abstracts of the 1st All-Union Symposium on Liquid-Crystalline Polymers, Suzdal (1982), pp. 48-50.
292. W. J. Toth and A. V. Tobolsky, J. Polym. Sci., B8, 289-295 (1970).
293. A. De Visser, J. Feyen, K. De Groot, and A. Bantjies, J. Polym. Sci., B8, 805-809 (1970).
294. A. De Visser, K. De Groot, J. Feyen, and A. Bantjies, J. Polym. Sci., 9A-1, 1893-1899 (1971).
295. A. De Visser, K. De Groot, J. Feyen, and A. Bantjies, J. Polym. Sci., B10, 851-857 (1972).
296. G. Hardy, F. Cser, A. Kallo, K. Nyitrai, G. Bodor, and M. Lengyel, Acta Chim. Acad. Sci. Hung., 65, 287-301 (1970).
297. J. Shiniski, J. Phys. Chem., 80, 88-89 (1976).
298. E. Hsu, S. Clough, and A. Blumstein, J. Polym. Sci., Polym. Lett. Ed., B15, 545-548 (1977); Am. Chem. Soc. Polym. Prepr., 18, 709-710 (1977).
299. A. Blumstein, Y. Osaka, S. Clough, E. Hsu, and R. Blumstein, in: Mesomorphic Order in Polymers and Polymerization in Liquid-Crystalline Media, A. Blumstein (ed.), ACS Symposium Series, No. 74, Washington, D.C. (1978), pp. 56-70.

300. Y. Tanaka, S. Kabaya, Y. Shimura, A. Okada, Y. Kurihara, and Y. Sakakihara, J. Polym. Sci., B10, 261-265 (1972).
301. H. Saeki, K. Jimura, and M. Takeda, Polym. J., 3, 414-418 (1972).
302. A. De Visser, J. Van den Berg, and A. Bantjies, Makro-Mol. Chem., 176, 495-500 (1975).
303. A. L. Smolyanskii and V. P. Shibaev, Vysokomol. Soedin., A21, 2221-2228 (1979).
304. Ya. S. Freidzon, E. G. Tropsha, V. P. Shibaev, and N. A. Platé, Makromol. Chem., Rapid Commun., 6 (1985)
305. H. Finkelmann, H. Ringsdorf, W. Siol, and J. Wendorff, Makromol. Chem., 179, 829-834 (1978).
306. H. Finkelmann, J. Koldehoff, and H. Ringsdorf, Angew. Chemie, Int. Ed. Engl., 17, 935-936 (1978).
307. V. P. Shibaev, H. Finkelmann, A. V. Kharitonov, M. Portugall, N. A. Platé, and H. Ringsdorf, Vysokomol. Soedin., A23, 919-924 (1981).
308. Ya. S. Freidzon, N. I. Boiko, and N. A. Platé, Abstracts of the 1st All-Union Symposium on Liquid-Crystalline Polymers [in Russian], Suzdal (1982), pp. 24-26.
309. Ya. S. Freidzon, S. G. Kostromin, N. I. Boiko, V. P. Shibaev, and N. A. Platé, Am. Chem. Soc. Polym. Prepr., 24, 279 (1983).
310. Ya. S. Freidzon, A. V. Kharitonov, and S. G. Kostromin, Abstracts of the 4th International Liquid-Crystal Conference of Socialist Countries [in Russian], Vol. 2, Tbilisi (1981), p. 219.
311. Ya. S. Freidzon, N. I. Boiko, V. P. Shibaev, and N. A. Platé, Abstracts of the 5th Liquid-Crystal Conference of Socialist Countries [in Russian], Vol. 2, Part 1, Odessa (1983), pp. 130-131.
312. A. M. Mousa, Ya. S. Freidzon, V. P. Shibaev, and N. A. Platé, Abstracts of the 10th International Liquid-Crystal Conference, York, United Kingdom (1984), p. D19.
313. P. N. Keating, Mol. Cryst. Liq. Cryst., 8, 315-323 (1969).
314. L. N. Lisetski, Abstracts of the 1st All-Union Symposium on Liquid-Crystalline Polymers [in Russian], Suzdal (1982), pp. 26-27.
315. V. V. Tsukruk, Ya. S. Freidzon, V. V. Shilov, and V. P. Sibaev, Polym. Commun., 26 (1985) (in press).
316. H. Finkelmann, D. Naegele, and H. Ringsdorf, Makromol. Chem., 180, 803-806 (1979).
317. G. Aguilera, H. Ringsdorf, A. Schneller, and R. Zentel, Proceedings of the International Symposium on Macromolecules, Vol. 3, Litografia Felici, Florence, Italy (1980), pp. 306-308.

318. V. M. Moiseenko, Candidate's Dissertation, Moscow State University, Moscow (1978).
319. R. V. Tal'roze, S. G. Kostromin, V. P. Shibaev, N. A. Platé, H. Kresse, K. Sauer, and D. Demus, Makromol. Chem., Rapid Commun., 2, 305-308 (1981).
320. R. V. Tal'roze, V. V. Sinitsyn (Sinitzyn), V. P. Shibaev, and N. A. Platé, in: Advances in Liquid Crystal Research and Applications, Vol. 2, L. Bata (ed.), Pergamon Press, Oxford, Akademiai Kiado, Budapest (1980), pp. 915-924.
321. V. P. Shibaev, R. V. Tal'roze, V. V. Sinitsyn (Sinitzyn), S. G. Kostromin, and N. A. Platé, Proceedings of the 28th International Macromolecular Symposium, Amherst, Massachusetts (1982), p. 814.
322. N. A. Platé, R. V. Tal'roze, S. G. Kostromin, V. P. Shibaev, and H. Kresse, Abstracts of Communications to the 27th International Symposium on Macromolecules, Vol. 2, Strasbourg, France (1981), pp. 978-981.
323. R. V. Tal'roze, S. G. Kostromin, V. V. Sinitsyn, V. P. Shibaev, and N. A. Platé, Abstracts of Communications to the 12th European Conference on Macromolecular Physics, Leipzig, GDR (1981), pp. 313-314.
324. R. V. Tal'roze, S. G. Kostromin, V. V. Sinitsyn, V. P. Shibaev, and N. A. Platé, Abstracts of the 4th International Liquid-Crystal Conference of Socialist Countries [in Russian], Vol. 2, Tbilisi (1981), pp. 215-216.
325. W. R. Krigbaum, H. J. Lader, and A. Cifferi, Macromolecules, 13, 554-560 (1980).
326. W. R. Krigbaum and H. J. Lader, Mol. Cryst. Liq. Cryst., 2, 87-95 (1980).
327. W. R. Krigbaum, C. E. Grantham, and H. Toriumi, Macromolecules, 15, 592-602 (1982).
328. C. Casgrande, M. Veyssie, C. Weill, and H. Finkelmann, Mol. Cryst. Liq. Cryst., 92, 49-55 (1983).
329. M. F. Achard, G. Sigand, F. Hardouin, C. Weill, and H. Finkelmann, Mol. Cryst. Liq. Cryst., 92, 111-118 (1983).
330. R. V. Tal'roze, V. V. Sinitsyn, V. P. Shibaev, and N. A. Platé, Abstracts of the 5th Liquid Crystal Conference of Socialist Countries [in Russian], Vol. 2, Part 1, Odessa (1983), pp. 132-133.
331. H. Pranoto and W. Haase, Mol. Cryst. Liq. Cryst., 98 299-304 (1983).
332. R. V. Tal'roze, V. V. Sinitsyn, I. A. Korobeinikova, V. P. Shibaev, and N. A. Platé, Dokl. Akad. Nauk SSSR, 274, 1149-1153 (1984).

333. H. Ringsdorf, H. Schmidt, and A. Schneller, Makromol. Chem. Rapid Commun., 3, 745-748 (1982).
334. H. Coles and R. Simon, Mol. Cryst. Liq. Cryst. 102, 75-82 (1984).
335. H. Bentack, H. Finkelmann, and G. Rehage, Abstracts of Communications of the 27th International Symposium on Macromolecules, Vol. 2, Strasbourg, France (1981), pp. 961-965.
336. H. Ringsdorf, H. Schmidt, G. Bauer, and R. Kiefer, Polym. Prepr., 24, 306-307 (1983).
337. R. V. Tal'roze, V. V. Sinitsyn (Sinitzyn), V. P. Shibaev, and N. A. Platé, in: Polymer Liquid Crystals, A. Blumstein (ed.), Plenum Press, New York (1985), pp. 331-342.
338. V. P. Shibaev, S. G. Kostromin, N. A. Platé, S. A. Ivanov, V. Yu. Vetrov, and I. A. Yakovlev, Polym. Commun., 24, 364-365 (1983).
339. S. A. Ivanov, I. Ya. Yakovlev, V. Yu. Vetrov, S. G. Kostromin, and V. P. Shibaev, Pis'ma Zh. Tekh. Fiz., 9, 1349-1353 (1983).
340. R. V. Tal'roze, V. V. Sinitsyn (Sinitzyn), V. P. Shibaev, and N. A. Platé, Polym. Bull., 6, 309-314 (1982).
341. R. V. Tal'roze, V. V. Sinitsyn (Sinitzyn), V. P. Shibaev, and N. A. Platé, Polym. Prepr., 24, 308-309 (1983).
342. R. Simon and H. Coles, Mol. Cryst. Liq. Cryst., 102, 43-48 (1983).
343. R. V. Tal'roze, I. A. Korobeinikova, Ya. S. Freidzon, V. P. Shibaev, and N. A. Platé, Abstracts of the 10th International Liquid Crystal Conference, York, United Kingdon (1984), p. D16.
344. N. A. Platé, R. V. Tal'roze, and V. P. Shibaev, Polymer Yearbook (1985).
345. V. N. Tsvetkov, Rigid-Chain Polymer Molecules [in Russian], Nauka, Leningrad (1985).
346. E. V. Anufrieva and Ya. Ya. Gotlib, in: Adv. Polym. Sci., 40, 3-68 (1981).
347. E. V. Anufrieva, Pure Appl. Chem., 54, 533-548 (1982).
348. V. N. Tsvetkov, Vysokomol. Soedin., 4, 895-897 (1962); 7, 1468-1482 (1965); Dokl. Akad. Nauk SSSR, 165, 350-363 (1965).
349. V. N. Tsvetkov and L. N. Andreeva, Adv. Polym. Sci., 39, 98-207 (1983).
350. V. N. Tsvetkov, E. I. Ryumtsev (Rjumtsev), and I. N. Shtennikova, in: Liquid-Crystalline Order in Polymers, A. Blumstein (ed.), Academic Press, London (1978), pp. 43-103.

351. V. N. Tsvetkov, E. I. Ryumtsev, I. I. Konstantinov, Yu. B. Amerik, and B. A. Krentsel', Vysokomol. Soedin., A14, 67-78 (1972).
352. V. N. Tsvetkov, E. I. Ryumtsev, I. N. Shtennikova, I. I. Konstantinov, Yu. B. Amerik, and B. A. Krentsel', Vysokomol. Soedin., A15, 2270-2282 (1973).
353. E. I. Ryumtsev, I. N. Shtennikova, N. V. Pogodina, G. F. Kolbina, I. I. Konstantinov, and Yu. B. Amerik, Vysokomol. Soedin., A18, 439-448 (1976).
354. N. V. Pogodina, V. N. Tsvetkov, and I. P. Kolomiez, Abstracts of the 4th International Liquid-Crystal Conference of Socialist Countries [in Russian], Vol. 2, Tbilisi (1981), pp. 203-204.
355. I. N. Shtennikova, T. V. Peker, G. F. Kolbina, S. V. Bushin, and E. V. Korneeva, Abstracts of the 4th International Liquid-Crystal Conference of Socialist Countries [in Russian], Vol. 2, Tbilisi (1981), pp. 227-228.
356. Ya. S. Freidzon, V. P. Shibaev, I. M. Agranovich, V. D. Pautov, E. V. Anufrieva, and N. A. Platé, Vysokomol. Soedin., A20, 2601-2607 (1978).
357. Ya. S. Freidzon, V. P. Shibaev, V. D. Pautov, T. K. Bronich, G. D. Shelukhina, V. A. Kasaikin, and N. A. Platé, Dokl. Akad. Nauk SSSR, 256, 1435-1438 (1981).
358. E. V. Anufrieva, V. D. Pautov, Ya. S. Freidzon, V. P. Shibaev, and N. A. Platé, Vysokomol. Soedin., A24, 825-829 (1982).
359. V. A. Kasaikin, Ya. S. Freidzon, and E. E. Makhaeva, Abstracts of the 4th International Liquid-Crystal Conference of Socialist Countries [in Russian], Vol. 2, Tbilisi (1981), p. 179.
360. E. V. Anufrieva, V. D. Pautov, Ya. S. Fredizon, V. P. Shibaev, and N. A. Platé, Dokl. Akad. Nauk SSSR, 278, 383-386 (1984).
361. T. I. Borisova, L. L. Burshtein, T. P. Stepanova, Ya. S. Freidzon, and V. P. Shibaev, Vysokomol. Soedin., B19, 553-557 (1977).
362. A. Yu. Grosberg, Vysokomol. Soedin., A22, 96-103 (1980).
363. H. Finkelmann, B. Lühmann, and G. Rehage, Colloid Polym. Sci., 260, 56-63 (1982).

GPSR Compliance

The European Union's (EU) General Product Safety Regulation (GPSR) is a set of rules that requires consumer products to be safe and our obligations to ensure this.

If you have any concerns about our products, you can contact us on ProductSafety@springernature.com

In case Publisher is established outside the EU, the EU authorized representative is:

Springer Nature Customer Service Center GmbH
Europaplatz 3
69115 Heidelberg, Germany

Batch number: 10682323

Printed by Printforce, the Netherlands